FRAME ASSY JIGS
FR.14 51E 600095 D
- 15 51E 600096 D
- 18 51E 600097 D
- 19 51E 600098 D
- 20
- 21 38E 101121 D
- 24
- 22
- 23 38E 101122 D
- 25 38E 101124 D
HYDRAULIC BAY DOOR
BAGGAGE DOORS
FWD.
AFT.
SET
CROSS MATCH
FEMALE JIG REF
101E 500047D
101E 700041D
MALE JIG REF
101E 102264D
101E 500046 D
101E 700040 D
ASSY JIG
51E 102246D
ASSY JIG
51E 500031D
REAR DOORS ASSY JIGS
PASSENGER
SERVICE
FR 41 REF PLATE
101E 600983C
FR 43 REF PLATE
101E 600984D
101E 700021 D
101E 700026 D
101E 700022 D
101E 700028 D
A.P.U. JIG REF
101E D
A.P.U. DOOR JIG REF
MALE
101E 700019 D
A.P.U. DOOR JIG REF
FEMALE
101E 704026 D
A.P.U. DOOR ASSY JIG
051E 700905 D
T0100048
T/P L/E JIG REF
101E
T/P F/S JIG R
101E
T/P C/S BOX
ASSY JIG
MASTER FUSE RING
101E 600003D
MASTER FUSE FRAME
101E 500015 D
PLUG FR 19
58E 102177 D
MAIN U/C LEG
JIG REF
101E 600038D PORT
101E 600039D STB
233E 600037D
101E 900328D
TAILPLANE ASSY JIG
CHASSIS-CONTROL
BOX-ASSY JIG
101E 700053
101E 700010D
TAILPLANE EQUIPPING
SITE
101E 200020D
101E 900023D
101E 900327D
U/C BAY UNIT
507E 600080D
CENTRE FUSE
TROLLEY UNIT
507E 600079D
CENTRE FUSE
TROLLEY CHASSIS
508E 600078D
SET WING
ASSY JIGS
BOXING SITE
TROLLEY
FROM 7
FIN ASSY JIG
FIN L/E REFS
101E 900020 D
101E 900021 D
101E 900022 D
SET
FIN L/E ASSY JIG
58E 600963-4-5D
600102 7D
101E 900333D
FAIRING MODEL
245E 600985D
SPINE FAIRING
ASSY JIGS
58E 600089D-90D-91D-92D
BOXING 6 SITE
TROLLEY
299E 501875D
299E 501865D
101E 501861D
233E 501640D
233E 501962D
101E 600603D
101E 600604D
101E 606060D
101E 601040 D
101E 601471-2D
101E 600605D
TO SET WING
ASSY JIG
101E 102543D
101E 102703D
051E 600982D
051E 600980-1D
COMPOSITE JIG REF
FLAP FAIRING No. 4
101E D
101E D
101E D
101E
TO SET
WING ASSY
JIG
I.C.Y MEDIA shewn thus:-
BUILD SEQUENCE thus:-
APPLICATION of MEDIA,
CROSS MATCHING etc thus:-
Site
HATFIELD
FILTON
BROUGH
MANCHESTER
PRESTWICK
AVCO
SAAB SCANIA
SHORTS
BRITISH AEROSPACE AIRCRAFT GROUP
HATFIELD HERTS ENGLAND
B.Ae 146 BASIC INTERCHANGABILITY
MEDIA IN RELATION TO BUILD SEQUENCE

www.avgeektv.com

Book design and layout by Darin Kirschner

Edited by Andy Martin

978-1-63405-027-2 (print)

978-1-63405-028-9 (e-book)

Published by LA JETÉE PRESS

www.lajeteepress.com

Printed and bound in China.

Art created by: Darin Kirschner

Author of Fighting to Be Heard hugging his favorite aircraft, the BAe 146 (this one owned by Cello Aviation).

Acknowledgements

This book wouldn't be complete without the help and support of numerous people.

To my wonderful wife Carrie who's been very understanding of my taking off or devoting time to work on this book, including frequent traveling for research. Few of my projects recoup the costs involved, but she's been supportive of me over and over again. My children, Braniff, Anika, Katia and Coraline - none of them understands my aircraft passion, but Coraline certainly shares it. My mom and dad for being supportive, and Charles Kennedy who supported bringing it to life. Lot's of love to Roger & Sue Curtis who've always supported my crazy endeavors.

David Dorman has not only been incredibly supportive but has connected me to many other key people involved with the life of the 146. David's enthusiasm for my project has been the inspiration to ensure I put the best BAe 146 book out there, and he's been absolutely key to much of the content. His lovely wife Jane was a wonderful host during my research trip in the UK which produced a considerable amount of new content never before published.

Peter Connolly, who arguably deserves a co-writing credit. He was a walking book of knowledge on the entire BAe 146 project, and devoted quite a bit of his time reading what I had wrote, offering corrections, clarity, and even new information on the BAe 146. His feedback were chapters in themselves. I've encouraged him to write about his adventures at British Aerospace in a separate book.

Former Squadron Leader Graham Laurie was kind enough to share information about his time flying the No 32 Squadron BAe 146-100 aircraft. Graham also introduced me to former Squadron Leader David Gale, who wrote a considerable amount about the RAF evaluation of the BAe 146 as a replacement for the No 32 Squadron Andover aircraft. David's insight and stories really bring the RAF operations to life.

John Stevens, Mark McArdle, Terry Snow, Stephen Morrison, and Andrew James at BAE Systems Regional Aircraft at Prestwick were instrumental in providing further operational information as well as never before published photos in this book. Further former BAE sales and marketing persons (including Asset Management Organization) Nick Godwin, Steve Doughty, Paul Stirling, Nigel Benson, and Phil Bolt provided amazing insight to the sales process, the near collapse of BAE, and their efforts to dig the firm out of the predicament it found itself in.

Derek Taylor, Clive Nicholson and Roger Pascoe, all former BAe reps stationed at PSA, who eventually became PSA employees, provided numerous photos, information, and actual on-the-ground performance details of the airline's operations which bring this book and story to life.

Richard Thomasson and Scott Becker not only helped proof read my draft book, but provided suggestions on where blanks needed to be filled in. Pilots Chuck Ross of PSA and Pat Doyle of Air Wisconsin provided detailed accounts of flying the BAe 146 while, David McLees who worked at BAe in Arkansas offered quite a bit of information covering operations in the South.

Maurice Gallagher was gracious with his time talking to me about Pacific Express, WestAir, and the industry in general. He is the founder of Allegiant Airlines based in Las Vegas, Nevada. Gary Ellmer, also with WestAir, Royal West, and BEX gave me a further insight to operating the aircraft at each airline. Jim Skinner and Kevin Govett, both formerly with Montex Oil, flew and maintained the only privately owned BAe 146 in the United States, and provided insider information on piloting and maintaining such an aircraft. John Sloan of Royal West delivered insight into the short-lived operations of the airline. Debbie Seracini for being patient and accommodating with my requests to essentially pull all 74 boxes on PSA from storage for my research at the San Diego Air and Space Museum. Vince Essex of Cello Aviation allowing me access to both aircraft, and sharing his Avro RJ garden summer home conversion.

Howard Guy and Greg Shilton of Design Q, who designed a number of VIP Avro and BAe 146 interiors, provided detailed information about design proposals generated for BAE Systems. Andrew Scott at London City Airport offered some wonderful photos including the Loganair 146 day at LCY. David Banmiller, formerly of AirCal who provided a detailed look from inside. Albert Boring for sharing insight and photos of AirCal's BAe 146 maintenance operations. Amber Biela at Air Wisconsin provided photos and information on

their BAe 146 operations. Ben and Phil of Juneau Projects produced an interesting futuristic artwork concept using a former tail section from a BAe 146 for Juneau Projects.

Kåre Halvorsen and Anne-Line Bjella-Fosshaug of the Accident Investigation Board of Norway. BAE Heritage Center (Farnborough) Barry Guess allowed me unfettered access to peruse their archives and provided many of the photos in this book. Mark Goodliffe, who organizes the restoration and maintenance of the world's only surviving Avro RJX (100) aircraft and was helpful in providing information and photos. Derek Ferguson, formerly of BAE Systems flight test crew was on the final Avro RJX aircraft, helped correct research and provided additional photos for the book.

Fellow avgeek friends Emma Conroy and Sandy Tweedie (who suggested the title to this book) who've supported my work. Sites like AVGEEKERY.COM which are a great resource for the aviation geek. Darin Kirschner is a fantastic designer, and thankfully designed the work you are now reading, along with Andy Martin who artfully edited my verbal drivel that wound up on paper. I must not forget Harsha, who fed me with amazing food and company during my work trips to Bangalore, India where I wrote a substantial amount of this book.

Tej Soni, my manager (yes, I do have a day job), has been incredibly supportive of my aviation projects, giving me the time to chase down much needed research, especially during my international work travels. Without his support, this book would have taken another 2-3 years to finish. I am forever grateful. Sanjay, Chindu, Shashi & Krishna, Sarbashish, Sarah, all true supporters.

Tom Harris and JP Joseph Santiago who wanted to launch their new aviation publishing company by making Fighting to Be Heard their first release. God help 'em! Emily Brubaker Harris for her role in helping these two chuckleheads (and me!).

Godspeed to Captain Daniel Burke, Boeing 747SP Lead Captain for Sands Aviation, Ted Brogan, an amazing avgeek who shared his love of junk food with me, and Brooke Watts.

Finally Jerry Jessop (Golden Fro) continues to pimp slap me ensuring my aviation enthusiasm never dies. Per Dolemite's instructions, no cotton drawers were worn during the production of this book.

David Dorman, former Marketing and PR with British Aerospace, the author, and Former Squadron Leader Graham Laurie.

PROLOGUE

FIGHTING TO BE HEARD

From the very beginning, the 146 project has been fighting to be heard. Airlines were dubious of a very short-range jet aircraft when turboprops were the norm. They were even more skeptical of four jet engines on a commuter jet when the configuration was associated with larger and longer-range transports. The BAe 146 would eventually prove to be the quietest jet aircraft ever built; it was ironic that Hawker Siddeley and subsequently British Aerospace were constantly fighting for airlines to listen to what it had to offer.

The roots of the BAe 146 began in 1959 when de Havilland was looking to build a Douglas DC-3 replacement. Starting off as the DH123, a 30-40 seat turboprop, the aircraft went through multiple iterations, progressing next to the DH126 twin turbojet, Hawker Siddeley HS131 twin turbofan, and the HS144 which was a rear twin engine jet design. But with no suitable indigenous engines available, the HS146 was born with foreign powerplants.

The BAe 146 arrived at a time when the industry was in turmoil, especially in the United Kingdom. It was then shelved when fuel prices skyrocketed, there was labor unrest, and the world's economy was faltering. But the British airliner industry persevered. After a very long incubation period, the 146 went into hibernation. Four years later British aircraft manufacturers were nationalized, forming British Aerospace. The BAe 146 programme launched in 1978 with the first flight following in 1981 and initial deliveries commencing in 1983. It went on to become the best-selling British jetliner ever built, and the best-selling regional aircraft (jet or prop driven) with more than 40 seats to come out of Britain. It also went on to start what became the regional jet revolution as propliners fell out of favor with passengers.

But the BAe 146 wasn't all sunshine and roses. It had a long gestation period, followed by a very rough entry into service due in part to lackluster reliability from a powerplant that wasn't built for the rigor of airline operations. While the engine was proven with millions of hours of service prior to installation on the BAe 146, it was originally a vertical-lift turbine for helicopters, namely the Boeing Chinook CH-47. It was also used in the U.S. M1A Abrams, giving the battle tank amazing speed across a variety of terrain. The AVCO-Lycoming ALF series turbine had never been used in a commercial transport and it took time for it to settle into the rigors of airline service. As a result, the BAe 146 acquired negative acronyms like BAe "Bring Another Engine" and "14-sick".

Initial sales of the 146 struggled, not gaining traction until the end of 1983 when a U.S. regional carrier placed a substantial order. It was exactly what BAe needed; a long established and successful carrier buying the type in large numbers. To further bolster the 146's credibility, that regional airline was downsizing from larger aircraft such as Boeing 727s and 737s. The rest is history.

Later in its life the BAe 146 and its derivatives began to face an uphill battle as the regional jet scene exploded on the market. Two engines were more attractive, burning less fuel, and less maintenance headaches, and the unique airports that the 146 originally opened up began to be served with newer, cheaper and more fuel-efficient aircraft. The days were numbered for the 146, and BAe's inability to move to a twin-engine design after multiple failed initiatives sealed the type's fate. Supporting an airliner is not simple, nor cheap, and BAe in its early days had sometimes looked for a way out of the commercial aircraft business, instead preferring military contracts or to be a supplier to other commercial manufacturers. The company used the events of September 11, 2001 to bring an end to 23 years of Britain's most successful airliner project.

I've followed the 146 for the past twenty years and have an infatuation with it from every perspective. I think it's a great looking aircraft, and its performance is still something to behold. At the time of writing, I've only ever flown on the BAe 146 twice: a round-trip from San Francisco (SFO) to Las Vegas (LAS) with charter operator TriStar airlines. I was ecstatic, but the only thing I remember was the approach into LAS. I thought I could run faster than our speed across the threshold of the runway! It seemed like we just floated downwards like a feather. And that was one of the most positive aspects of the aircraft: it was difficult for pilots to make a bad landing.

The British Aerospace 146 is in my opinion the most over-engineered aircraft most airlines just didn't need. Practically every salesman, engineer, mechanic or pilot I've encountered used the same phrase to describe the construction of the 146: a brick shithouse. Sure, it had amazing STOL capability, could land on unpaved fields, didn't need noisy and complex thrust reversers, and had steep approach and takeoff capability. But most airlines (and airports) didn't need and couldn't justify an aircraft with such characteristics. In fact, carriers were moving away from four engine aircraft and the industry was headed towards a two-engine future, even for large airframes.

There have been plenty of books, articles, and magazines on the BAe 146, but I was always been left wanting more, feeling that many efforts were very 'dry' with information, and lacked the story that should be told. After acquiring a large collection of BAe 146 documents, brochures, photographs, and more from a fellow enthusiast, I had so much inside information I felt compelled to produce a book. I hope I've provided you with an insight into perspectives never-before discussed. Unlike most aircraft books this was never meant as a technical manual, but instead a story told from the front lines, weaving history with perspectives from former pilots and airline personnel who flew the BAe 146, the engineers who maintained the aircraft, and those that designed, built, and sold the 146. A testament to one of the best aircraft ever built.

RJ115
RJ100
AVROLINER
LF507
RJ85
AVROLINER
LF507
FADEC
RJ70
AVROLINER
LF507
FADEC

CONTENTS

FIGHTING TO BE HEARD

Framing for the mockup of the HS146 began before the program was terminated.

CHAPTER 01

British Airliners - The Beginning of the End

When Hawker Siddeley began researching a jet concept designed for short sectors, it wasn't aware that it would be developing the last airliner to be built in the United Kingdom. Nor did it realize it was establishing a brand-new aircraft category: the regional jet. The project marked the beginning of the end for the UK as a manufacturer of complete commercial aircraft, even before the aircraft's final configuration had been agreed upon. While Hawker Siddeley was engaged in concept development for its new airliner, it was also securing the future of Britain as a supplier to other civil aircraft manufacturers, notably becoming a risk sharing partner in the Anglo-French Airbus Industrie consortium.

Until the 1970s, regional routes were served by turboprop aircraft flown by commuter airlines which operated alongside a small subset of larger jets and major carriers. The British Aircraft Corporation was first to market with a small short-range jet, the BAC One-Eleven. With a seating capacity of 89 it beat the Douglas DC-9-10, another similarly configured rear engine aircraft, to market. Hawker Siddeley, having developed turboprop aircraft, was looking to move into jets and began designing smaller short-range jetliners as far back as 1959. Over time and unable to settle on a specific configuration, the project began to grow in size. In 1967, the company settled on a similar configuration to the BAC One-Eleven with an 89-passenger capacity. With widebody aircraft like Boeing's 747-100 series, the McDonnell Douglas DC-10, and the Lockheed L-1011 commencing operations, the 'hub-and-spoke' system was beginning to take shape and smaller aircraft feeding larger airliners in the United States sparked a new model for air travel. Nevertheless, Hawker Siddeley did not really grasp the concept at the time, nor did it see a need to feed passengers onto flights operated by larger aircraft in major cities.

Hawker Siddeley was building an aircraft that it believed state owned carriers or those operating in a regulated market would purchase. What the firm did not foresee was the wave of deregulation that would ripple out from the U.S. Prior to 1978, the U.S. market was heavily regulated by the Federal Government, with each airline required to apply for approval not just for each route (and aircraft), but also the ticket pricing for interstate travel (journeys between more than one state). This dictated the shape of the market, although regional airlines were exempt from Federal regulations and pricing provided they were operating as intrastate carriers (flights between points within a single state). But even intrastate carriers were subject to local approvals including route and pricing controls, which were similar to the Federal government's rules. Other markets around the world followed regulatory rules similar to those in the US. The regulated market Hawker Siddeley was using in its business model would begin to disappear even before the HS146 would make its first test flight.

At the time the majority of intrastate carriers flew small turboprop aircraft, although a very limited number flew jets too. The Boeing 737 and 727 (along with the Douglas DC-9 series 30 and 50) were too large for many markets, while smaller aircraft were perceived to be too small to feed hub systems for larger airlines. The hub-and-spoke model was just beginning to emerge in the U.S., but wasn't widely practiced in the rest of the world which focused more on point-to-point service.

Early trijet design for the Hawker Siddeley airbus project.

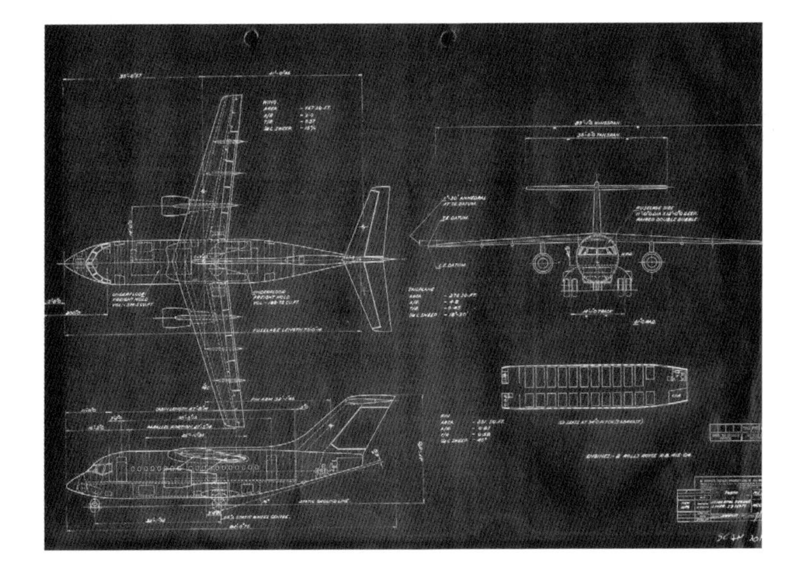

Twin jet concept for the Hawker Siddeley airbus project.

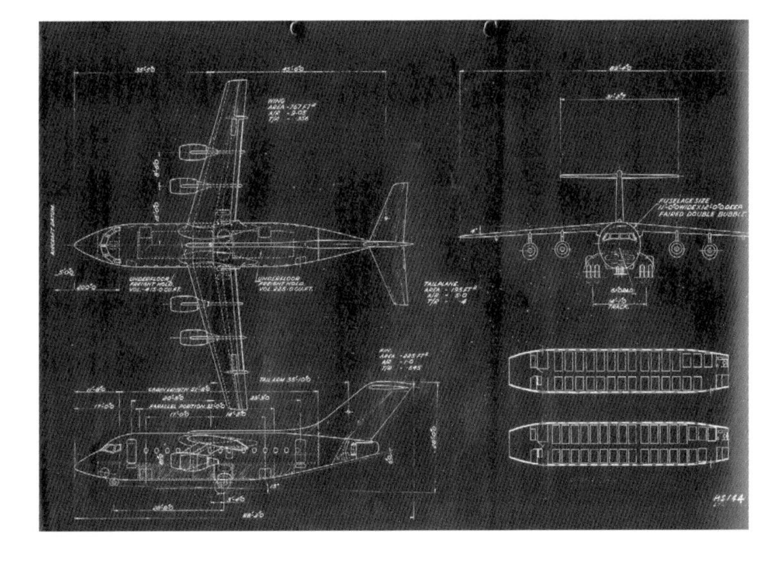

Revised design concept that would become the HS146 aircraft.

In 1980 the Civil Aviation Act (CAA) was passed in the USA. The changes enabled the market to act in the best interest of airlines, and to take into consideration the benefits to multiple carriers serving the same route. The CAA also transferred pricing controls from government to the airlines. The United Kingdom and Europe were still regulated markets and member states controlled their own airlines, which often included subsidized flag carrier airlines. It wasn't until 1984 that the European Council adopted a directive that would essentially deregulate markets between member states, introducing competition.

By then the BAC One-Eleven was an ageing aircraft with fuel and maintenance costs that were becoming uncompetitive in comparison to newer Boeing and McDonnell Douglas designs. Hawker Siddeley pursued a new-build aircraft to fit a perceived gap in the market between turboprops and small narrow body jets. The design goals for its next generation of airliner were becoming defined: long service life operating on very intensive short haul regional routes, with engineering simplicity and reliability. Having an APU for ground power, air stairs, and access to the baggage holds without baggage loaders meant that the aircraft could fly to any airport and be serviced without the need for additional equipment. It was initially pitched as the perfect self-contained aircraft to operate into small airfields in remote areas. The irony was that regional airlines that operated into such fields couldn't always afford the purchase price and typically used second (or third) hand aircraft.

As Hawker Siddeley further refined its regional jet concept, it became clear there was no power plant available that gave the aircraft the needed lift. Hawker Siddeley approached several engine manufacturers, but was really banking on Rolls-Royce to provide the powerplants it needed given the UK firm was already supplying engines to BAC, and to competitor Fokker for its regional jet. But financially struggling Rolls-Royce was unable to divert resources from the moribund Concorde project in order to develop an engine that met Hawker Siddeley's needs. Rolls-Royce eventually declared bankruptcy after it took on more projects than it was capable of delivering, specifically the RB211 that would power the American built-Lockheed L-1011.

Rolls Royce, liquidated and then nationalized by the British government as Rolls-Royce Ltd in 1971, narrowed its scope of work to support established financially viable programs, which meant it could not accommodate the HS144 . Hawker Siddeley's proposed air bus projects were originally configured with a low wing and twin rear mounted engines but with the company facing difficulties sourcing suitable power plants, the firm found itself investigating alternatives that would require it to modify its aircraft design.

While Rolls-Royce was trying to catch up with the RB211 program for the Lockheed L-1011 and its partnership with SNECMA for the Olympus 593 engine to power Concorde, Hawker Siddeley found itself in the opposite situation. It began talking to U.S. based manufacturer AVCO Lycoming, as it was the only company that had a potential off-the-shelf solution. The T55 was in the early stages of being converted from a vertical-lift helicopter engine into a powerplant for commercial aircraft and possible use on the Hawker Siddeley aircraft. However, the thrust required by the HS144 could not be achieved with just two modified T55s, and instead four were required to meet the lift needs.

The team at Hatfield felt it was worth pursuing the AVCO Lycoming engine, even though its low thrust output would necessitate a major design change and mean the HS144 would no longer be a twin-engine aircraft with rear-mounted power plants. Mounting four engines was no easy task as underslung wing-mounted jets introduced a risk of foreign object damage (FOD) ingestion. Moving the wing up above the fuselage would reduce the likelihood of FOD ingestion considerably, while giving the aircraft STOL (Short Take Off and Landing) performance capabilities. But installing four engines caused concern over perceived operating and maintenance expenses, leaving the Hawker Siddeley sales team with

British Aircraft Corporation was working on a similar concept, the QSTOL, but in widebody form.

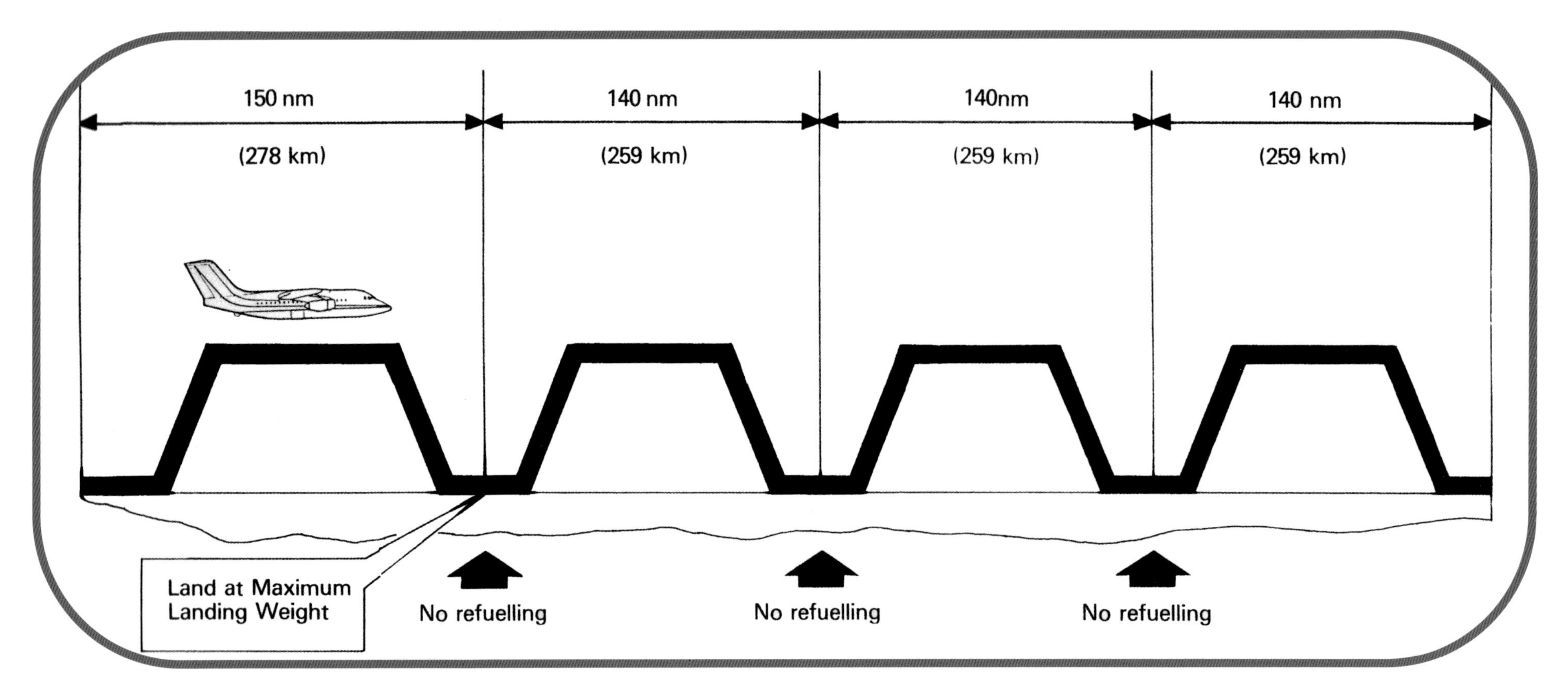

Ultra-short haul concept of route operation.

quite a challenge to convince airlines of the new aircraft's economic benefits. Nevertheless, the firm proceeded with the new configuration and design, which became known as the HS146, a high wing four-engine jetliner that was to eventually become the BAe 146.

AVCO Lycoming began testing the ALF502 geared turbofan for fixed wing aircraft in 1971, with its first application being the Canadair Challenger CL600 executive jet. AVCO Lycoming sold Hawker Siddeley on the idea that four ALF502 engines would equal two higher thrust motors, with its unique design still offering two-engine economics. At the same time the firm also began offering a fan engine derivative of its helicopter engine for smaller fixed-wing aircraft such as the HS125 business jet. Hawker Siddeley's selection of an unproven engine (at least in a high-utilization airline environment) produced by a manufacturer with no track record in the airline business was a high-risk strategy as events were to prove.

The Hawker Siddeley 146 (HS146) would be a short range 'air bus' to serve small markets that could benefit from a jet aircraft. It was designed to be fueled before the first flight, and then like a bus, carry out a series of short sectors (trips) between cities before needing to be refueled again. Each sector was expected to cover 100-150 miles, very short and regional in nature. Launched as a joint venture in August 1973, the HS146 was a £92 million (pounds) partnership with Hawker Siddeley putting up 50% of the development costs and the British government covering the remaining 50% with repayable loans. In U.S. dollars adjusted for inflation in 2018, it amounted to over $1.7 billion, with the British government on the hook for half the committed amount. The aircraft was designed for maximum flexibility, fast turnarounds, low operational costs and future growth; and it would be the quietest aircraft to fly into and out of noise sensitive airports. The HS146 was supposed to support everything a regional airline could throw at it, while offering growth derivatives for future needs and market expansion

The HS146 was initially designed with a modest 71-passenger, 5-abreast, 33-inch pitch cabin but offered flexibility to accommodate up to 88 passengers, 6-abreast at a 31-inch pitch. A larger derivative seating up to 102 passengers, 6-abreast with a 31-inch pitch was also planned as a later development. Hawker Siddeley's estimates suggested most airlines would prefer the smaller version. Powered by four AVCO Lycoming ALF-502H engines, the HS146 would be capable of flying up to 1700 miles. It also possessed, through an ingenious design combining a high wing and a petal-tail brake, the ability to land on unpaved fields making it attractive to operate into remote airstrips that needed air service. The high wing and lack of thrust reversers reduced the risk of damage from FOD ingestion, while reducing engine complexity, maintenance costs and noise.

The Hatfield Design Office, led by Bob Grigg, was responsible for the overall design assisted by the Manchester and Brough offices. Hatfield was where final assembly would occur for the HS146, while responsibility for assemblies and parts would be spread out across facilities all over the United Kingdom. The aircraft's price was set at $4.4 million (USD), 10% higher than the Fokker F.28.

The HS146 was not designed to be a long-range airliner, hence its early characterization in sales literature as an "air bus" (and even "bus jet"). Short sectors and fast turnaround were a primary design aim and operational principle. But the HS146 also opened up the possibility of regional airlines flying longer distances with higher passenger loads, increasing the reach of their route networks without the need for a partner airline, or to buy larger and costlier aircraft. This didn't initially work in the United States due to regulation, but in 1978 the market would change.

Following extensive internal studies and talking with nearly 120 airlines that expressed interest in the jet, Hawker Siddeley launched the HS146 just before the worldwide oil crisis hit in October 1973. Engineering work was in full swing by late 1973, driving towards a first flight planned for December 1975. A full-scale wooden mockup was built to show to potential customers, complete with seats from a Boeing 747-100 to demonstrate how spacious the cabin would be for passengers. The ultimate goal was to achieve certification in early 1977 with deliveries of the HS146-100 to customers shortly after. The larger series 200 aircraft was expected to make its first flight around the time of certification of the series 100. Hawker Siddeley estimated that roughly 30% of sales would be for the larger series 200 aircraft.

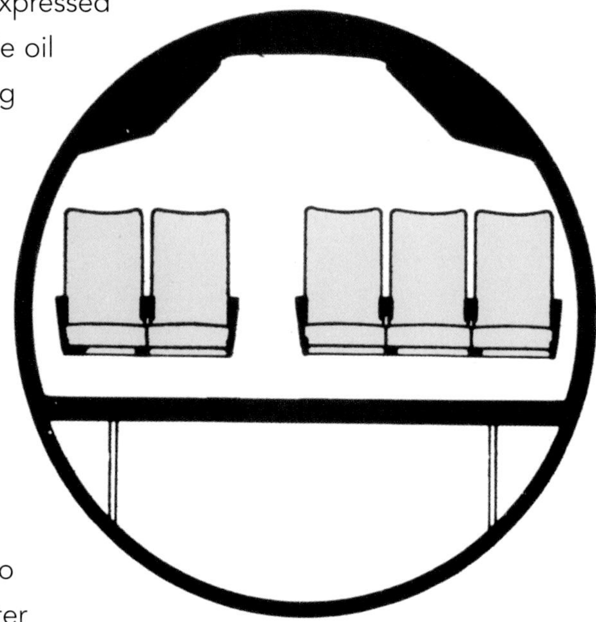

Luxury 5 abreast (B747 seats)
Double 44.5 in. (1.13 m.)
Triple 65 in. (1.65 m.)
Aisle 18 in. (0.45 m.)

In October 1973, a crisis was born. OPEC (Organization of Petroleum Exporting Countries) raised the price of oil nearly 70% and cut production in response to the United States supporting Israel in the Yom Kippur War. Less than a year after Hawker Siddeley unveiled the HS146 design the price of oil increased by nearly 400% compared to when the project was launched. Oil started at the equivalent of $3.60 for a barrel in 1972, quickly rose to $4.75 in 1973 and hit $9.75 by 1974, and inflation

skyrocketed overnight, driving up costs across the board. Oil prices continued to rise reaching $37.42 a barrel. There could not have been a worse time to launch a four engine jet aircraft. And it would be 1986 before oil prices declined to a low of $14.44 a barrel.

The oil price increases became a bittersweet gift for Hawker Siddeley, enabling it to re-evaluate the HS146 project. Internal studies determined that the aircraft's original development costs had been grossly underestimated. The original forecast of £94 million (of which the British Government would contribute 50% in repayable loans offset against aircraft sales) climbed to over £150 million in development fees due to inflation, world events and other unforeseen circumstances. This raised the break-even point of the project from 250 to 400 planes, making it economically unviable in the eyes of manufacturer. Adding insult to injury, Hawker Siddeley was responsible for any excess in the development of the aircraft beyond the initial amount it had agreed with the British Government. As a result, Hawker Siddeley needed to find an additional £56 million to move forward with the HS146 and cover the aircraft's development. With the British Government unwilling to provide additional funding, Hawker Siddeley moved to end the project swiftly, without consulting the Government.

During July 1974, the board of Hawker Siddeley began considering recommendations for the termination of the HS146 project, which claimed it was no longer economically viable. This came in complete contrast to its position during the previous year when the firm's view was that it was the right aircraft for a growing market and would be a lucrative program. British politician Anthony (Tony) Wedgwood Benn, the Minister responsible, wrote a letter to the Chairman of Hawker Siddeley summarizing his feelings: "I must tell you frankly that the Government are not at present convinced by your arguments that continuing expenditure under the contract is no longer justified." Hawker Siddeley was trying to use every reason possible, from lack of airline interest to the increase in oil pricing which resulted in inflation, to justify cancelling the project. The company's perspective was largely kept out of view of the union and the workers involved, while Benn had to come to terms with the ultimate loss of upwards of 8,700 jobs by 1978 if the HS146 was canceled. The aerospace industry as a whole had yet again become a political football.

But Hawker Siddeley wasn't being honest; it had in fact just turned down an order for eight Trident aircraft valued at £24 million from North Korea. John Rowan, divisional organizer of the engineering union Technical, Administrative and Supervisory Section (TASS) said he was told the reason the business had been rejected was that the production line would be filled with HS146s. Hawker Siddeley refused to comment or acknowledge the claims. On July 29, Hawker Siddeley and Chairman, Sir Arnold Hall maintained then firm's position, which was to cancel. Benn asked Hawker Siddeley if it had consulted the unions on the decision, prompting Sir Arnold to respond: "Oh, they have nothing to contribute." Benn countered: "Maybe not, but I will have to consult them before I can give you a measured reply."

October 1974 began with a very public battle to save the HS146 project. Hawker Siddeley had been playing its cards close to its chest. It had been hinting that the HS146 might not continue development, but in public its position was based on rumor and conjecture. Two days before parliamentary elections took place, Hawker Siddeley notified its work force that the HS146 program had been terminated. The move roiled the labor unions and angered the British Government which felt taken. Tony Benn disclosed he had been battling for three months to save the now-cancelled HS146 air bus project. Just two days later, after the Labour party won the election, Hawker Siddeley sent a letter to the Minister informing him that they were terminating the project, citing soaring costs that caused the price of a HS146 to jump from £1.7 million to nearly £5 million per aircraft.

The Rolls-Royce fiasco, still fresh in everyone's minds, had resulted in the previous Conservative Government nationalizing the company not only to save jobs but also to preserve a strategic national asset. Benn now found himself pitched into a similar unfolding debacle, with another private sector project gone wrong. The political football was now in play, with Hawker Siddeley making its view clear that the project was too great a risk for the company and its shareholders. The Labour Government, with its intimate links to the ever-powerful Trades Union movement, was determined that the interests of the work force were not going to be ignored nor was a project of national importance to be so lightly discarded.

Governments of both political persuasions had become increasingly involved in the Aerospace Industry. The creation of larger groups in the manufacturing sector and the importance to

Ease of serviceability in the field was one of the selling features for the HS146.

HS146 HATFIELD UNIONS NEWSLETTER No.15

Coffin leads workers' air-bus march

by ANDREW ADAMSON
Echo Industrial Reporter

THOUSANDS of aircraft workers marched through Hatfield today carrying a coffin and placards to demonstrate their support for the Hawker Siddeley HS 146 jet.

Most of the 4,700 workers from the Hawker Siddeley factory downed tools to walk to White Lion Square in the town centre — losing about two hours pay.

Mr Stan Davis, works convener, said: "We want to make it absolutely clear that everyone here believes in the HS 146 and wants work on it to continue. We are virtually a one industry town and if anything happened to Hawker Siddeley it would have a very serious effect on the trade of Hatfield."

The town centre meeting was addressed by Mrs Helene Hayman, Labour MP for Welwyn Hatfield, and officials of the shop floor and white collar unions.

A House of Commons motion, sponsored by Mrs Hayman and backed by five other Labour MPs, demands immediate action to rescue the airbus. They say the government should fulfil the Labour manifesto commitment to take the aerospace industry into public ownership quickly. They are also worried about reports that British Airways may instead buy the Dutch-German Fokker F28 version.

Hawker Siddeley management has decided to stop work on the new air-bus, due to fly at the end of next year, because it says the project is no longer commercially viable.

Redundancies have been delayed for a month.

Some work is going ahead and the unions are using the time to mount a political campaign for government intervention to save the project.

ECHO 30-10-74

Hawker workers at Commons

by ANDREW ADAMSON Echo Industrial Reporter

HAROLD WILSON was being pressed by Hawker Siddeley workers this afternoon to save the HS146 bus-stop jet.

A deputation of men and women working on the plane at Hatfield and other factories were hoping to see Mr Wilson at Downing Street.

Last night MPs including Mr Raphael Tuck of Watford, Mrs Helene Hayman of Welwyn Hatfield, Mr Robin Corbett, of Hemel Hempstead, and Mr Ivor Clemitson of Luton East, discussed the situation with shop stewards.

Mrs Hayman pleaded the plane's case in her maiden speech in the Commons last night.

She said it was a necessary replacement for aircraft that were going out of usable condition. It would be a tragedy if, as had been reported, British Airways purely and simply because of a unilateral management decision, made perhaps on political grounds, or on commercial grounds, had to go abroad to buy a foreign plane, which was considered inferior to the HS146.

She did not intend to go into that point because she believed she would be accused of being controversial. But on that basis, she added, the country's balance of payments position would be made even worse and the British aircraft industry run down.

She called for immediate Government action to ensure the 20,000 jobs involved in the project into the 1980s were safeguarded.

"We are not talking about 200 redundancies at Hatfield and we are not talking simply about the Hawker Siddeley group. There are firms throughout the country which are dependent for their livelihood upon the health of the aircraft industry."

She said everyone in the country had reason to be proud of the industry's performance in the past.

"We cannot allow a decision made out of spite, a decision made out of a very narrow commercial definition of what is viable and what is profitable to prejudice the future of 20,000 jobs for our people and one of our greatest industries."

ECHO 31-10-74

WORKERS START BATTLE TO SAVE BUS-STOP JET

THE CAMPAIGN to get public and Government support to save the Hawker Siddeley HS146 plane got under way yesterday with a march through Hatfield.

About 3,000 workers downed tools, pens and typewriters to go to a meeting in the town's shopping area.

Mrs Helen Hayman, Welwyn Hatfield MP, told workers and passing shoppers: "It is not going to be an easy battle to win. We have got to persuade a lot of other people."

She warned that the Government had a lot of other calls on its money and had to be persuaded that the HS146 was worth spending on.

Mr Ernie French, divisional organiser of the Amalgamated Union of Engineering Workers, said the effect of cancellation of the plane would be felt not only by people at Hawker Siddeley but by everyone working in the town, by every shop keeper and every businessman.

Experience had shown that if there were redundancies among the staff they would not end there.

In 18 months there would be redundancies of ten times the number among the shop floor, he predicted.

The company last week agreed to delay redundancies for four weeks and allow limited work on the plane to go ahead.

Although the Hawker Siddeley board says the plane is not a commercial proposition, workers are trying to get the Government to step in.

ECHO 31-10-74

MPs CALL FOR HS146 HEARINGS

By JOHN HILTON

THE CONTROVERSY over the Hawker Siddeley HS 146 project took a new turn today when Tory and Labour MPs called for a select committee to inquire into the affair.

But in the meantime they emphasise work should continue on the airliner.

They have tabled a Commons motion sponsored by Mr David Price, Tory MP for Eastleigh and a former Junior

He is backed by Conservatives Mr Victor Goodhew, St Albans, Mr David Madel, South Bedfordshire — Labour's Mrs Helene Hayman, Welwyn Hatfield and others.

If the Government agreed to a Select Committee the MPs would urge the need for a debate in the Commons.

Mr Goodhew said today that with the Government's nationalisation plans for the aircraft industry it seemed absurd that there should be any question of dropping the HS 146 simply because there appeared to be a hiatus between the private sector

for the Government to make," he said. It would be a pity to miss a potential market for at least 400 aircraft.

After attending yesterday's meetings between workers representatives and MPs at the Commons Mr Madel believed part of the trouble was a lack of information coming from the company.

"A Select Committee would bring everything out into the open," he said.

He looked for an early meeting between Industry Minister Tony Benn and company chairman Sir Arnold Hall.

them of orders from the state-owned national airlines BEA and BOAC, meant that too many vital decisions ended up being lowest common denominator compromises fashioned around domestic and insular considerations. Promising projects like the VC-10 and the Trident ended up being emasculated with neither achieving many sales outside the UK market as a consequence.

The following two months heralded the start of a very heated public debate on the future of the HS146 programme. The public, hearing of the latest aerospace mess, was already feeling burned by previous crises and Government interventions in the aerospace industry involving Rolls-Royce and the Concorde project. It was a familiar tune for British taxpayers, who watched as their pounds continued to be offered as welfare for unprofitable aerospace programs. Yet, just a month earlier Hawker Siddeley had announced an incredibly profitable quarter, begging the question of why the project was in trouble, and why the state should save it.

Minister Benn began a battle with Hawker Siddeley Chairman Sir Arnold Hall, releasing to the press letters that he had sent to Sir Arnold decrying the decision. The position each side took was pretty simple: Hawker Siddeley said that the sudden increase in oil prices, coupled with inflation and consequent increased labor costs meant it could not offer fixed priced aircraft year over year. The Government insisted that the HS146 project should continue in an effort to protect its investment and the wider aerospace industry, along with jobs in Britain. In an interview Mr. Benn cited that "inflation wasn't just an industry issue nor a British issue, but for all countries and industries it is present. Inflation is a problem no doubt, but it's not unique to the British."

There was recognition that the HS146 would be the only British civil aviation project for the foreseeable future, and it meant a lot to Britain's aerospace industry: jobs, technology, exports - everything was on the line. Workers at the Hatfield factory engaged in a 'work-in', refusing to stop work on the HS146. Employees went as far as 'seizing' engineering drawings and tooling to prevent Hawker Siddeley from halting the progress, while a public war of words was exchanged between Benn and Sir Arnold. Efforts to engage tripartite talks between the British Government, Hawker Siddeley and the labor unions, failed. Marches on Parliament began in November, along with the distribution of a report to Parliament justifying saving the HS146. Calls came not just from Hawker Siddeley workers, but from the British Government claiming that nationalizing the aerospace industry in an effort to save it from itself might be worthwhile. After all, the government was the largest customer for aerospace products in Britain, and there was strong belief it should have a seat on the board and a say in the decisions.

Publication of the letters that a Hawker Siddeley spokesman said "takes my breath away" intensified the bitterness of the confrontation between the government and Hawker Siddeley. The documents listed five Hawker Siddeley arguments to terminate the project, which the

HS146 HATFIELD UNIONS NEWSLETTER No.19

Bilateral talks on future of HS 146

TIMES 5-11-74

Bus-stop jet talks hold-up: Benn tries to clear confusion

TALKS about the future of the Hawker Siddeley HS 146 bus-stop jet involving the Government, unions and the company are expected soon.

But the form they will take was confused today following a statement in the House by Industry Secretary Mr Tony Benn.

Sir Arnold Hall, chairman of Hawker Siddeley has agreed in principle to tripartite talks, but Mr Benn told the House he had sought them without success.

Some sources indicated that there was a problem over the agenda for such a meeting. The company is apparently not prepared to discuss the terms of the contract between the Government and Hawker Siddeley.

I understand it considers the decision to terminate the contract under which the development costs — £92 million at last year's prices — would be shared between the Government and Hawker Siddeley is final.

But Hawker Siddeley's board is prepared to consider other ways of keeping the 70-100 seat short haul airliner project going.

Mr Benn told the Commons that his intention now was to have a series of bilateral talks.

Mrs Helene Hayman, Labour MP for Welwyn Hatfield asked Mr Benn to make as much effort as possible to ensure there were not redundancies while the review of the project was taking place.

Mr Benn told her: "I would be deceiving the House and the workers if I gave the impression it was within my power at this moment to safeguard jobs in a company which has taken action which I am advised is in breach of contract."

Mr Benn said the Government was still considering the future of the plane and the house would be kept informed.

He told Mr James Johnson, Labour MP for Hull West, that it had not been only the workers and managers in the industry who had confidence in the aircraft.

Until the beginning of July he had had every reason to believe the company shared the feeling.

"One of the reasons I have not felt able to agree to what the company say is that the arguments presented by the firm were not self-evidently accurate.

"We are anxious to look at it together with others concerned."

Mr Benn told Sir Derek Walker-Smith, Tory MP for Hertfordshire East, that he had sought very hard without success to get tripartite talks with the company and the unions.

"Inflation creates serious problems for the aircraft industry. On the other hand it is not confined to the British industry and not confined to this country alone.

"It would be dangerous for anyone in this country to assume the present level of world inflation made it necessary for us automatically to destroy projects upon which the long-term health of British industry depends."

Mr Benn added that these were the reasons why the Government wanted to go into the matter in greater detail.

Mr Benn told Mr Raphael Tuck, Labour MP for Watford, that legislation to bring the aircraft industry into public ownership would be introduced as soon as practicable.

Mr Tuck said the HS 146 was the only civil aircraft this country had which would fly in the 1980s. If the chairman of Hawker Siddeley was allowed to "wriggle out," the long-term result would be the loss of 20,000 jobs, perhaps hundreds of millions of pounds of exports and the finest design team in the country.

Britain would then be forced to purchase civil aircraft from abroad.

■ About 3,000 Hawker Siddeley workers, many of them from Hatfield are expected to take part in a mass lobby of the Commons tomorrow.

government would not accept. They also expressed surprise that the project was apparently proceeding according to plan up to the beginning of July, and yet was described as courting disaster at the end of July. In one of the six letters addressed to Sir Arnold Hall, dated September 12, 1974, Benn said: "I must tell you frankly that the Government are not at present convinced by your arguments that continuing expenditure under the contract is no longer justified, especially the emphasis you have on the way you think inflation may affect the UK in relation to other countries. I note what you say about unilateral termination and I would ask you most earnestly not to take precipitate action in the way." The letter repeated a request for a joint government, company and union meeting and referred to the importance of further talks.

Tony Benn had set out candid reasons why the termination proposals were not justified based on Hawker

Minister Tony Benn.

Siddeley's position on July 22. The firm said that technically the airframe and engine were proceeding well with no insoluble problems, and that in spite of a few months of slippage, costs were well within estimates and envisaged contingencies. Additionally, the letters confirmed that the results of updated surveys broadly supported market and sales forecasts made in July 1973. But Hawker Siddeley argued that no airline seemed anxious to be the first customer for the HS146. Tony Benn said that was not unknown or unusual, and claimed that Hawker Siddeley had only recently become pessimistic about its sales forecasts.

Tony Benn asserted that it was a 'matter of conjecture' whether Hawker Siddeley concerns over cost escalation were relevant. Benn's first letter continued by saying: "Cancellation of a project of this magnitude, amply and adequately backed by a Government launching contribution which allowed it to go forward, would be a very serious thing and would necessarily raise questions of national interest." He also pointed out that the HS146 was Britain's only substantially new civil aircraft project and the Government could not view its cancellation without serious concern, if the country was to remain a leading aircraft manufacturer in a market which that was likely to continue to grow. Benn had difficulty persuading Hawker Siddeley to attend any three-way discussions, but his concern to keep the debate going was clear.

With thousands affected by the decision, workers at Hatfield staged numerous public protests aimed at saving the HS146. These took place both locally and nationally, including a mass lobbying of Parliament. At stake were not just jobs, another aircraft project, or national exports, but what could be the beginning of the end of Britain's role in designing and manufacturing its own commercial aircraft.

Unlike Concorde, which was a prestige aircraft with no real market beyond a handful of destinations due to skyrocketing oil prices and environmental pressures, the HS146 was expected to sustain thousands of jobs and have a long future supplying small aircraft to regional and national airlines. Additionally, unlike Concorde, the HS146 would be solely a British venture and not built by a co-operative consortium. Recouping investment would be faster with fewer partners to share the revenue, and the coordination and spread of work would be less complex. Hawker Siddeley always felt that it could manage the entire project without partners.

The fight escalated further on November 6 with a war of words shared in a meeting between Sir Arnold Hall and Tony Benn. The debate had been quite heated in public but was as

much more bitter behind closed doors. Sir Arnold stated he had no obligation to discuss the contract with Tony Benn, while Benn argued that maybe a court appearance was in order. Sir Arnold wasn't budging, with Benn continuing on "Quite frankly, if I worked for you and was treated by you in the way you have spoken of these people, I would be suspicious of everything you intended to do." To drive his point home further Benn continued "You have wrecked this aircraft, and I am advised you're in breach of contract." The differences between the two were ideological, personal and generational so unsurprisingly there was little common ground to build on. However, in the final analysis, it was Hawker Siddeley's decision to make and the Government did not have the power to override it. It is not hard to see how this confrontation added weight to the view of Benn and his supporters that the only way forward was nationalization, and that quickly became the direction of travel.

The fight to save the HS146 and the associated jobs went on for nearly four years. Although Rolls-Royce reconsidered supplying a power plant, its interest was muted by its existing substantial commitment to

the rival Fokker F.28 project. Companies including Boeing and Aerospatiale recognized what was happening and began a recruitment effort aimed at luring engineers away from Hawker Siddeley while the aircraft was on hold. Simultaneously, they studied the market that the HS146 was poised to enter. Hawker Siddeley had already completed labor-intensive market research, and there was a risk another established aircraft company could swoop in and win most sales by building a similar aircraft.

Then there were competitors such as Fokker and the F28 Fellowship jet aircraft. The Dutch manufacturer was very vocal in its opposition to the HS146 on multiple occasions throughout its development, stating that the market wasn't big enough for more than one short haul airliner and pointing to the F.28' substantial UK content. Fokker distributed marketing material stating that four engines were a costly expenditure and that two were more economical, but their cries that the HS146 was nothing more than a job-creation scheme fell on deaf ears, and the project was unaffected. The HS146 was the ideal aircraft to replace the ageing and inefficient Fokker 28 aircraft which was quickly headed for obsolesce, and Fokker was aware of the threat.

The British aerospace industry would ultimately be nationalized and consolidated by the Labour Government's 1977 Aircraft and Shipbuilding Industries Act. A newly formed company, British Aerospace, was created and began operations on April 29, 1977. The HS146 wasn't out of the woods though. Newly designated members of the board of British Aerospace took over, and their initial reaction was to not to restart the HS146 project based on anecdotal information suggesting it was a terminal project. Instead, it was decided to make further redundancies at Hatfield while continuing to sort out which aircraft British Aerospace would move forward with. For more than a year, meeting after meeting of the new board took place, with lobbying from all sides. It wasn't until July 10, 1978 that the Minister of State at the Department of Industry, Gerald Kaufman, announced in the Commons that the 146 would in fact continue. Thereafter, British Aerospace was finally given the official green light to move forward with production, sales, and marketing of the aircraft.

Fokker emerged from the shadows again and filed formal complaints with the European Economic Community (which would become the European Union) claiming that the aircraft was a 'duplicate' of an existing aircraft, and that the British Aerospace 146 should not be allowed to move forward or be built. The EEC ruled against Fokker citing there was insufficient evidence to support their allegations.

The fight to move the aircraft from a concept and wooden mockup for prospective customers to production still had a long way to go. British Aerospace was in for a rude awakening, as the very markets it had originally designed the aircraft for were changing. The worldwide markets de-regulated, with the ripples arising from the change spreading throughout the airline industry, and that would directly impact sales of the BAe 146.

The finished and fully painted full size mockup of the HS146.

CHAPTER 02

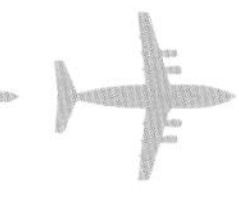

More from Four: Introducing the British Aerospace 146 (BAe 146)

In 1978, following the nationalization of the British aircraft industry, the 'new' conglomerate British Aerospace (BAe) began actively remarketing the 146 to airlines. The aircraft's technical specifications remained largely unchanged from the original Hawker Siddeley design, showing confidence the newly formed BAe had in the original aircraft. It also betrayed a reluctance to commit no more investment than was absolutely necessary, something that would undoubtedly have raised questions if BAe had really stood back and examined the project against the revised market conditions. With the recent deregulation of the U.S. market, widebody aircraft taking hold, and the expansion of hub-and-spoke systems of airlines, BAe began to refer to the 146 as a 'feeder-liner' in its sales literature, an airliner suited to providing jet service to growing markets in preference to turboprop aircraft. The BAe 146 was marketed as the self-contained airliner that could get in and out quickly, safely, and feed larger destinations from smaller (and underserved) markets. BAe felt that the market would be for around 1,500 aircraft in the 70-120 seat range, and its sales projections for the 146 were 400 aircraft with breakeven expected at 250 units. This contradicted Hawker Siddeley's previously revised revision of 400 aircraft being the break-even point.

Despite deregulation of the airline route and regulation system in the United States (with the United Kingdom and Europe to soon follow), BAe was still primarily focused on state-owned and remote operations in Africa and Asia (specifically Southeast) as the primary market for the aircraft. BAe was also rebuilding its sales force which had been run down to skeletal levels between 1974-1978. In comparison to Fokker, which had an experienced and unified sales, marketing, and customer support team, BAe was back to square one. Ironically, BAe had two very good sales and marketing teams, one at Weybridge working on the One-Eleven, and the other at Woodford supporting the HS-748 turboprop. There was some resentment and bitterness from both that the BAe 146 was selected as the firm's main future civil aircraft programme. BAe's approach was to have local product-centric teams, and thus a new team was to be built for the BAe 146 at Hatfield. This made it challenging for the firm to have a coherent position to the market, with multiple and sometimes contradictory sales campaigns being run, each with their own messages and themes. BAe would change how it marketed, sold, and supported the aircraft many times throughout the life of the 146 programme, and each reorganization took its toll.

BAe approached the market knowing that ageing, less efficient, and noisier aircraft such as the Fokker 28, the BAC One-Eleven, plus older propliners such as the HS-748, Fokker F27 and Douglas DC-3 would be ideal targets for replacement.

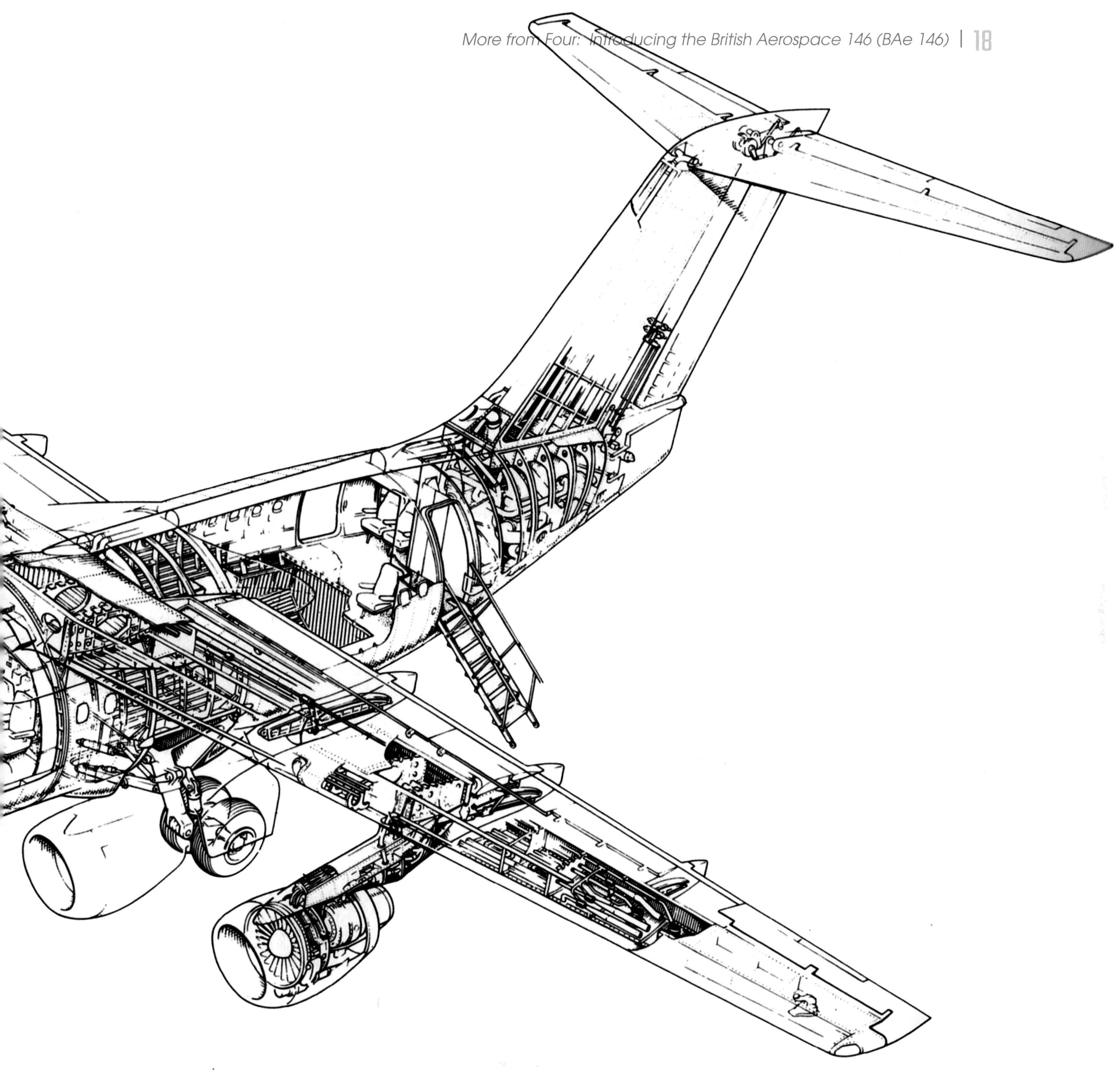

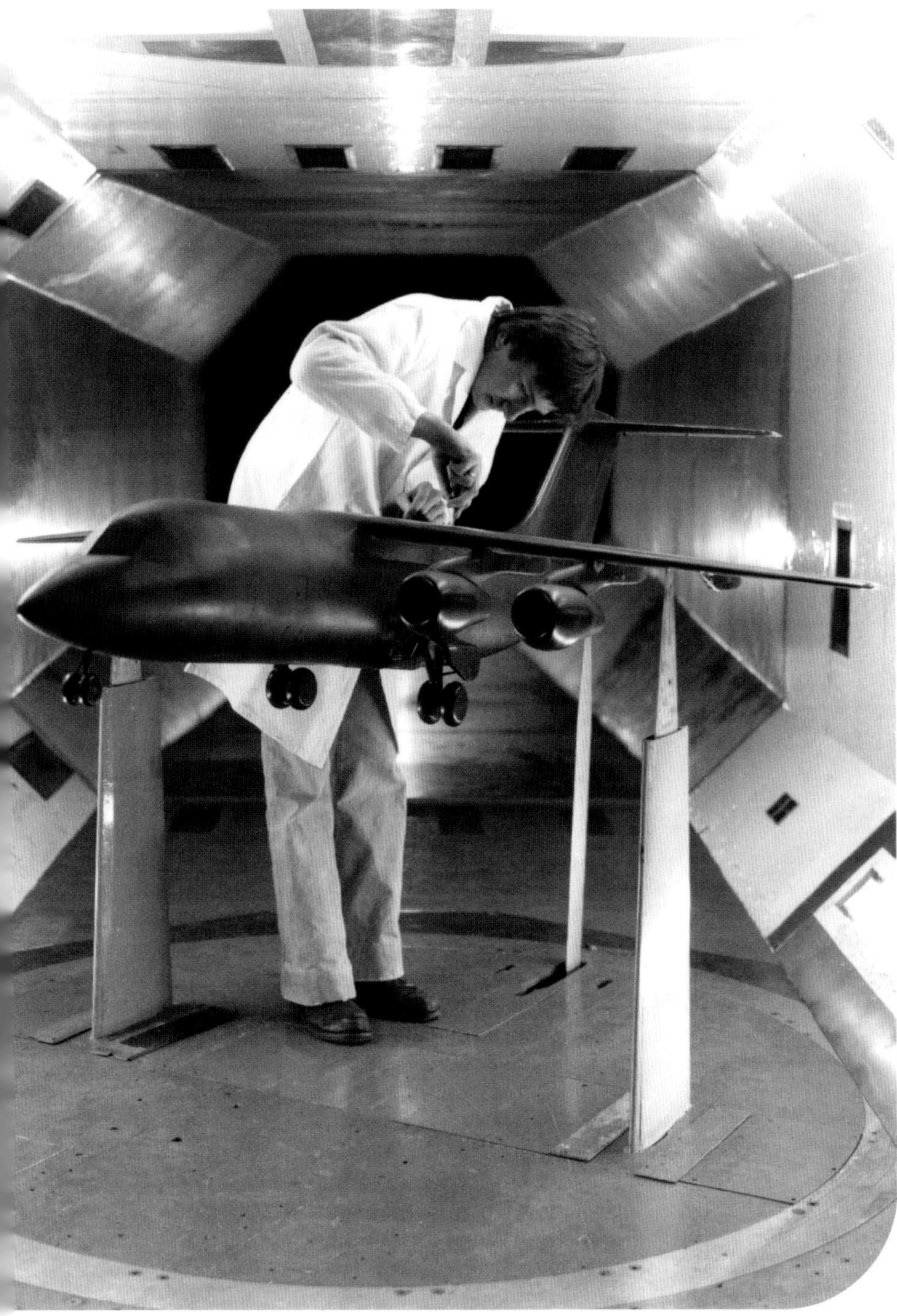

Wind tunnel testing of the HS146 using a 1/10 scale model.

And BAe wanted the 146 to be that replacement aircraft. BAe claimed that airlines operating current jet aircraft on ultra-short stage lengths (e.g. 150nm) could operate the 146 with a similar load with as much as 20% lower costs. On stage lengths under 250 miles the BAe 146, which was slower than some other jets, would not even impact block times. What BAe didn't account for was the sudden onset of 'hush kit' installations that lower the noise profile of jet aircraft already flying with prospective 146 customers, extending their operations without necessitating the need or expense of new airframes. The firm also failed to realise that many carriers operating older aircraft couldn't afford new-build planes. BAe would have to shift gears and target smaller commuter airlines that were growing and ripe for transition to jets. BAe couldn't simply offer a jet aircraft to those customers – it had to be one that offered the operating economics of a turboprop, which commuter airlines were accustomed to.

Further fallout from industrial strife in 1979 resulted in the Labour party losing control of the UK government following the election of Margaret Thatcher and the Conservative Party she represented. Shortly after Thatcher's win the Labour Party appointees within BAe were booted out, replaced by private sector individuals who were more in tune with the new government's drive to privatize the company. The success of the BAe 146 would become tied to the company itself, without further influence or money from the British Government.

Aircraft Overview

The BAe 146 was built as an air bus in the literal sense: to transport people across short distances many times a day, in markets where commercial airline jet service had not previously been economically viable, or where demand for more than 100 seats did not exist. The BAe 146 marked the beginning of a regional jet for airlines: larger than typical turboprops, but smaller than the smallest Boeing or McDonnell Douglas aircraft at the time. The BAe 146 was truly aimed at establishing a brand-new category of 70-100 seat jet aircraft in market that had not been widespread before. It was quiet, flexible, self-contained and ultra-short haul; it would have the benefits of operating into small and short airfields, including those surrounded by difficult terrain, and unpaved fields, but with jet performance. There are many suggestions that the 146 was designed as a direct replacement for ageing Vickers Viscount or Douglas DC-3, and there is documentation that supports these views. The reality was that the BAe 146 was being promoted as the "Swiss Army Knife" of airliners: an aircraft with so many attributes it would fit into a virtually unlimited variety of market segments. Of course the cliché "jack of all trades, master of none" could also apply as well.

Competitive jets, such as the Boeing 737-200 and McDonnell Douglas DC-9-10 could not match the BAe 146 for operating costs. Nor did those legacy designs possess either the STOL capabilities necessary for steep approaches and to access to unpaved landing fields, or the reduced-noise factor required to operate into sensitive city centers. The previous British-built short haul airliner, the BAC One-Eleven, was a faster yet aging and fuel thirsty twin-engine aircraft and was struggling to maintain its relevancy when faced with high 1970s oil prices. Competing newer Boeing and McDonnell Douglas aircraft offered better fuel and maintenance efficiency, and were less expensive to operate than the BAC One-Eleven. They were also part of growing families of aircraft. Multiple attempts to improve the One-Eleven with updates including new engines failed to win over airline customers. And Rolls-Royce, still recovering from receivership, was not able to invest in developing further engine programs to improve the economy and noise-footprint of the aging BAC jet.

The Dutch aircraft maker Fokker offered the primary competition to the BAe 146. Price wars were waged over sales campaigns, with Fokker initially offering the F27 Friendship turboprop and later the F28 Fellowship jet which entered service barely a decade before the 146. The 65-seat F28 Mk1000 was smaller than the BAe 146-100, but the Mk4000 variant which increased passenger capacity to 85 in 1977 was to become the most direct competitor to the 146 in the early days.

BAe cancelled planned derivative versions of the BAC One-Eleven including the Two-Eleven and Three-Eleven which were both still in the design stage. For the jet market the firm bet everything on the BAe 146 which was much further developed, showed more market promise, and had much improved political sensitivity given past struggles and its production line sited in important marginal electoral constituencies.

The 146 was built for quick turnarounds of less than 30 minutes and was designed to enable loading and unloading of passengers from both the front and rear port side of the aircraft. Baggage was loaded via waist level cargo doors a mere 2 feet 2 inches from the ground, negating the need for additional ramp equipment and staging time. The aircraft's design, with podded engines under the wings which were accessible from the ground, allowed most power plant maintenance to be performed without lifts or platforms. Every ground activity that was eliminated or enhanced could play to the quick turnaround advantage being sold by BAe, as well as the ease of maintenance.

The fuselage was designed for flexibility. With two planned capacity increases and further development beyond what was originally envisioned, the BAe 146 was initially designed to carry 71-88 passengers and four crew members (two pilots, two flight attendants) in its original short-bodied configuration (-100). The proposed 'long body version' (-200) had an increased length of 93 feet, supporting a maximum capacity of up to 102 passengers.

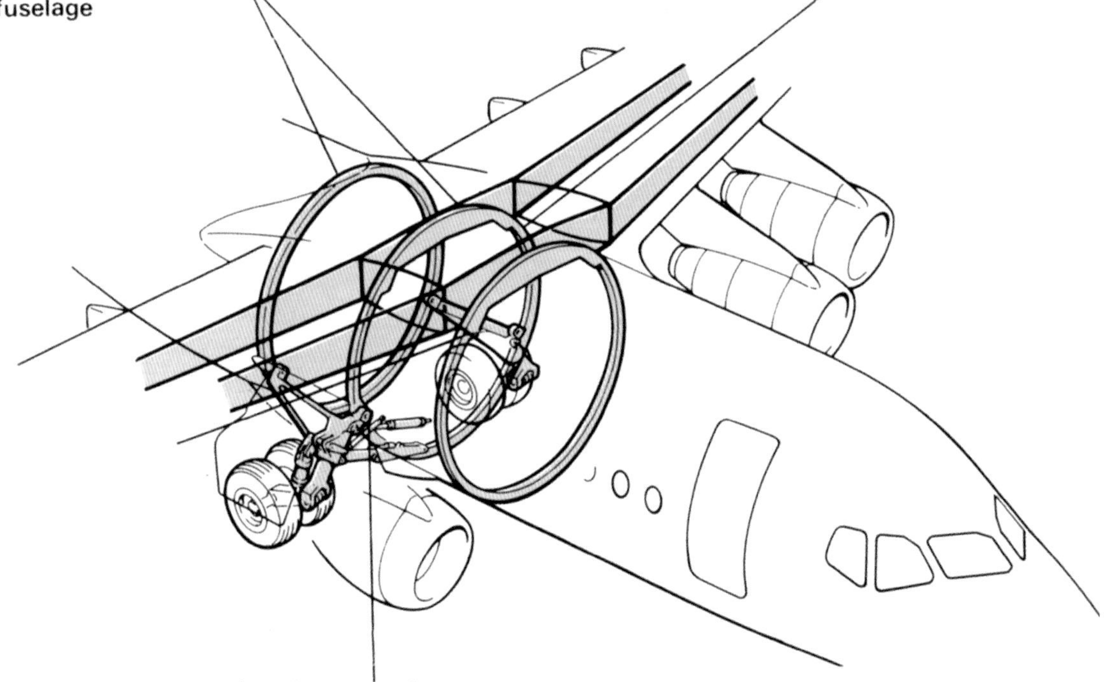

The high-lift wing with a 15-degree sweep and a span of 86 feet provided 832 sq ft of gross lift area. The wing was optimized for short haul performance and low speed handling. It enabled nearly unobstructed views of the ground for passengers, even those sitting next to the engines. The minimal amount of sweep minimized bending loads and allowed the use of a simple structure. The flaps extend across 80% of the wingspan, enabling low-approach and take-off speeds, and excellent maneuverability at four settings: 18, 24, 30, and 33 degrees. A computer controls operation of the flap positions set by the pilot and stops flap deployment in the event of a malfunction. The wing sweep limited the aircraft to a maximum speed of Mach 0.75 (High Speed Cruise) with operations generally occurring at Mach 0.65-0.70, slower than its Boeing and McDonnell Douglas rivals which often cruise between Mach 0.75 and 0.82. With the 146 typically operating over such short distances, a faster cruise speed was deemed unnecessary and would have come at the expense of greater fuel burn. The Vmo (maximum operating speed) of the -200 would be reduced by a further 15 knots, albeit permitting an increase of 5,000 lbs in maximum zero-fuel weight.

The relative locations of the T-tail and the high wings are similar to those in many low-wing aircraft. No tendency to deep stall (also known as super stall) was discovered during testing, but a stick pusher was deemed necessary to improve the nose down response of the aircraft at low speeds. At the rear of the fuselage an 'petal' air brake induces drag in order to slow the 146 down. This, combined with the large steel wheel brakes (or optionally, later to become standard, carbon brakes) and powerful lift dumpers allowed the 146 to stop on a runway. These features obviated the need for thrust reversers, further reducing engine complexity, lowering noise on arrival, and reducing aircraft weight – all of which all added up to lower fuel consumption and in theory, maintenance costs. The petal air brake at the rear of the aircraft allowed the BAe 146 to not only reduce its speed quickly, but also enabled it to serve airports with steep approaches.

The BAe 146-100 brochure listed the aircraft as having maximum ranges with a 14,000 pound payload of 71 passengers and baggage of 1,100 miles at max cruise speed (up to 1,250 miles with optional extended range pannier fuel tanks). Operating at the slower cruise speed, range would be extended to 1,550 miles (or up to 1,800 miles with optional pannier fuel tanks which extended range).

The take-off runway requirement was approximately 3,400 feet at sea level in ISA conditions, or 5,000 feet at an altitude of 2,000 feet above sea level and ISA +20 C conditions.

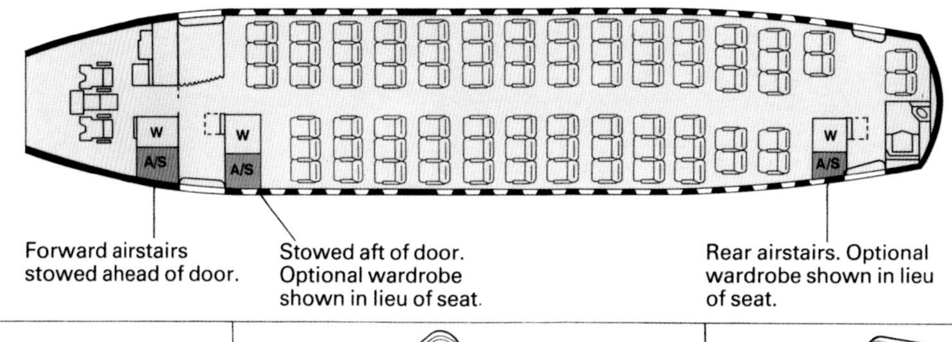

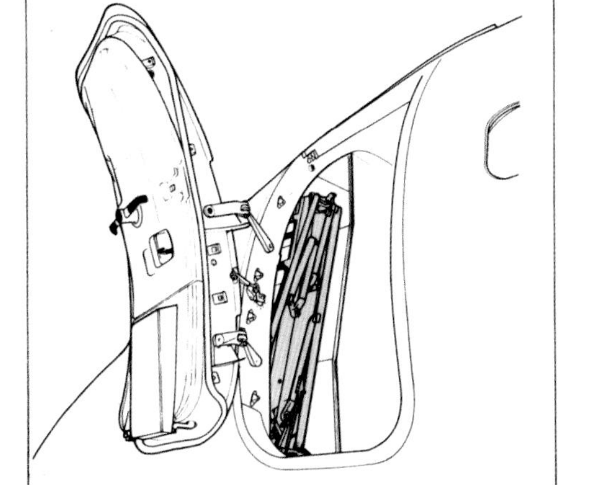

Integrated air stair operation and locations onboard.

Powerplant

Leading the way in reduced fuel consumption was the AVCO Lycoming ALF-502H engine (originally H for Hawker Siddeley). The engine core (T55) was used in a variety of applications and had gained more than three million service hours from military usage. It provided enough power at reduced noise levels to give the BAe 146 the competitive advantage in the short haul sectors it was designed for. The ALF-502 was deemed a mature power plant, which was another factor in its selection. In fact, the BAe 146 version of the ALF-502 turbine had 10% lower thrust rating than the military version, giving growth potential for the BAe 146 derivatives. The ALF-502H was a geared turbofan, the type of engine that is now powering the latest Airbus A320 "Neo" aircraft. The technology had been considered by engine makers for larger aircraft in the past, but never caught on.

When the BAe 146 program was restarted, the ALF-502H that was originally planned to be used was replaced with the newer ALF-502R3, which was capable of generating up to 6,900 pounds of thrust by the time the aircraft commenced its first test flights. The ALF-502R3 featured a seven-stage axial/single stage centrifugal compressor, eliminating the need for four more small erosion-sensitive axial stages and enabling the use of a shorter, stiffer shaft.

A two-stage air-cooled turbine is coupled directly to the compressor shaft; a two-stage uncooled fan-drive turbine and a "folded" reverse-flow annular combustor gives a cool engine exterior for improved fire zoning characteristics.

The fan module comprised a single stage fan with an additional core-engine supercharger stage. The single stage planetary helical reduction gear transmits power from the turbine to the fan, while the accessory drive (from the power producer) is externally mounted on the lower fan shroud to ensure easy access to engine-driven accessories. These features provided the ALF-502R3 with 6,970lbs of thrust and a bypass ratio of 5.6-5.7. Each engine weighed less than 1,300 pounds.

What made the AVCO power plant welcome, and later during initial operations essential, was its modular construction. Each section could be removed and/or replaced, allowing faulty modules to be swapped out to be worked on in a shop. Replacing modules rather than having to pull the entire engine improved on-field serviceability

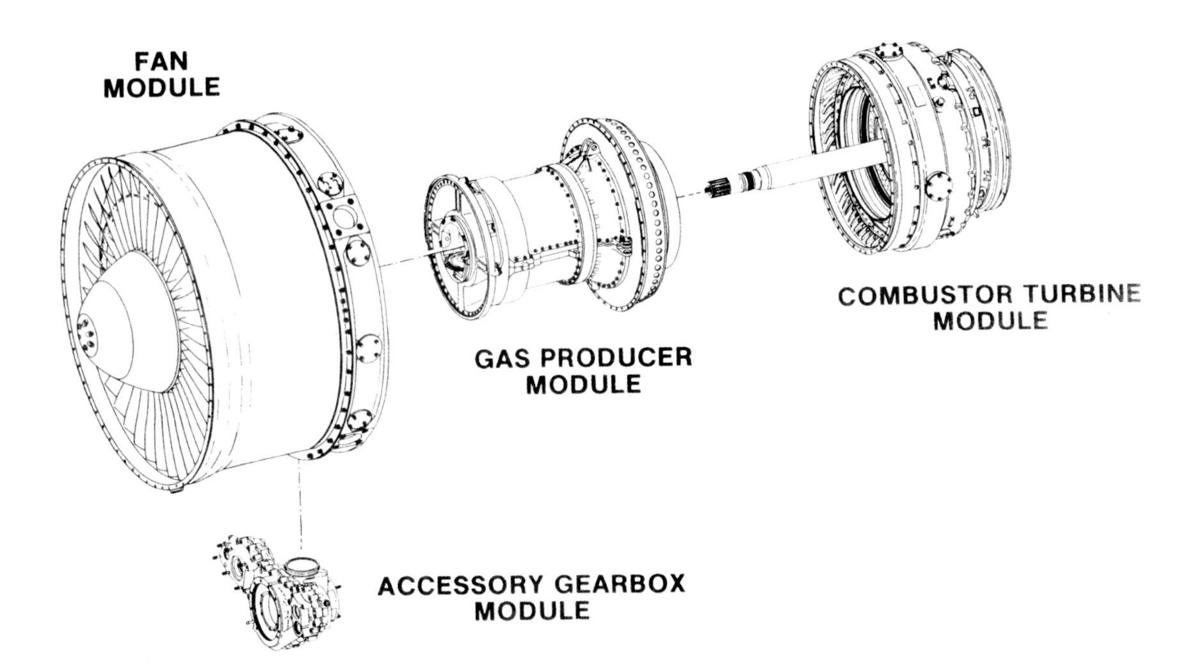

The simplicity of the Avco-Lycoming engine core.

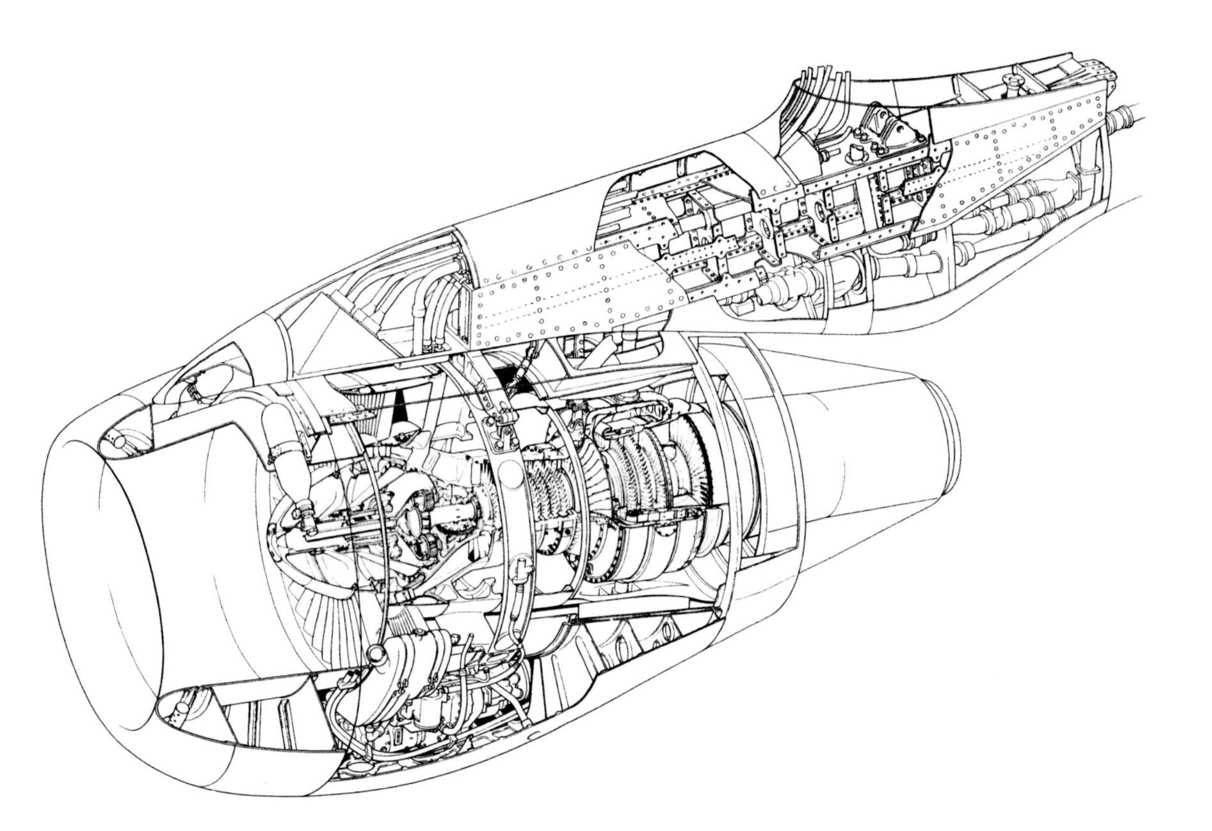

and offered a cost benefit to airlines as it also reduced spares inventories. Each engine was installed in identical nacelle, and access was provided through two large hinged doors and a rearward sliding gas generator cowl. Smaller inset doors allowed oil level checks and top offs without opening the main panels. A simple hoist system was designed to support rapid engine changes.

Additional operational maintenance was provided in three forms: condition monitoring (action only when problems arise); on-condition (scheduled on-aircraft inspection with periodic shop visits); and hard time (TBO – time between overhauls). This is again where the simplicity and flexibility of the power plant came into play. One of the primary selling points to airlines was that if the BAe 146 lost an engine on takeoff or during flight, it still retained 75% of the available power which resulted in minimal climb gradient restrictions. With a two-engine aircraft, the loss of an engine leaves only 50% of the available power remaining and almost certainly requires emergency procedures be followed with an immediate return to the airport. The BAe 146 could conduct ferry flights with only three engines operational, although in practice this rarely happened.

The other primary selling point was that the engine and consequently the aircraft was so quiet that not only could it operate into city centers, but it could also fly into noise-abatement controlled airports. This gave one airline in the United States a competitive advantage out of noise-sensitive airports in California. The BAe 146 was forecast to produce a noise footprint of 96dB on approach, a sideline of 87dB, and a flyover of 84dB. The chart produced at the time showed the BAe 146 compared to similar sized competitors (DC-9/MD-80, 737, 727) demonstrated that it was almost always the quieter aircraft.

Fuel & Hydraulics Systems

BAe marketed the 146 as having the benefit of four engines with the economics of a two-engine aircraft. While this was mostly true, there were some areas where there was some 'hocus-pocus' going on. In the majority of airliners powered by two, three or even four engines, each power plant generally drives an electric generator as well as a hydraulic pump. The BAe 146, instead of having four hydraulic pumps (one per engine) and for electric generators (one per engine), was equipped with only two of each. The outboard engines (#1 and #4) carried the electrical generator pumps, while the inboard engines (#2 and #3) carried the hydraulic pumps. Hydraulic pumps, housed in the 'bulge' in lower part of the engine nacelles, manage the power to flaps, air brakes, landing gear, and brakes. It could be argued that a regional jetliner should not need the four-engined complexity and redundancy built into a long-haul airliner. The short sectors the BAe 146 was designed for, and modularity of the engine, reduced its reliance on redundancy but operational resilience was also being considered. In the event of a failure of both hydraulic pumps, the electrical generators are able to take over (and vice versa). This is achieved through the use of electrically driven hydraulic back-up pumps. A standby hydraulic generator is also automatically activated if both engine driven generators fail and the APU generator is unavailable.

Fuel capacity for the BAe 146 initially totaled approximately 3,099 gallons (U.S.), accommodated in two primary wing tanks and a center wing tank. Later options provided for extended operations with the addition of dual 155-gallon pannier fillet tanks in the wing root, bringing capacity up to a maximum 3,409 gallons (U.S.). Average fuel burn was approximately 600 gallons per hour, around 20-30% less than 737-200/300 series and the MD-80 aircraft. The tank design ensured adequate natural water drainage and the structure is also protected against corrosion by well-proven methods and treatments.

Fuel flows to the outer wings via gravity, where fuel pumps send it to each engine. The system incorporates a pressure refueling system that enables the tanks to be filled from empty to full in under 12 minutes with automatic cut-off at preselected loads. This is accomplished from a single point on the right wing. The system has two AC driven pumps in each wing, supplying both engines and greatly simplifying the overall system, yet meeting all requirements for an independent feed to each engine. No crew management is required under normal flight conditions. Any one pump can feed both engines on its respective side, or all four engines if required. In the event of total electrical failure, a hydraulically powered fuel pump maintains the fuel supply. The tanks are protected against over pressurization by a mechanical sensor which automatically cuts off supply when the tanks are full.

Passenger Accommodation

The BAe 146-100 and -200 offered a variety of passenger capacity options for airlines depending on the seat and aisle widths chosen. This allowed airlines to install five or six abreast seating, with a 33-inch pitch, declining to 29 inches at maximum passenger capacity. The aisle width ranges from 16 inches at maximum capacity to 20 inches in more luxurious configurations. A customer could choose to use similar seats to those in the rest of their fleet such as the DC-10, 737, or the L-1011, enabling almost unlimited combinations of capacity and comfort to be provided.

The proposed 'long body version' (the -200) increased the fuselage length to 93 ft. (an additional four frames – two forward of the wing and two aft of the wing), with passenger seating increased to 102 at 6-abreast and a 31" seat pitch. Standing room inside the 146 provided for up to 6 ft. 7" in the center, except under the wing where there was a modest 3-inch reduction in height (6 ft. 4"). Compared to other narrow body jets, the 146 was more spacious than the Douglas DC-9 and Fokker F.28 series, but slightly narrower than the 737 series.

BAe 146 customers were also offered flexibility and configurability – various galley, lavatories, and integrated air stairs for remote field operations could be installed on request. The standard 146 layout offered both a single front and rear lavatory on the left side of the aircraft, along with small galleys on the right side of the aircraft next to the service doors which were suitable for short haul operations during which only drinks and a small snack were likely to be served. But operators could take some liberties in how their aircraft were configured, with options of integrated front and rear air stairs; Galleys (up to 4 positions), both for long range (3 positions) or commuter applications (1-2 positions); Toilets (up to 3: 1 up front, 2 in the rear of the aircraft); and finally storage and wardrobe closets at the front and rear of the aircraft.

The 146 provided a of interior options so each operator could add or subtract equipment in lieu of additional (or reduced) seating. BAe expected most customers to select higher density 6-abreast layouts with minimal galley and amenities. But operator-selected options soon started to cause significant difficulties to the production flow, demonstrating that theoretical offers made in a brochure could prove rather more difficult to incorporate in reality.

Using an engine bleed system from all four engines (maximum system integrity), cabin climate circulation originates from the luggage bin level of the aircraft, with air recycling occurring at floor level. Each passenger has an individually controllable air outlet above their seat. On-ground cooling performance is provided by the Garrett AiResearch GTCP 36-100M APU and air conditioning packs. On a +38-degree centigrade day (100 degrees Fahrenheit) with 45% relative humidity, the cabin temperature could be reduced from an acquired 45 degree centigrade (110F) to 28 degree centigrade (82F) within 30 minutes. With a full load of passengers and crew, the cabin could be held at 28 degrees centigrade indefinitely, maintaining passenger comfort whilst at cruise altitude. Depending on flying conditions, altitude and ISA conditions, the temperature can be maintained as low as 21 degree centigrade (69F)

The air from the engine pylons is pre-cooled and then supplied to the air conditioning packs through pipes in the dorsal spine; high temperature air does not enter the pressurized zone. There are two separate cold air units that permit separate temperature control for the cabin and the flight deck. These packs are located in the aft bay behind the pressure bulkhead with external access allowing simple maintenance or removal of the entire units. Cabin odors are eliminated by positive extraction of air from galleys and lavatories. When on the ground, the APU provides power for the air conditioning packs without the engines running. During flight, unlike conventional aircraft systems that feed in hot and cold air, pass it through the cabin, then dump it (replacing with new air), 40% of the air in the 146 can be re-circulated through filters. All of this is operated by a Normalair-Garrett cabin environmental system.

In the event one of the air conditioning packs becomes unserviceable, a ram air ventilation system will permit the aircraft to be dispatched, and the APU can even be used up to 20,000 feet to handle cabin air conditioning. Pressurization of the cabin gives sea-level conditions up to an altitude of 15,000 feet, followed by 8,000 feet cabin altitude while the aircraft is cruising at its maximum 30,000 feet flight altitude. Finally, anti-icing is achieved by hot-air bled from the engines, with no aerodynamic penalty.

Baggage and small cargo can be accommodated in two underfloor compartments. The one at the front of the aircraft is just over 13 ft. long and 3 ft. 5" tall with a capacity of 265 cubic feet while the rear compartment is slightly longer, but the tapering fuselage reduces its capacity to just 252 cubic feet. With the cargo doors so close to the ground and easily accessible, there is no need for baggage loaders. On the -200 model, the underfloor cargo volume is increased by 30% with a total capacity of 674 cubic feet.

The aircraft features two full-sized entrance doors front and rear to support loading simultaneously on the port side, whilst the starboard side features two smaller doors for servicing of galley equipment. All four doors are marked for emergency evacuation as required.

Reduced Economics & Fuselage Construction

The fuselage was built of an aluminum-copper alloy, with naturally aged L109 alloy used for the pressurized portion of the cabin. Working in favor of the BAe 146 was the commonality between the majority of major parts of the airframe. This reduced the requirements for operators to have a large parts inventory, as well as lowering the

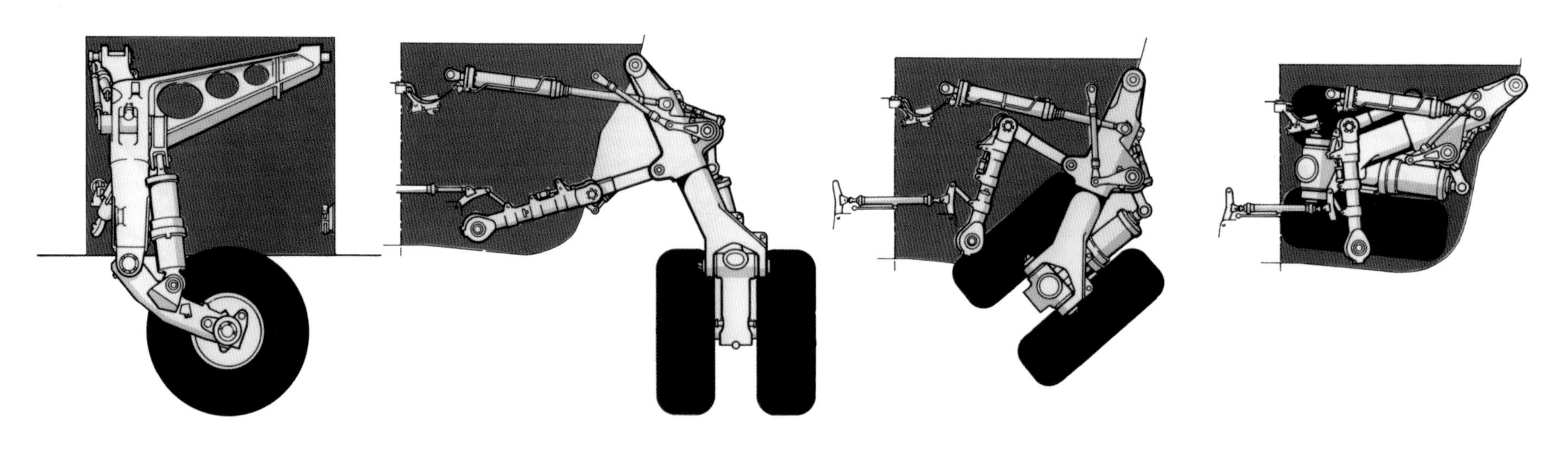

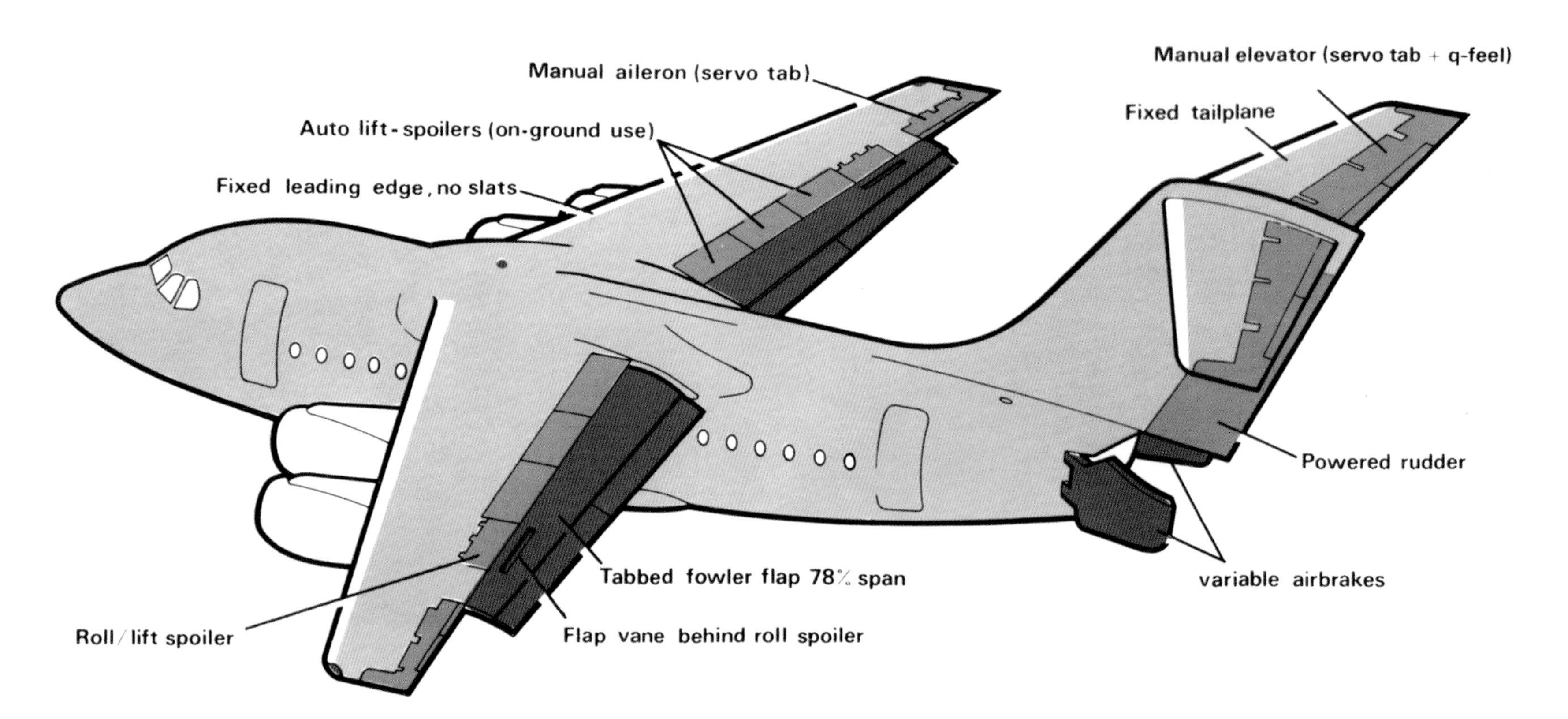

overall cost of operating the aircraft. Keeping the design simple contributed to reducing costs included no stringer/frame cleats on fuselage, and a simplified wing/fuselage joint with only a single joint in the top wing skin.

Maximum standardization of parts (wherever consistent) to keep costs low also contributed to the simplicity of the aircraft design. Examples include all six lift-dumpers and wing spoilers being identical, as well as all four engine pods being interchangeable. Flap tracks and many landing gear elements are non-handed, and seat rails positioned so that all floor panels in a parallel section are identical. The airframe was designed for long service life (proven today with examples now fighting fires), with conventional and widely-available materials used at moderate stress levels. The entire airframe structure was designed around a fail-safe philosophy with duplication of load paths and bonded crack-stoppers where appropriate. Fail-safe fuselage frames can contain a skin crack well in excess of one frame pitch in length due to extensive use of Redux metal-to-metal hot adhesive bonding, which reduces the number of rivet holes and improves fatigue resistance (eliminating more than 100,000 rivets). Aluminum/copper alloy fuselage construction, with frames that contain an inner ring to absorb bending loads, and a notched outer ring to handle shear loads, result in simpler and cleaner construction.

These well-proven protection techniques ensure freedom from corrosion in the most adverse operating environments. This is accomplished via bead or shot blasting large areas of aluminum parts with Alochrom treatment (chrome acid) followed by application of an epoxy-based primer rich in leachable chromate.

Fatigue testing was planned to show 80,000 flights crack-free, followed by a further 40,000 flights to demonstrate the fail-safety of the structure for a total of 120,000 flights. The landing gear structure was demonstrated by testing to 200,000 simulated landings. All of this coupled with the efficiency of the ALF-502 power plant meant that the operating costs per flight on short stages were up to 20% below the other short haul jets of the period. The operating cost per seat mile was also 10-15% below twin-turboprops then in operation.

Flight Deck and Avionics

The flight deck of the 146 was designed with the aim of providing an 'ideal' environment. Accommodating two pilots plus one jump seat position, the spaciousness offered was second to none. The view from the seats was panoramic, extending to 20-degrees forward below the horizon to aid landing, as well as 40 degrees above and

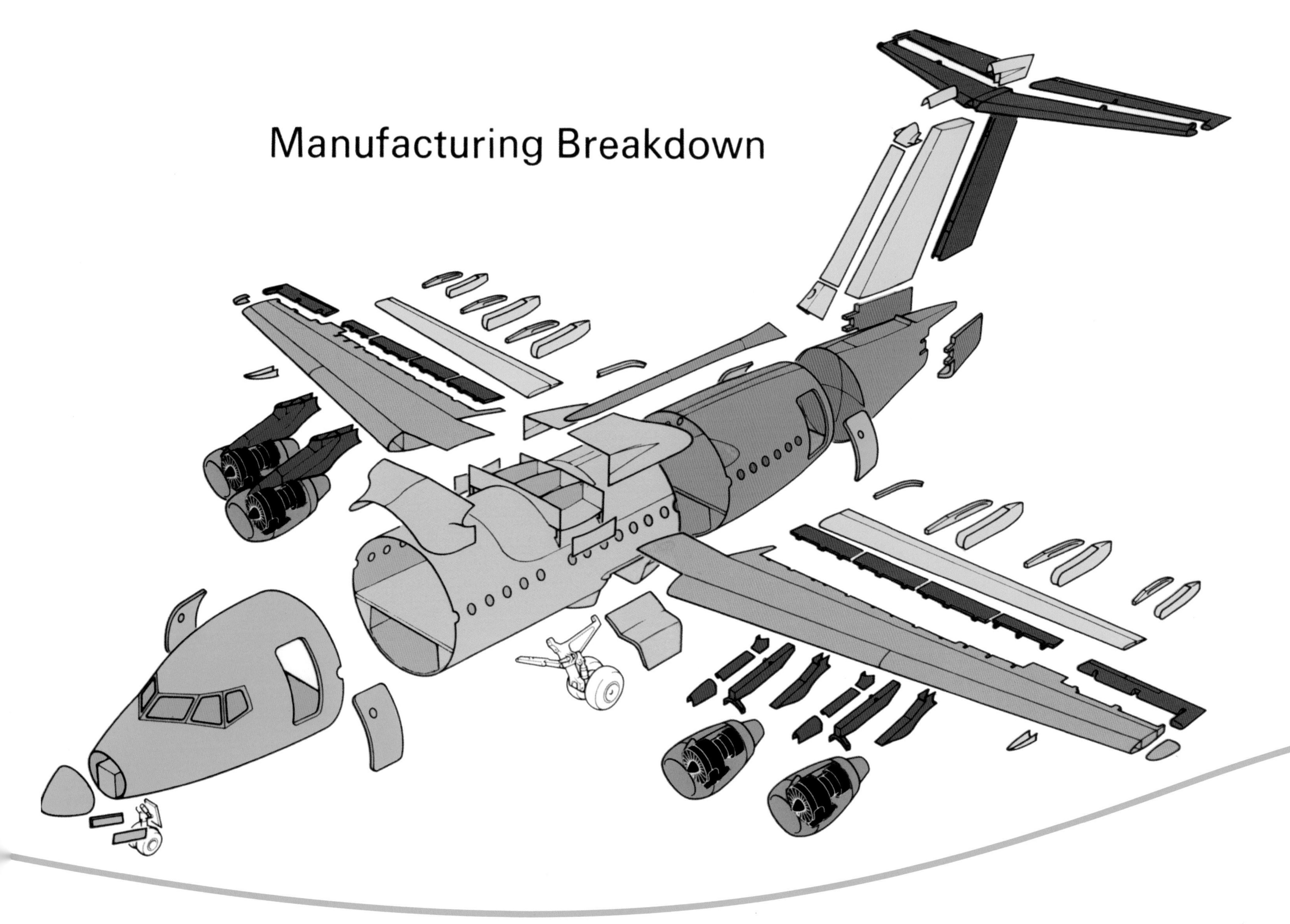

BRITISH AEROSPACE Aircraft Group Divisions

Hatfield/Chester

Weybridge/Bristol

Manchester

Kingston/Brough

Scottish

Other suppliers

Shorts

Avco Aerostructures, USA

Saab, Sweden

Dowty Rotol

Avco Lycoming, USA

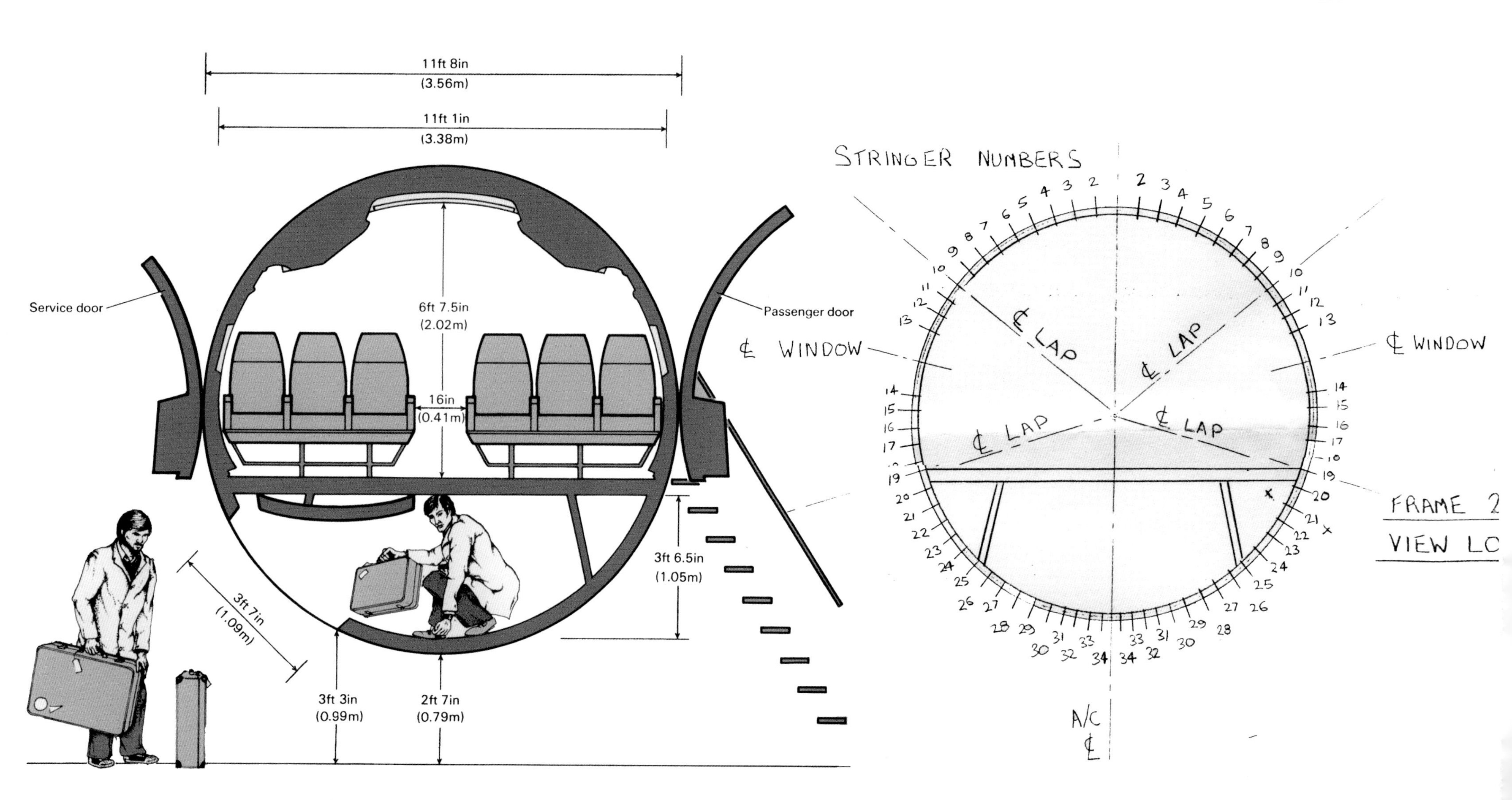

Fuselage frames and stringer numbers

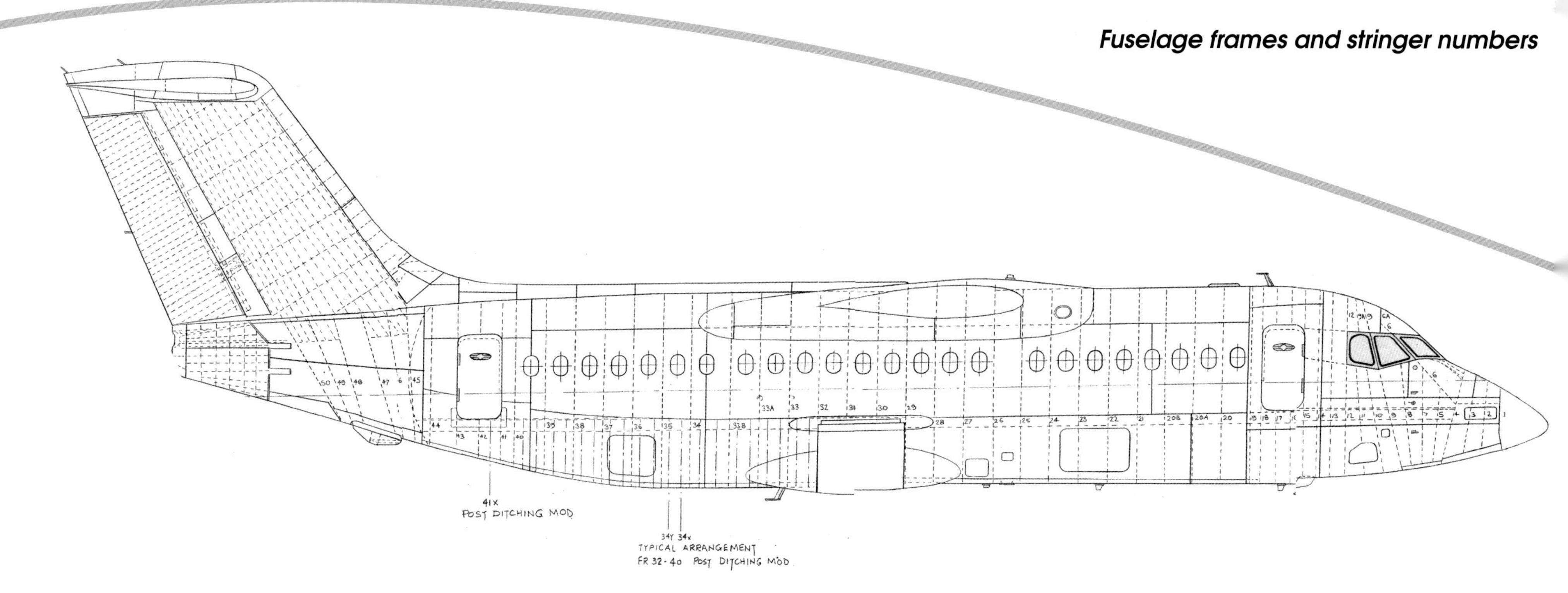

Typical assembly procedure

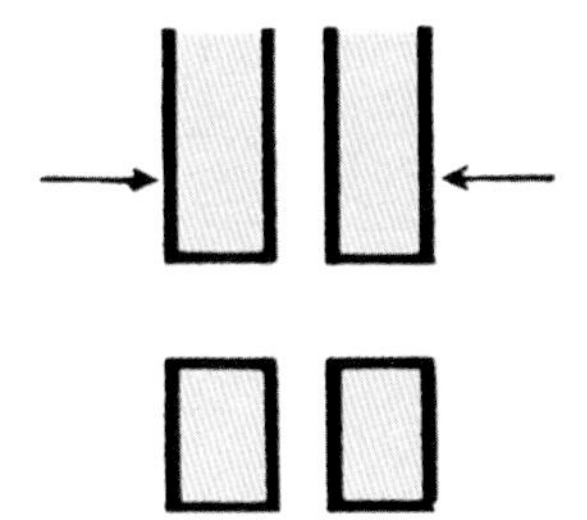

Initial assembly – drill in situ

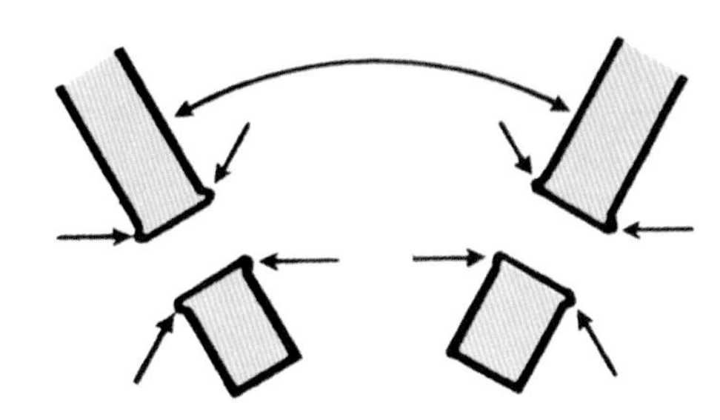

Disassemble–de-burr holes to enhance fatigue life

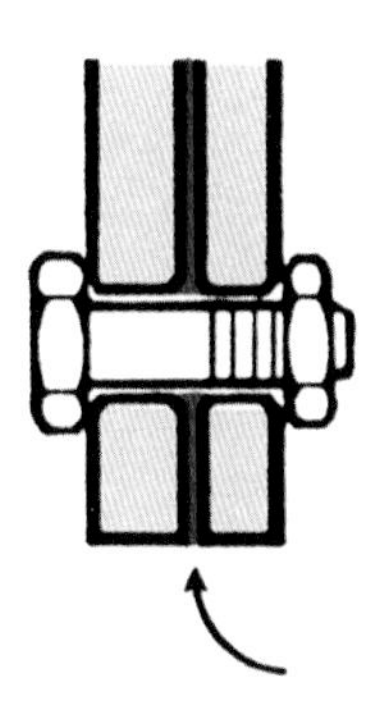

Re-assemble – wet with Thiokol interlay:-enhances fatigue life, precludes corrosion

30 degrees below the large side windows. The windows were angled so that they did not need to be open for direct vision. The aft side windows opened for ventilation while on the ground (as well for escape in an emergency) and the panoramic view also eliminated the need for the 'eye-brow' windows some airliners had in the past.

The wide body gave the design team a clear 40" of spacing between the two direct instrument displays, just 2 inches less than the 747. There is plenty of room for both pilots to work, with ample storage on each side of the seats. The primary flight instruments follow a standard "T" layout but in addition, the bar of the "T" leads naturally to the key engine instruments in the center block.

The TMS (Thrust Management System) of the 146 offers four-engine handling with single-engine simplicity. A computer controls four throttle actuators, giving the crew the ability to select automatic synchronization of the engines or to control them to preset figures of N1 RPM (fan speed); N2 RPM (core speed) or TGT (turbine gas temperature). Synchronization can be selected during acceleration or in steady flight using

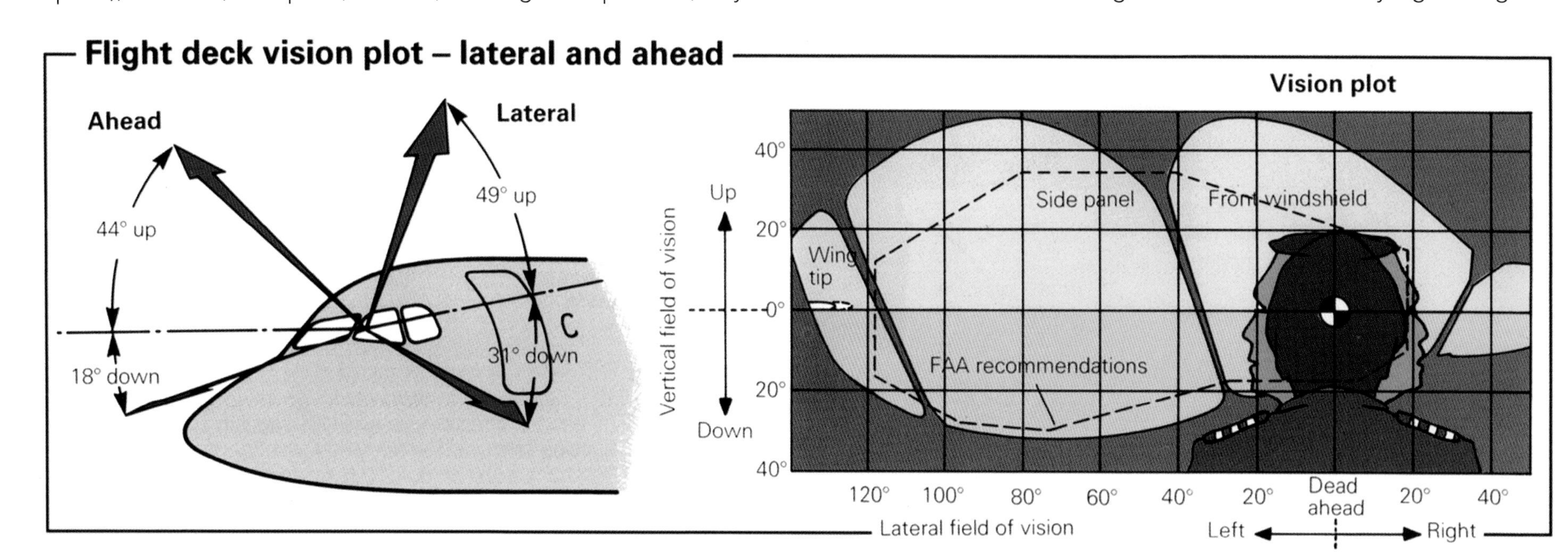

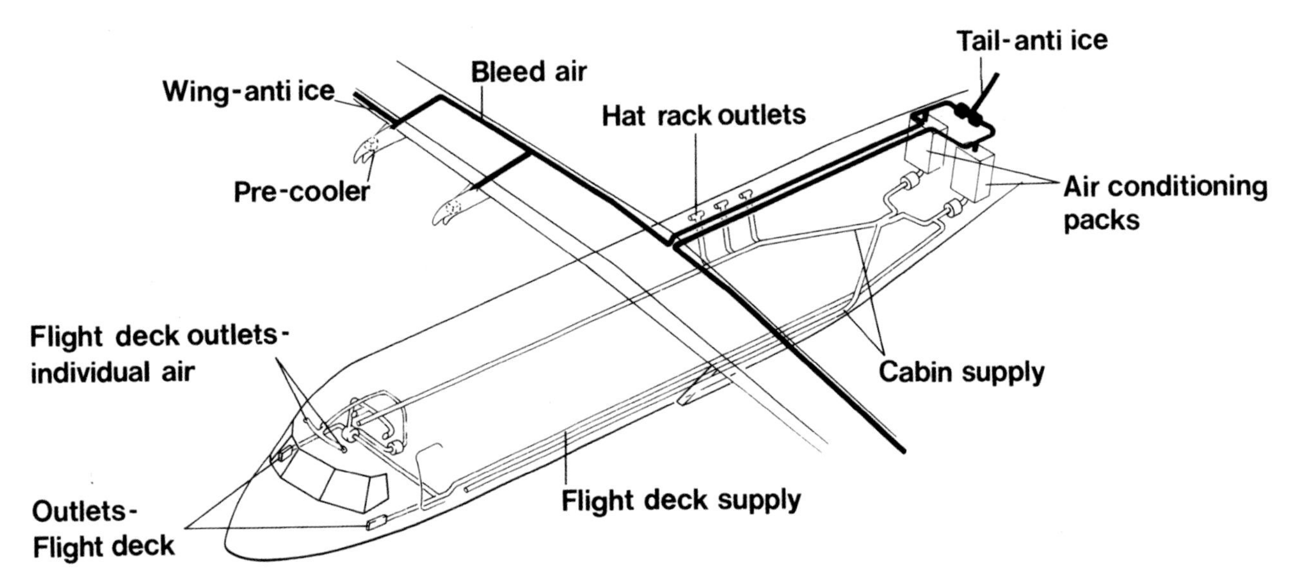

Advertisement for the BAe 146 in trade magazines

In two years, when the 146 enters service, its nearest competitor will be 20 years old...

BRITISH AEROSPACE 146

Only the new British Aerospace 146 – the first jet conceived specifically for feeder-line and local service operators – meets the needs of the 1980s with the technology of the 1980s. Its competitors are aircraft designed in the 1960s with structures, systems and engines which are now 20 years old – and with economics to match . . . In the 1980s, the new 80-100 seat British Aerospace 146 will make the difference that means success.

BRITISH AEROSPACE 146

New-technology fanjets and advanced aerodynamics enable the 146 to improve profit margins on even the shortest sectors, and to operate from short, hot and high strips or noise-sensitive urban airports.

Wide cabin and jet speeds enable the BAe 146 to bring intercontinental standards of comfort to feederline operations whilst carrying significantly more passengers than its closest competitors.

Economy of maintenance is ensured by a long-life structure, modular engines, and simple, accessible systems.

BRITISH AEROSPACE
unequalled in its range of aerospace programmes
AIRCRAFT GROUP
Kingston upon Thames · England

engine #1 or #2 as the master. The slave actuators are prevented from exceeding any of the engine limitations during synchronization.

The BAe 146 incorporated a flight director and simplex autopilot, providing a wide range of control modes including coupled approach. The standard autopilot offered the following modes: Independent yaw damping, including turn co-ordination; Airspeed, Pitch attitude, and Altitude hold settings; Heading select and hold, with turn control, as well as VOR and ILS coupling. Optional facilities included Vertical speed hold, Altitude acquire, Category 2 equipment, and finally VLF/Omega area navigation system coupling.

The standard BAe 146 included a full complement of communication and navigation equipment to meet requirements for IFR operation. The avionics were located in an air-conditioned bay, accessible via an external access door near the front of the landing gear, and an internal access door in the floor of the cockpit. Standard equipment included VHF communication and navigation radios, weather radar, ADF (automatic direction finder), DME (distance measuring equipment), ATC transponders, radio altimeter, GPWS (ground proximity warning system), compass, and altitude alerting system in addition to the flight data recorder. There were additional options to extend the BAe 146's capabilities including the addition of high frequency communications systems, VLF/Omega area-navigation systems, and SelCal (selective calling) systems.

Assembly of the first airframe

Electrical power was provided using 'proven off the shelf components' and included two separate power channels designed so that a fault in one channel wouldn't affect the other channel. Primary 115/200V 400Hz constant frequency AC power from engine-driven 40/50kVA integrated drive generators on outboard engines; 28V DC supplies derived through transformer/rectifier units; Electrical (28V DC) engine starting system enabling starting from the most widely-available ground facilities such as conventional AC or DC power trucks. Optional lead-acid batteries could provide independence via an internal start capability. The majority of components were located in the avionics equipment bay in the forward fuselage, and the equipment was modular to facilitate rapid trouble shooting and replacement. The APU could provide additional source for electrical power and engine starting.

The Master Warning System would signal any system failure via two lights: yellow (amber) or red to note the severity of the problem detected. These lights flashed on the main control panel in front of the pilots, with the corresponding system or area of detection displayed in the overhead panel.

Flight Controls

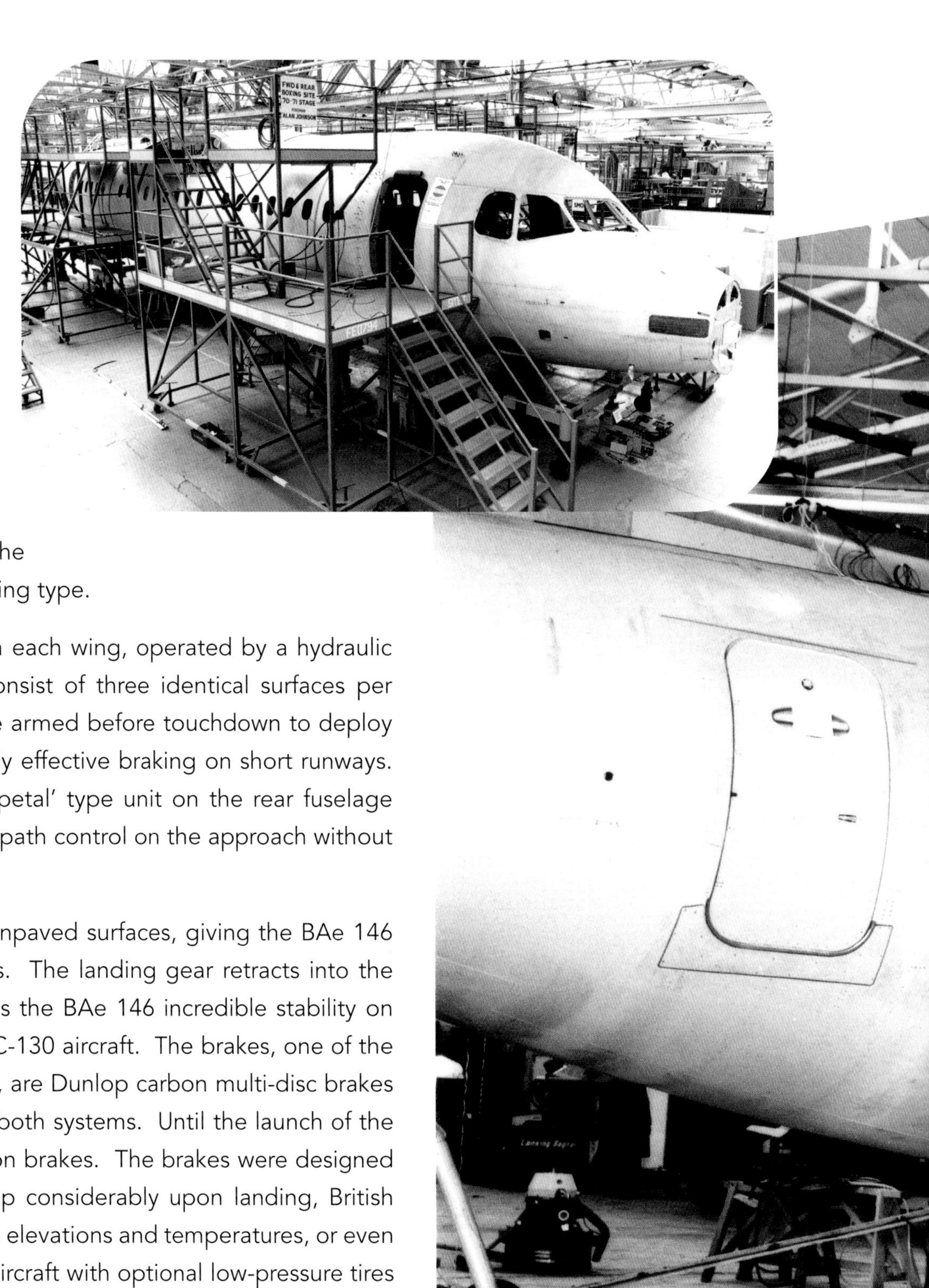

The BAe 146 elevators and ailerons are manually operated whilst the rudder is powered by two hydraulic jacks, and the spoiler on each wing by a single jack. This reduces weight and complexity (and cost) versus typical cable operated controls. Input circuits were designed to permit continued flight and landing even after a control jam or disconnect. This meant that the Captain's control yoke was connected to the left side controls of the aircraft (and the First Officers to the right side), with spring struts at the intersection in the center of the flight deck. In the event of a jam, the unaffected pilot can break out the spring strut of the affected side and retain full control of the aircraft. The horizontal stabilizer is of the fixed, non-trimming type.

Secondary controls include single tabbed Fowler flaps on each wing, operated by a hydraulic motor driving a screw jack system. The lift dumpers consist of three identical surfaces per wing, each one with a hydraulic jack. The spoilers can be armed before touchdown to deploy automatically on wheel spin-up, thus ensuring immediately effective braking on short runways. Finally, the speed brake is an infinitely variable hinged 'petal' type unit on the rear fuselage powered by a single hydraulic jack, allowing precise flight path control on the approach without disturbing wing lift.

The landing gear was designed for both for paved and unpaved surfaces, giving the BAe 146 flexibility to be used in a variety of operational situations. The landing gear retracts into the fuselage where it is fully enclosed. The wheel track gives the BAe 146 incredible stability on the ground – the track is wider than the larger Lockheed C-130 aircraft. The brakes, one of the three key items required to stop the aircraft upon landing, are Dunlop carbon multi-disc brakes with a duplicate hydraulic supply and anti-skid facility on both systems. Until the launch of the BAe 146, only Concorde had used the much lighter carbon brakes. The brakes were designed for up to 1,000 landings. Because the brakes heated up considerably upon landing, British Aerospace offered optional cooling fans for use in extreme elevations and temperatures, or even for expedited turnaround of the aircraft. Equipping the aircraft with optional low-pressure tires gave the BAe 146 unpaved and rough landing strip capability and the high-mounted engines reduced the likelihood of debris ingestion.

Hawker Siddeley envisioned growth and derivative versions of the 146 at the time of launch, including increased capacity, capability to fly steeper takeoff and approach gradients, and military derivatives. Finally, the aircraft had maximum design weights of up to 74,000 pounds at take-off (72,000 pounds at landing), and a zero-fuel weight of 61,000 pounds.

First -200 series in production

Bringing the 146 to Life

Having settled the design, BAe then determined where components for the BAe 146 would be built in order to start up the production line, and began marketing the aircraft to airlines for real. Manufacturing was spread amongst five BAe locations, and three risk-sharing partner locations:

- Production of the cockpit, doors and radome, as well as final assembly of the aircraft would take place at the Hatfield division
- Center fuselage, main landing gear doors, and wing root would be built by the Filton/Bristol division
- Rear fuselage would be built at BAe's Chadderton facility, part of the Manchester division
- Vertical stabilizer and associated flaps would be provided by the Brough division
- Engine pylons and tail plane, elevators and wing spoilers would come from Prestwick division in Scotland
- Engine nacelles would be built by Shorts in Northern Ireland
- Engines would be supplied by AVCO Lycoming in the U.S.
- Wing assemblies would be sourced from AVCO Aerostructures, U.S.
- Flaps, rudder and horizontal stabilizer would come from SAAB in Sweden

BAe, unlike Hawker Siddeley, sought risk sharing partners to participate in the construction of the 146. In 1979 the firm signed a contract with SAAB-Scania to produce the tooling for some major components, as well as for manufacturing. The initial contract for $20 million covered 20 sets of tail planes, rudders, ailerons, elevators and spoilers. Delivery for the first production set was due for May 1980, one year before E1001 would roll out of the factory for its official public unveiling.

Prior to the termination of the HS146, Aerospatiale had been given a contract to produce wings for up to 12 aircraft, along with engineering drawings, at its Nantes factory. When production resumed as the BAe 146, AVCO was selected to produce the wings. AVCO Aerostructures division (part of AVCO Lycoming) signed a risk-sharing agreement with BAe to produce the wings for the 146 at its plant in Nashville, Tennessee. AVCO was also manufacturing parts for other aircraft such as the Lockheed L-1011 TriStar and the Lockheed C-130 Hercules. The wings and the engines produced by AVCO accounted for nearly 40% of the total content of the 146 aircraft. A bit of jingoism had been forming in the United States, and BAe thought this would help the BAe 146 obtain a foot hold in the country given the number of parts and jobs that would be created locally. Short Brothers at Belfast were awarded a contract to build approximately 20 sets (4 pods x 20) of engine pods for the BAe 146, worth an estimated $6.5m USD.

BAe continued to employ a manufacturing process that used Redux adhesive bonding of components (e.g. stringers attached to the skin), carried out in an autoclave. The process used very high temperatures and considerable pressure to ensure the adhesive cured properly to provide the strength needed for the airframe. Additionally, the firm used integrally machined components extensively in the load bearing areas of the aircraft. This process minimized the number of riveted joints, which in turn reduced weight and improved resistance to fatigue. Fewer joints subsequently reduced the inherent risk of corrosion problems. The same machines that produced the wing spars for the Airbus Industrie A300 were utilized for the 146. Hatfield undertook the manufacturing of major load bearing frames in the fuselage including the wing center section ribs and spars.

The First 146

On January 1, 1980 in accordance with the British Aerospace Act, a public company was registered on behalf of British Aerospace PLC. Preparations were made to sell stock to the public, enabling the government to liquidate its holdings in the firm. Construction of aircraft E1001 was underway, having begun early in 1980. As the parts fabrication progressed, the pieces of the fuselage were assembled and began to form the first aircraft. It was not happening at a record-breaking pace, but the programme was moving forward nonetheless.

June 4, 1980 was an historic day. The BAe 146 received its first order, for three aircraft (two -100s, one -200) plus an option for three more from LAPA – Lineas Aereas Privadas Argentinas SA. The airline was an operator of short haul commuter services between Buenos Aires and La Plata in Argentina. LAPA flew small Swearingen Metro IIs and its commitment to the BAe 146 indicated it was going to expand. Unfortunately, the order from LAPA lapsed after it was unable to get approval for the intended routes. Shortly afterwards the war in the Falkland Islands occurred, and the order would have almost certainly been cancelled as a result of political pressure.

Sales of the BAe 146 were off to a slow and troubling start. The only airlines to have initially shown any interest were very small regional carriers, not the blue-chip airlines that would have established the 146 as a formidable aircraft. Outside of one U.S. based airline that also placed an order (Pacific Express), the BAe 146 was struggling to gain traction in the market. Many smaller carriers neither had the cash nor the assets to secure financing for jet aircraft, and this began to prove to be an Achilles heel for BAe. These airlines often bought used aircraft for a fraction of new aircraft pricing, putting additional pressure on BAe. The cliché "Build it and they will come" wasn't holding true – BAe was building an aircraft, but no one was buying.

Towards the end of the summer 1980, with the first production wings from AVCO Aerostructures in place, the first aircraft was taking shape. In another part of the still largely empty factory at Hatfield, aircraft E1002 was also coming together, and parts for E1003 were slowly arriving. On February 4, 1981, the British government sold 51% of the shares it was holding in British Aerospace Plc. This marked the beginning of the government's divestiture, turning BAe into a publicly traded company that would eventually have to live and die by its business model.

While BAe highlighted how easy it was to perform engine maintenance, most operators pulled the engines off to work on them..

BRITISH AEROSPACE
146

Both pages:
Fabric samples and design concepts for factory supplied interior. Clients often supplied their own materials.

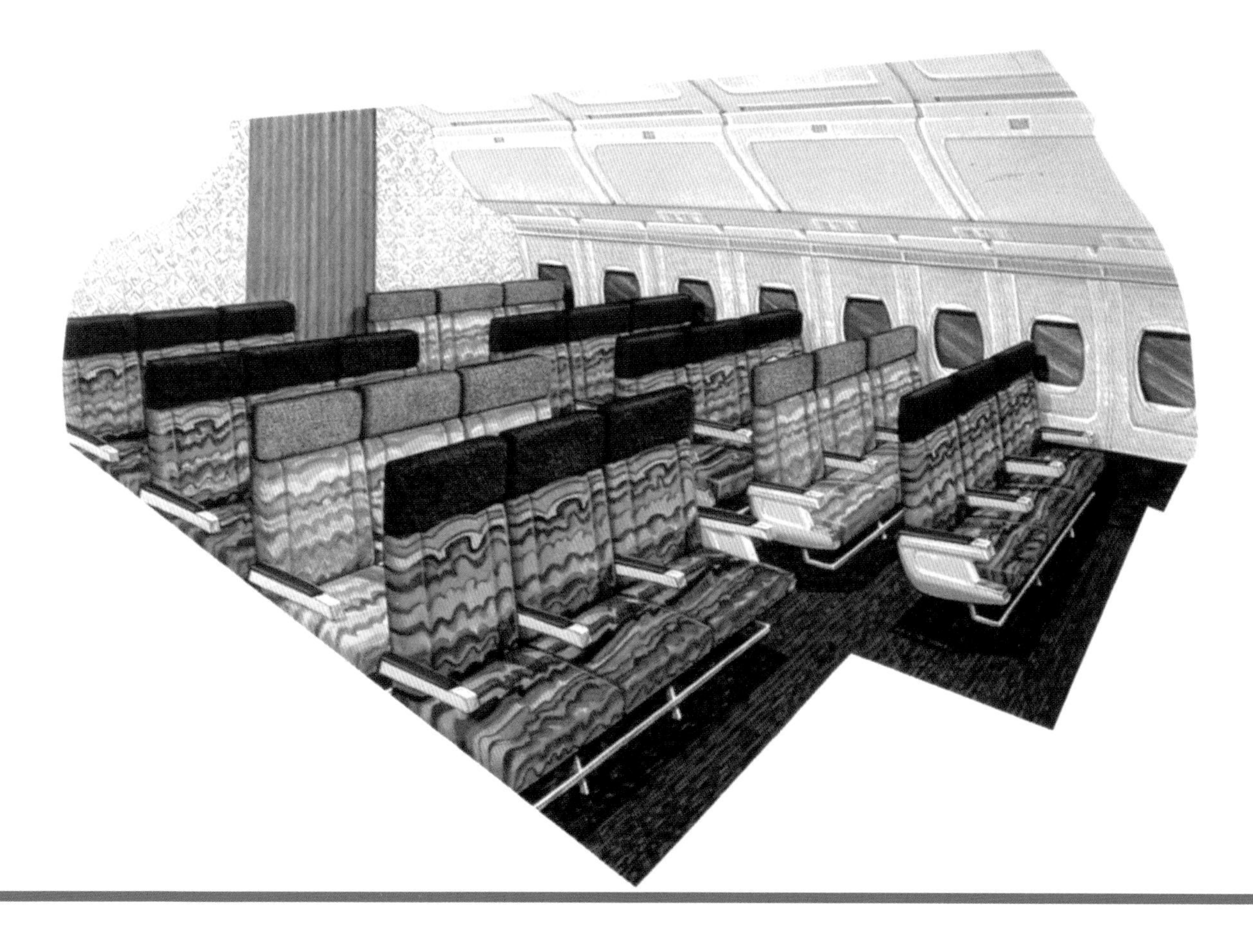

EXIT

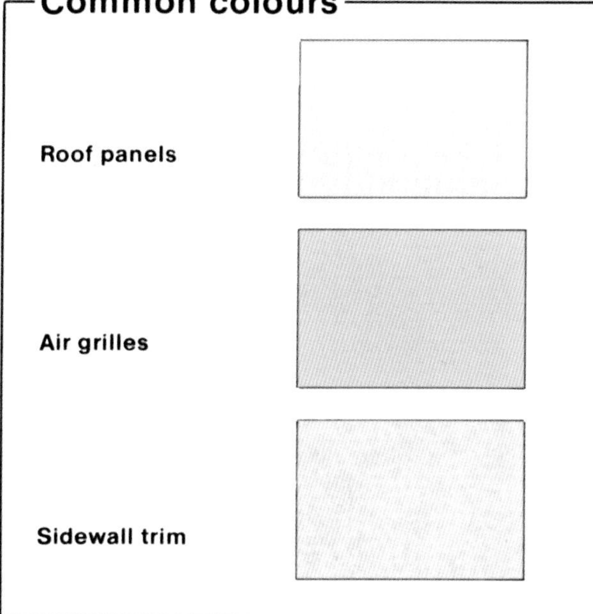
Common colours
Roof panels
Air grilles
Sidewall trim

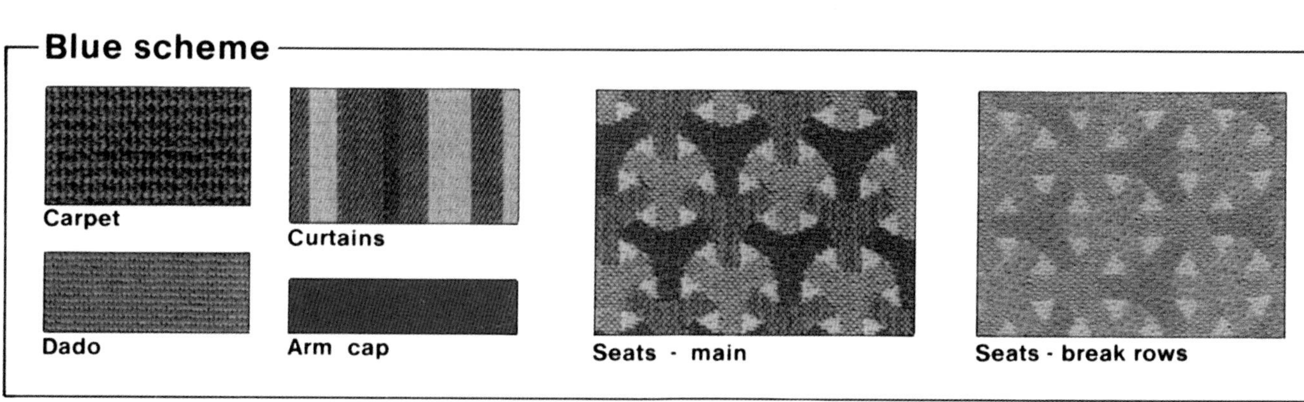
Blue scheme
Carpet
Curtains
Dado
Arm cap
Seats - main
Seats - break rows

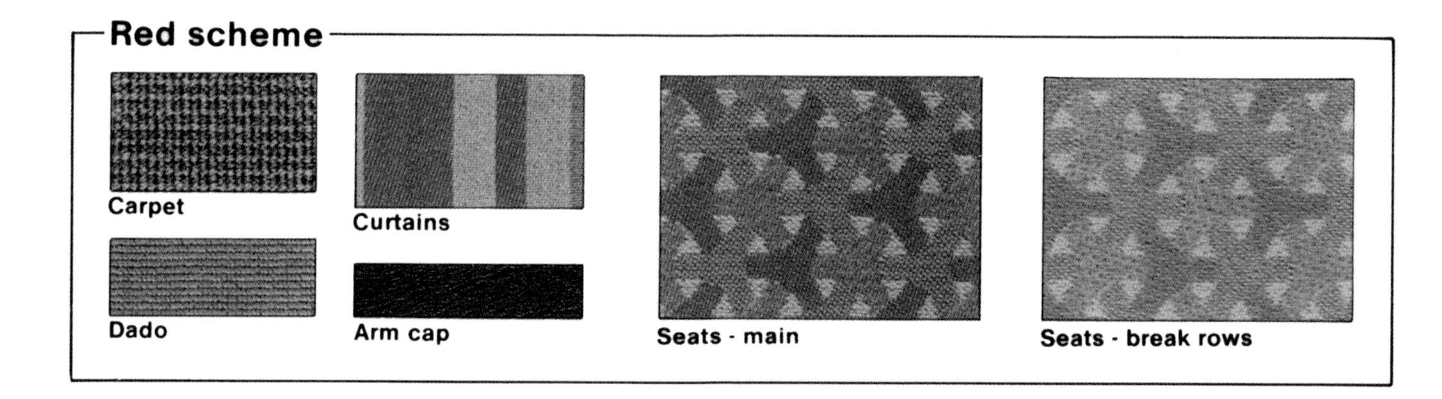
Red scheme
Carpet
Curtains
Dado
Arm cap
Seats - main
Seats - break rows

By March 1981, the first aircraft rolling down the assembly line was nearly completed. Flaps and flap tracks had been attached to the wings, along with the engine pylons and spoilers. Engines had already been delivered by AVCO Lycoming and were awaiting installation on the prototype 146, E1001. The undercarriage doors along with the rudder and fin had been added to the airframe. But with two cancelled orders, and a third placed by an airline that was on shaky ground (Pacific Express for six series -200 aircraft), the BAe 146 did not have a credible order from a mainline carrier.

E1001 was on schedule to roll out in May as originally planned, after which its structural and functional tests would commence. These were the final tests before the 146 could take to the air for the first time. Simultaneously E1002 and E1003 were being assembled from parts arriving from different facilities around the United Kingdom. The center section for E1003, delivered from Filton, had all the necessary wing-box structure along with the hydraulic systems in position, so that the wings could be attached quickly, showing how sub-assembly construction could work

Cabin mockup for sales and marketing of a series 200. Note the 3-2 seating arrangement.

in practice. Aircraft E1002-1007 had been earmarked to be -100 series, BAe believing that the smaller derivatives would be the dominant sellers.

E1001 was registered G-SSSH, a registration idea that came from an internal competition to signal in any language the 146's quietness. Surprisingly four identical entries were received with the winners being Mike Cull (Wind Tunnels), J.A. Barker (Laboratory), L.J. Underwood (Design) and D.F. Cook (Technical Publications). Each received portraits of the 146 signed by Chief Test Pilot Mike Goodfellow. There were plenty of other good suggestions, but unfortunately they had already been assigned to other aircraft by the UK CAA.

With the assembly of E1001 nearing completion, it was subject to electrical and instrumentation fitting, followed by the first "power-on" of the aircraft and completion of the flying controls. Aircraft E1002 entered the paint shop in May 1981 and was also subject to vital pre-flight structural stiffness tests. E1003 wasn't far behind in having the remaining aspects of its airframe completed.

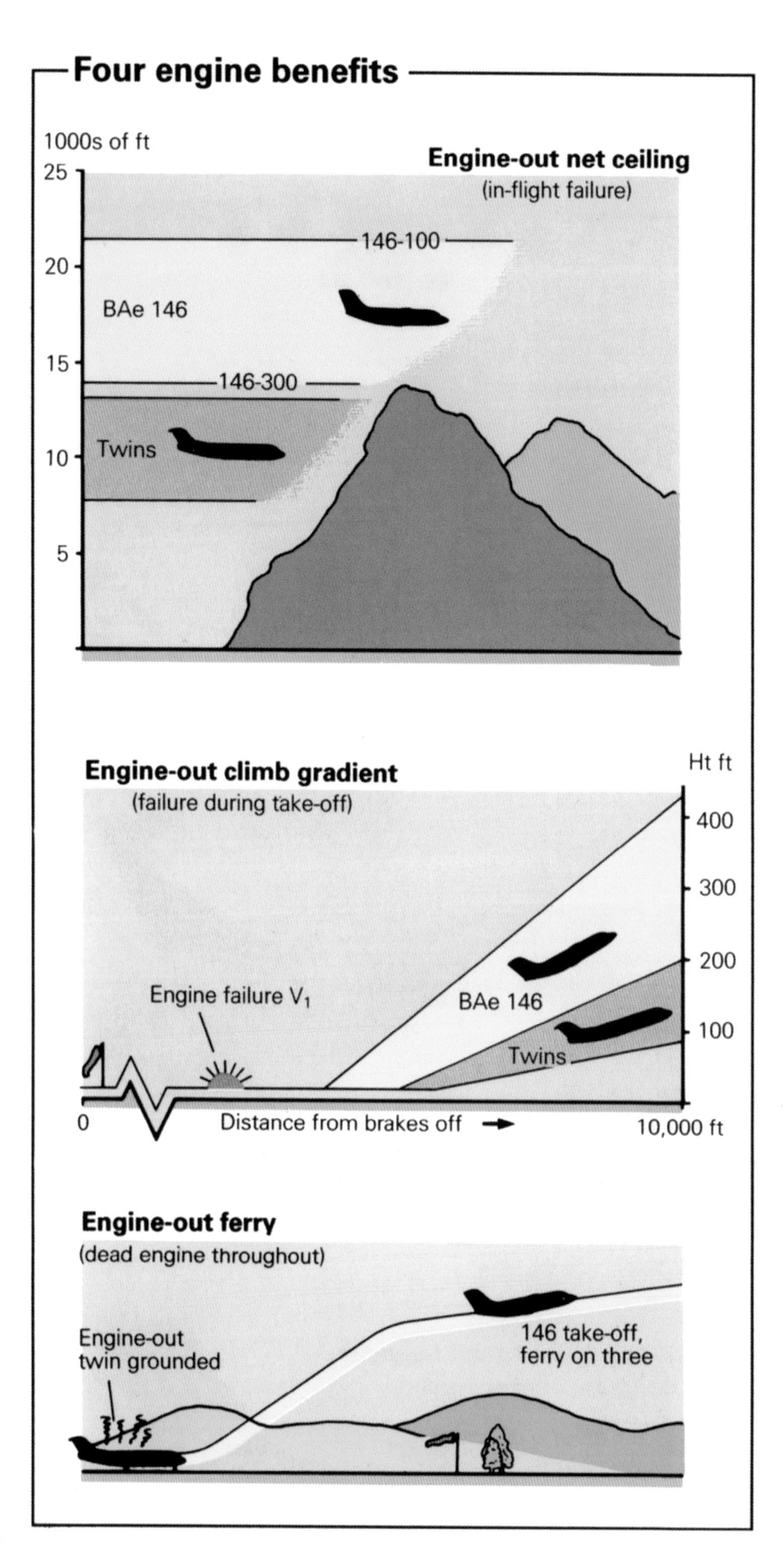

Extoling the virtues of four engine performance.

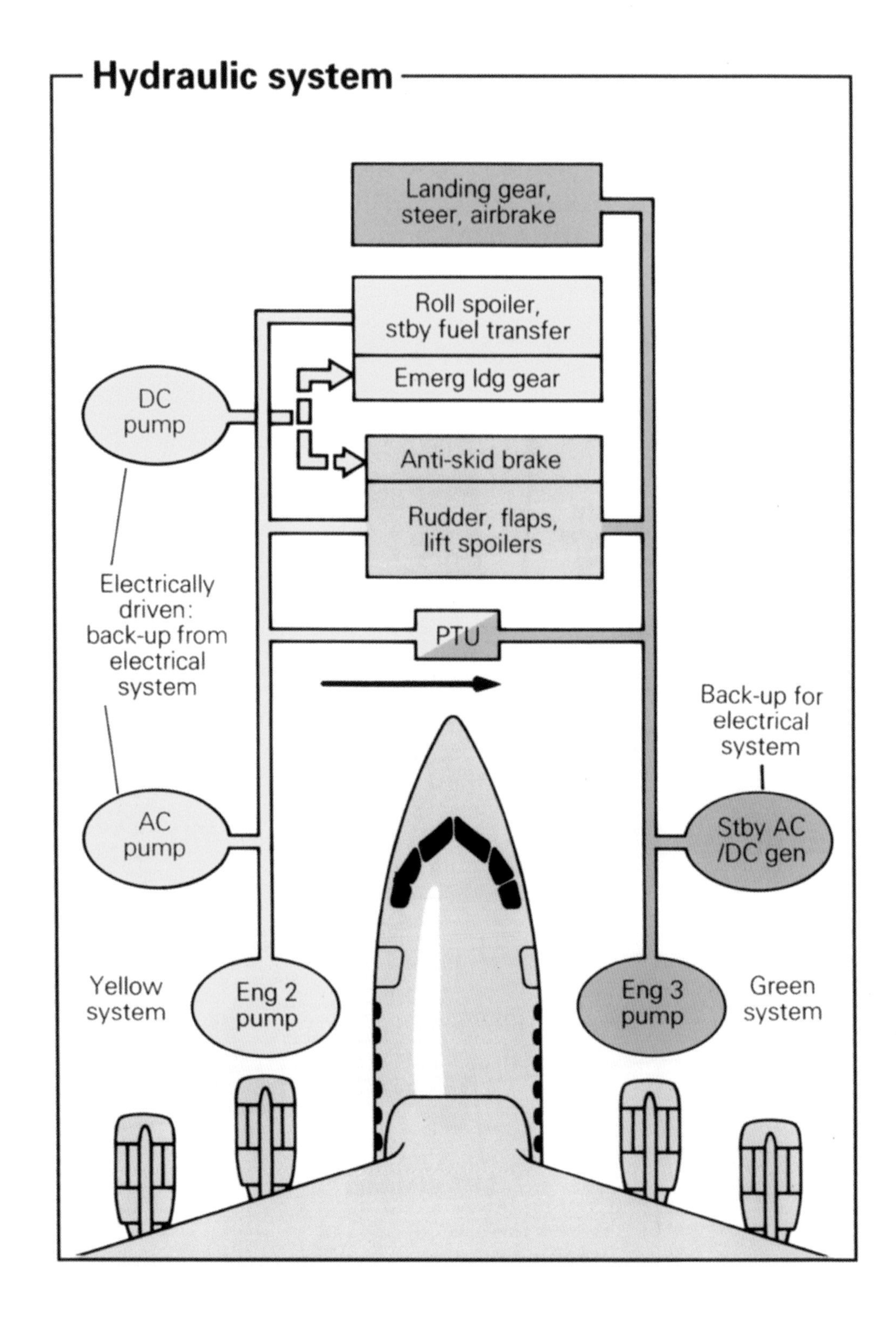

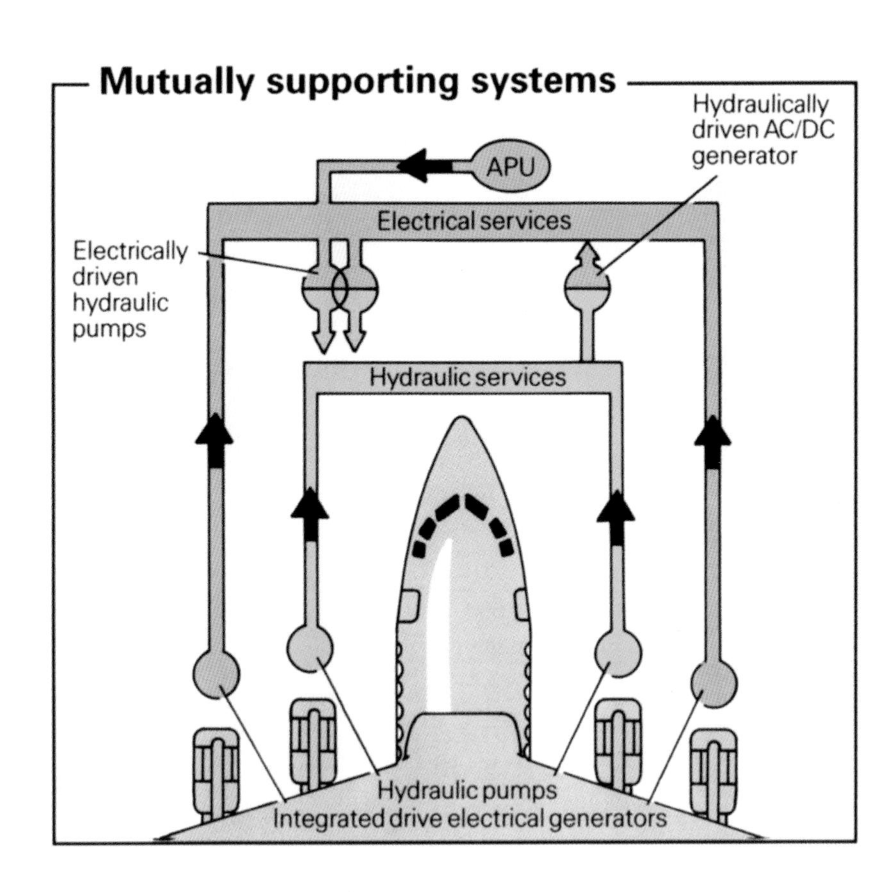

BAe still believed that the biggest market potential was in Asia and the Third World, so the sales teams for both regions were larger than those covering North and South America. The salesman were out pursuing customers, with the general response "we have a big order just around the corner" frequently being uttered to management. In reality, the BAe 146 was facing an uphill battle: airline maintenance departments were not fans of four engines from a cost and perceived complexity perspective; Flight Operations were dubious of block time performance because of the lower cruising speed; and large airlines tended to be very conservative with every aircraft purchase. This put BAe at a very large disadvantage, as it had to convince every department at every level to secure a sale. Lacking an existing customer base for the aircraft made overcoming these reservations difficult, but BAe was scrambling hard for customers, and hoping for a roll-out day miracle.

146
BRITISH AEROSPACE
G-SSSH

CHAPTER 03

Hey, Look Me Over – The First BAe 146 aircraft

May 20, 1981, on a dark and cloudy day aircraft E1001 was rolled out of the assembly hall at Hatfield. It was towed to a static display area, where the press, VIP's from manufacturing partners, and other contributors were waiting. Employees on duty were permitted to leave their posts at 11:45am to attend the event. Many lined the grass along the north taxiway to watch the first BAe 146 being moved into its final position in front of the flight hangar on the north apron. More than 4,000 people were in attendance as the Central Band of the Royal Air Force played "Hey, Look Me Over".

The aircraft was subsequently moved inside the flight hangar as onlookers sought relief from the threatening rain, while Chairman Sir Austin Pearce gave a speech introducing the BAe 146 to the world. Sir Austin stated the BAe 146 "rests here at Hatfield with the men and women who fought off all sorts of objections to produce the 'plane you see here today." Sir Austin also addressed the unmentioned elephant in the room: government subsidies. In his speech he continued by stating "I would like to correct one other impression – this aircraft has not been built with Government money. We have had no free gifts, no hand-outs in developing this aircraft. We have put our own money on the line." This was only partially true, given the nationalization of the industry and the loan supporting the programme. However, the Government had already begun the firm's transformation to a public company with a sale of 51% of the stock it held just a couple months earlier, and the sale of the remaining shares by 1985.

The roll out received significant press coverage from the major television and radio networks, in addition to print media. To ensure proper coverage across the Atlantic, BAe used the event to announce a sales commitment from the first U.S. customer, regional carrier Air Wisconsin, which placed an order for four aircraft and four options worth $70 million. Air Wisconsin would be the launch carrier for the larger -200 series aircraft. U.S. based media were quick to point out that the small jetliner was designed to capture a market ignored by U.S. aircraft manufacturers. Air Wisconsin President Preston H. Wilbourne said "Air Wisconsin selected it because of its advanced technology design, good fuel efficiency, low operating costs, and the fine passenger amenities which it affords."

The airline, based in Appleton Wisconsin, was growing rapidly and it wasn't far from Chicago, a hub that it served intensively. Tom Helm, Lead Pilot of Air Wisconsin brought the BAe 146 to the airline's attention. The aircraft would be the airline's first foray into jets, with deliveries expected to begin in March 1983. Preston Wilbourne, in conjunction with local businessman, sought jet service and the BAe 146 would serve that role. Air Wisconsin planned to operate its aircraft with 100 seats and would go on to become one of the biggest and longest running operators of the type, flying all three passenger variants.

In fact, work had already begun on the first -200 series airframe, E2008, that it was originally envisaged would be delivered to Air Wisconsin. However, the rear fuselage was damaged during Vmu (minimum unstick speed) take off trials at Bedford on September 29, 1982 and was kept as a demonstrator in the BAe fleet.

Air Wisconsin's commitment was the first order that came to fruition, and it became the first U.S. airline to operate the aircraft. This initial sale suggested BAe's belief that the series -100 would be the more popular derivative was already becoming questionable. British Aerospace was struggling to achieve any more sales, with the only firm commitments being for the series -200 aircraft while three series -100 aircraft were on the assembly line without customers. Hawker Siddeley's original forecast that the series -100 would be the most desired size was not panning out.

Testing and Proving of the 146

Prior to the first flight of any aircraft, significant testing occurs on the ground to simulate flight and operational profiles. A large-scale trials program was put into place for the brand-new 146 aircraft. A large rig known as an "iron bird" was built featuring an exact replica of the flying controls and hydraulic systems, which was used for testing everything from the flaps and rudder to the elevators. Portions of the rig also provided resistance that replicated inflight loads. It was not a flight simulator in the traditional sense, but comprised a series of gears connected to cables and pulleys. The landing gear was cycled repeatedly to ensure it functioned as expected; wings were pushed beyond their service limit to ensure the aircraft would remain airworthy; the fuselage (four test units) were submerged into a water tank to ensure pressurization could be maintained over tens of thousands of (simulated) flights, well beyond the designed service life of the airframe. By the time of the delivery of the first commercial version, over 20,000 simulated flights had been completed. The trials continued after the 146 entered airline service and by the year 2000 the test specimen had been subject to 180,000 "flights".

Before the first flight, the flight crew trained in the mock simulator. This enabled them to understand how the aircraft was supposed to fly, and what they could expect when they flew the real aircraft. The flight test crews rehearsed the tasks scheduled for the test flight on the eve of the 146's first trip aloft, right down to going over the actual flight path using a BAe 125 aircraft. In parallel, E1001 was used for ground tests including taxiing and turning, powering up and down as well as heavy braking. In fact, during one of the taxi tests the first prototype 146 became airborne briefly, its lift capabilities proving to be extremely good.

First Flight

With a crew of four headed by Divisional Chief Test Pilot Michael Goodfellow, along with Deputy Chief Test Pilot Peter Sedgwick, Assistant Flight Test Manager Roger de Mercado, and Senior Instrumentation Engineer Ron Hammond, the first BAe 146-100 taxied onto the runway September 3, 1981. Once the fog cleared and gave way to the sun, E1001 made its first test flight, lifting off the Hatfield runway at 11:54am BST after a 17 second take-off roll. The aircraft lifted off at a take-off weight of 64,000 pounds including 14,000 pounds (~2,100 gallons) of fuel. Ballast was used to obtain an accurate weight and center-of-gravity position.

A BAe HS.125 chase plane followed E1001, visually monitoring the performance. After the landing gear of the BAe 146 was retracted for the first time, chase plane pilot Peter Tait remarked "I think we have ourselves a pretty aeroplane", noting that he could now see the smooth lines of the all-new feeder liner. Climbing through 10,000 feet and turning to the North-East with a 95-minute flight ahead of them, the crew began running through the detailed tests that were planned for the initial trip aloft. These included cycling the landing gear several times as well as evaluating the feel and effectiveness of the flight controls. Pilot Mike Goodfellow noted that the control characteristics were "very responsive", and that engine handling and fuel flows were noticeably low.

Prior to the 146's first ever landing, the flight test team made two low passes at 300 feet over a large crowd at Hatfield. Making a perfect touchdown at 1:29pm, Pilot Mike Goodfellow summed up the flight as "Remarkably stable, very responsive, and delightfully quiet." The first flight marked the beginning of a lengthy test program aimed at achieving a British Certificate of Airworthiness, a prerequisite for the aircraft to be delivered to and operated by airline customers. All of the flight test results would be submitted to the CAA (Civil Aviation Authority) for review and approval. Those same results would also generally satisfy the U.S. FAA (Federal Aviation Authority) and lead to rapid approval for the 146's commercial use in the USA following type approval in the UK, assuming no adverse issues were uncovered during the testing cycle.

Three aircraft carried out the bulk of the development and certification program, which required a minimum of 1,000 flying hours. Aircraft E1001 was used for handling trials including control responses, trim change assessment, low speed behavior, and stability checks. Structural damping was validated progressively to allow the airspeed limitations to be increased to those planned for the flight manual. By the end of 1981, aircraft E1001 was excelling in every aspect it was being tested. Stall speeds at all flap settings provided good low speed handling without undesirable flight characteristics, coming very close to predictions. The first series -200 was also coming together, with plans envisaging being certificated and delivered shortly after the first series -100 had been handed over. The first series -200 joined the latter stages of testing as part of the CAA and eventually FAA certification.

The third and final test aircraft had calibrated engines and nozzles for accurate thrust measurement. One of the primary tasks it was engaged in involved capturing readings at level cruising speeds, as well as confirming fuel flows, climb rates, and minimum runway lengths for takeoff and landing. Another task assigned to the third test (G-SSCH) specimen was the demonstration of the 146's low noise, one of its strongest selling points. Accurate noise footprint data was to be made available to airport authorities worldwide, because in some cases, the 146 would become the first jet aircraft to serve an airfield.

By mid-November 1981, aircraft E1001 had accumulated approximately 80 hours of flying and was demonstrating outstanding serviceability. The aircraft was then grounded for installation of further test equipment, which was required to begin the next phase trials that involved expanding the flight envelope. Flutter inducers were fitted and used in conjunction with the extensive strain gauging already installed, so that speed could be increased systematically to the

Taxiing out to the runway for the first flight.

full design limit. Over the course of 36 flights, the aircraft was flown without problems through the center of gravity range. The landing gear and brakes were tested in over 100 landings during 'touch and go' exercises. The early aircraft were fitted with carbon brakes but because they proved to be delicate (they broke frequently), production aircraft until 1985 were fitted with steel brakes instead. In addition to the general testing that took place during flights from Hatfield, the 146 landed at Stansted where CAA staff admired and photographed the aircraft. It was also flown to Brough and the adjacent airfield at Holme on Spaulding Moor to enable the staff that produced the fin and flaps to see their handiwork.

In January 1982, aircraft E1001 had accumulated almost 114 hours of flight time, 59 flights and 133 landings. The second aircraft, E1002, made its first flight January 25, 1982, and was focused on systems evaluation. These included confirming the performance of the hydraulics, anti-icing, avionics, auxiliary power unit, electrics, as well as testing the thrust management system. The aircraft was also assigned to tropical trials in warm and humid environments. Performance was measured in order to prepare data for operating manuals and to ensure the aircraft achieved the targets guaranteed to customers. If an aircraft failed to live up to the performance marketed to and guaranteed to airlines, BAe would have to pay financial penalties – which could be substantial – in the form of cash, support, or a combination of the two to its customers. And if the aircraft failed to meet its guarantees before delivery took place, customers could refuse to accept airframes until suitable remedies were agreed. Each sales contract was different, but there was always potential for financial compensation should the 146 not meets its design targets. It was therefore imperative that BAe tested every parameter, and ensured the aircraft met or exceeded what it had proposed to its customers.

At this point in the testing phase the fuel consumption figures had been found to be very low, which was seen as paramount to the success of the aircraft. The first series -200 had its wings and fin attached, and progress on the airframe was being expedited so it could join the flight certification program as soon as possible. By January of 1982 the second 146, E1002, was approaching its first test flight just as E1001 completed 100 of the expected 1,000 hours of flight testing. Very few issues arose during those first 100 hours, and far more data than was anticipated had been collected as a result. Additionally, two major milestones were passed as 1981 drew to an end. The first came when the CAA Chief Test Pilot flew his first two assessments of the aircraft while logging the longest flight to date: four hours and 30 minutes. That flight saw the aircraft loaded to its forward center-of-gravity limit and was followed by an evaluation at full aft loading. The 146 has a wide center of gravity range allowing baggage to be loaded

First Rotation.

without the need for complex weight calculations. It was incredibly important to prove the aircraft's behavior across all loading configurations, which could expose differences in pitch control between the extremes. The further aft the center of gravity, the less stable the aircraft becomes and acceptable control response can govern the center-of-gravity range that is approved in the certificate of airworthiness.

The second major milestone was passed with the initial exploration of the stalling behavior of the 146. Plenty of wind tunnel testing had taken place using models which enabled the stall characteristics to be refined, but the first air tests were a key moment in the development of the aircraft. The stall had to be clearly signaled and there needed to be plenty of advance warning. Equally of concern was that the stall occurred at the right airspeed, as takeoff and landing performance data is related to the stall speed. Earlier British "T" tail aircraft including a BAC One-Eleven and a Hawker Siddeley Trident had been lost during stall testing. To avoid such a tragedy repeating, it was decided that the rear fuselage of the prototype 146 would be reinforced to support the mounting a stall recovery parachute. This installation was an appropriate precaution to take during low speed handling trials and the parachute replaced the rear clamshell tail brake during this phase of testing only. The parachute could pull the tail up and push the nose down if it were to be deployed following in an unrecoverable deep-stall, allowing the pilots to regain control of the aircraft after which the chute would be jettisoned. Alternatively, it would allow enough time for the pilots to bail out using their own parachutes. The 146 was also temporarily equipped with a nose probe that enabled the precise measuring and recording of angle of attack and sideslip.

The tests revealed the 146 was very close to predictions for stalling at all flap settings, as well as with flaps stored in a clean configuration. Handling was good and no undesirable characteristics appeared. A second round of testing followed after a static pressure source was added, and the stalls were repeated to confirm indicated air speeds. Results again showed that the 146 came close meeting the expected performance and the parachute was thankfully never deployed.

Testing continued, and so did manufacturing of additional airframes at Hatfield. By the beginning of the summer of 1982, the assembly line had an additional six airframes coming together and BAe was gearing up to produce approximately three aircraft per month. Aircraft E1005 was being outfitted with a completed and furnished airline interior in preparation for its role as a sales demonstrator while E2008, the first series -200 painted in Air Wisconsin colours, made its first flight on August 1, 1982.

Test aircraft E1002, painted in light and dark shades of blue in the same pattern as the orange and yellow on E1001, began tropical testing late in the summer of 1982. On August 25, E1002 left Hatfield for Torrejon airfield near Madrid, 2,000 feet above sea level and hot enough to meet the test requirements. The focus was primarily on performance, with a series of takeoffs planned to assess field length requirements and climb gradients under varying engine thrust and gross weight configurations. Trials included some with simulated engine failures at critical points which measured performance during emergencies. Once these tests were completed E1002 headed to Sharjah on the Persian Gulf, an airport noted for its combination of high temperatures and humidity. While in the United Arab Emirates, the aircraft's systems including their ability to keep the cabin cool on the ground, were tested. From there, the aircraft visited Sondre Stromfjord on the west coast of Greenland for ground tests in temperatures of minus 40 degrees Celsius. The engines and APU were started at dawn to examine the ability to get the aircraft up and running after being exposed to below freezing temperatures, and to establish the overall systems reliability. Warming the cabin to an acceptable temperature within 30 minutes after sitting overnight was another design requirement.

To further test the aircraft's abilities, including a three-engine ferry (from a remote location) as well operations from unpaved landing strips, E1002 was flown to Iceland where a natural gravel strip was used to give the

E1001 arriving back to Hatfield after its maiden flight, petal air brake fully deployed.

landing gear a thorough work out. During one flight, the undercarriage was dropped to allow icing tests to be carried out, resulting in two inches of ice accumulating on the gear. Once testing had reached 3,000 hours, all of the results were submitted to the CAA and FAA to demonstrate that the 146 could enter service without any restrictions.

Divisional Chief Test Pilot Michael Goodfellow receiving the logbook to E1001 after the first flight.

Aircraft E1003 (painted in a yellow/brown BAe livery) was used to continue to extend the scope of the flight trials, this time in Granada, Spain. These were focused mainly on noise measurement. More than 17 hours of testing took place including operating the aircraft at weights above the expected maximum for the BAe 146-100 (80,750lbs), as well as examining flyover, approach, sideline and departure noise. The trials confirmed that the BAe 146 was the quietest jetliner in the skies at that time.

The final requirement ahead of CAA certification was for a production 146 to fly 200 hours simulating real-world airline operations. This was done using the fully furnished and outfitted fourth aircraft, which was flown on a series of routes carrying volunteer (non-paying) passengers, mainly recruited by BAe and the airline performing the proving flights. The fourth aircraft, E1004 registered G-OBAF, entered the trials fleet carrying the airline livery of UK-based carrier British Air Ferries (BAF). It was used to assess flight deck workload and was also deployed on the critical route proving flights using BAF's own pilots. This aircraft, prior to entering test program, appeared in the static park at the Farnborough Air Show where visitors saw a completed aircraft, including a fully outfitted interior. BAF went on to present British Aerospace with a Letter of Intent (LOI) to purchase up to ten aircraft, but this was never converted to a firm order.

In September 1982, UK regional airline Dan Air purchased two series -100 aircraft, and placed options on a further two series -200s. With a 31-inch pitch and a 3-3 layout Dan Air would operate the -100s with 88 passengers and up to four flight attendants (more than the minimum required). The aircraft were initially scheduled to operate from the airline's Gatwick base, and their use was subsequently extended to Newcastle, serving Berne, Zurich, and Dublin. Dan Air became the first UK carrier, and only the second airline globally, to order and take delivery of a BAe 146. The order also marked the first sale of series -100 aircraft.

The third test (E1003) aircraft, painted in Dan Air's colours, showed off how agile the 146 was during

HRH Prince Charles during an evaluation of the BAe 146.

the flying display at the September 1982 Farnborough, where it flew alongside Air Wisconsin's series -200. In October 1982, it then went to Cranfield for water ingestion trials, a prerequisite for certification. This was aimed at demonstrating that during heavy rain or flooding, while taxiing, takeoff, and landing, the aircraft did not create a water pattern from its nose wheels or main gear that resulted in significant water being directed (and ingested) at the engines. The high mounted engines meant the 146 easily passed this test. E1003 was subsequently used for Crew Workload evaluation trials as well as Avionics certification trials. It was later used for crew training for Dan Air, Air Wisconsin, Air Mali, the RAF, TABA, and AirPac.

By this time British Aerospace's Sales Director, Johnny Johnstone, was under a lot of pressure to generate orders from the seemingly stalled sales campaigns. The aircraft was mere months from being certified to fly paying passengers, and yet there had been only two firm orders, both from small carriers. BAe had targeted Asia and Australia as the regions appeared to offer many strong prospects for the 146, so the firm dispatched aircraft E1005 (G-SCHH) on its first demonstration tour (the 'Far East Tour') that began on October 24, 1982. A total of 14 staff, consisting of flight crew, a sales team, and engineers were put through the wringer, getting only one day of rest per week as the aircraft covered a total of 60,000 miles visiting 15 countries over a 50-day period. It certainly proved to be a good way to demonstrate the reliability of the airframe, its systems and its engines. During the tour, prospective customers not only got to view the aircraft to see what it could offer but also experienced the 146 demonstrating its prowess operating out of airfields that were often incredibly challenging. Hot and high, short runways, dangerous approaches through jungle and mountains – the 146 conquered everything with flying colors. Some locations, however gave the aircraft and crew a bit of hassle. Upon arrival in India for example, customs wanted to apply tariffs to all the spare parts carried in the hold.

The 146 was the first jet airliner to land and take off at many of the airfields it visited and often used only half the available runway length, demonstrating to customers that it was the right aircraft for the job. Reliability was fantastic, fuel consumption below expectations, and the small noise footprint impressed many civic authorities and ministers concerned about additional noise pollution in urban areas.

Even more amazing was that G-SCHH had only taken its maiden flight one week prior to leaving the UK and had accumulated a mere 10 hours and 10 minutes of flight time before taking off from Hatfield. British Aerospace was taking a big gamble taking such a new and fresh aircraft on the road. It took off from Hatfield at 1:20pm, with fuel stops at Brindisi, Larnaca, and Bahrain (due to fog at Dubai). After making a few more stops along the way, Thai Airways was the first airline to have a look on October 26. The tour was scheduled to allow a one or two day stop in most locations where it was shown prospective airlines and officials. However, G-SCHH spent the majority of its time in Japan and Australia, from where large orders were anticipated.

Arriving in Okinawa Japan on October 28, the aircraft began demonstrations to South West Airlines. The very next day BAe received favorable comments about its low noise profile, something of great

E1001 fitted with probe for testing.

concern to residents living near Japanese airports. Heading up to Tokyo (Narita) the aircraft demonstrated its performance capabilities to a large turnout of reporters who noted the short take off roll, steep climb out, and quietness. Firstly, it was shown to the pilots and executives at TOA Domestic Airlines, and the next day to All Nippon Airways' (ANA). Prince Philip, the Duke of Edinburgh, was in Japan for an event as the President of the World Wildlife Fund and as an accomplished pilot (who would later receive conversion training and type approval on the BAe 146) he flew the aircraft from Tokyo to Sapporo ahead of further demonstrations. He flew the aircraft again, from Sapporo to Kushiro, the following day and then took it back to Tokyo. Towards the conclusion of the Japanese sales tour, the Japan Civil Aviation Bureau measured its noise on multiple approaches and take-offs, noting it produced some of the lowest recordings ever.

G-SSHH E1002 painted in blue joins the flight test program.

The BAe 146 was put through its paces on the tour and became only the second jetliner to land at Baguio, Philippines high up in the mountains. The aircraft performed admirably, and a bird strike to the number three engine went unnoticed until it was discovered during a routine inspection back in Manila. While being demonstrated to Mount Cook Airlines in New Zealand, the 146 became the first jet to ever land at Mount Cook airfield. It touched down in a 40kt cross wind, which required Mount Cook's older HS748 turboprop aircraft to divert to another airfield 25 miles away.

Although the BAe 146 was impressing airline executives with its performance, many found the interior to be underwhelming. The worlds quietest jet aircraft on the outside was deemed to be too noisy inside. Being an early build aircraft, the materials used in the cabin of E1005 were not quite ready for "prime time", and the interior suffered from 'quirks' and 'kinks' in quality. From the seats to the wall panels and the lack of sound proofing, these elements cast a shadow over the 146 which was already facing an uphill battle to win favour and convince customers that a four-engine regional jet made sense. Many Asian operators placed a premium on passenger comfort and the overall customer perception of the airline. While prospective customers seeing an early prototype was not a definitive reason why the 146 failed to gain an Asian customer in the first seven years, these issues certainly didn't help the marketing effort. Indeed, BAe mistook the warm reception and genuine astonishment at the aircraft's performance and low exterior noise levels for real interest in buying the 146. The firm was going to learn the hard way that the ability to operate into challenging airports was rarely going to be enough to justify an order. Another paradox was that noise levels were simply not an issue in most Third World countries where BAe saw target markets. Airlines were more interested in day in day out high utilization, with economics and reliability being key factors.

BAe took test aircraft number four, painted in a hybrid-BAF livery, on a 21,000 miles sales tour of Africa, demonstrating its hot and high performance as well as its unpaved airfield performance. BAe 146-100 G-OBAF was equipped with a six abreast 72 seat interior and left on the 18-day trip on the February 6, 1983. Arriving in Cairo, it performed for Egypt Air and the Egyptian Air Force, and also conducted a photo flight around the pyramids. The aircraft was later demonstrated to oil industry executives on remote air strips. G-OBAF left Egypt ahead of further demonstration flights in Sudan carrying Sudanese government personnel, followed by visits to Nairobi

All three test aircraft, with G-SSCH (E1003) to the left wearing a short lived dark red and ochre paint livery.

G-SSHH

Left: One of many route proving flights over the Christmas 1982 holiday break in order to meet CAA requirements.

Right: Onboard the route proving flights which carried BAe employees and their families.

Bill Batchelor

Bill Batchelor

and Zimbabwe where a quick flyby to demonstrate the noise profile (or lack of) was missed because no one heard the 146 coming. The 146 was shown to Air Zimbabwe and the Zimbabwe Air Force, and was taken on flights around one of the country's most popular tourist attractions: Victoria Falls. The 146 showed its prowess further by visiting airports that were usually served by Vickers Viscounts, some of which were hot and high with short runways.

Control of the aircraft was handed over to Air Zimbabwe pilots, enabling them to experience all the aircraft's capabilities including dual out engine performance. After Air Zimbabwe completed its tests, Air Force pilots were given a turn at the controls and allowed to fly into some of the local airfields. Most of the flight crews' previous experience had been with very old Douglas DC-3 aircraft and helicopters, but they transitioned to their first jet aircraft quickly. The aircraft carried two flight attendants from British Airways as well as two from Air Zimbabwe who provided first class service on the very short trips aloft. The aircraft was shown to Air Botswana where its hot and high performance was demonstrated along with a touch-and-go at the mining airstrip at Selebi Pikwe. At the conclusion of the tour, the 146 had visited 11 countries in Africa and carried 912 passengers. Many of the issues that surrounded the Far East Tour were experienced again during the African tour. With the exception of the short-lived Republic of Mali operation and similarly to the outcome of the Far East Tour, it would be nearly seven years until the 146 obtained a proper customer base on the African continent.

On November 16, 1982, the first 146-200 painted in launch carrier Air Wisconsin's livery departed for Casablanca, Morocco, in North Africa. It was there specifically for noise measurements, in an environment that was perfect for accurate testing: blue skies, little to no wind, and temperatures in the 70s (F)/23C+. International requirements for noise certification of an aircraft are very stringent, and suitable weather conditions are rarely prevalent in the U.K. The team from British Aerospace set up a base station in the old Casablanca air terminal. The conditions had to include a nominal head wind of not more than 10 knots, with relative humidity between 20-90%, air temperature between 36-95 degrees F (2C-24C), and a cloud base above 6,000 feet. Noise measuring sites were positioned underneath and to both sides of the flight path, enabling a determination of the

Employee ticket for route proving flights over the Christmas holiday weekend in 1982.

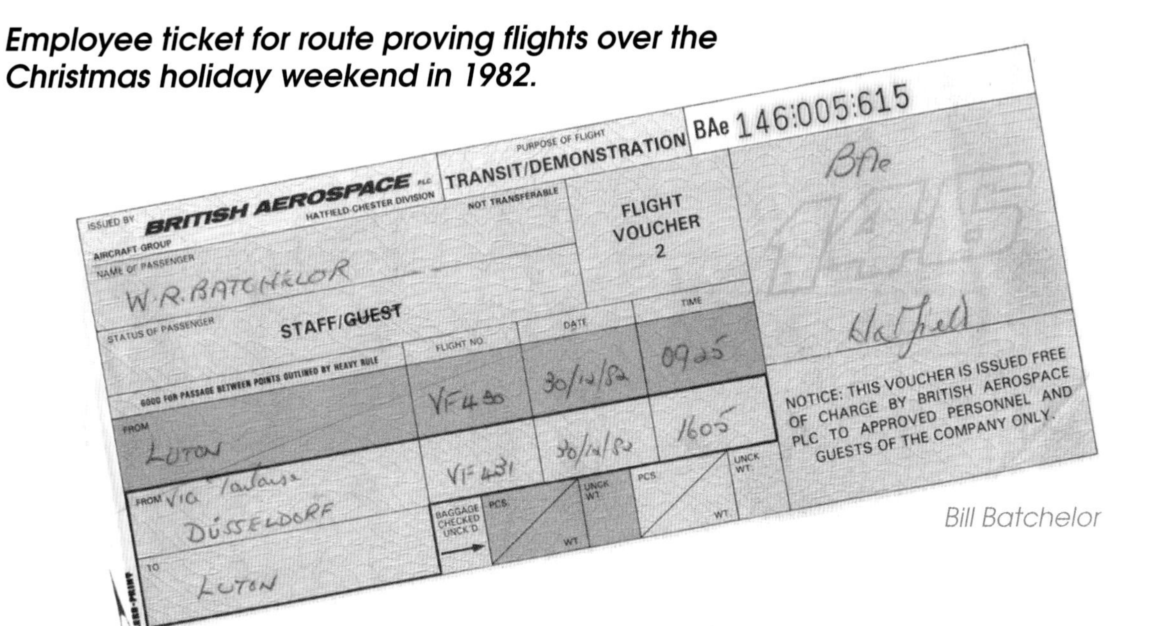

ISSUED BY BRITISH AEROSPACE PLC
HATFIELD-CHESTER DIVISION
AIRCRAFT GROUP
PURPOSE OF FLIGHT: TRANSIT/DEMONSTRATION
BAe 146:005:615
NOT TRANSFERABLE
FLIGHT VOUCHER 2
NAME OF PASSENGER: W.R. BATCHELOR
STATUS OF PASSENGER: STAFF/~~GUEST~~
GOOD FOR PASSAGE BETWEEN POINTS OUTLINED BY HEAVY RULE

	FLIGHT NO	DATE	TIME
FROM LUTON	VF430	30/12/82	0925
FROM VIA Toulouse DÜSSELDORF	VF431	30/12/82	1605
TO LUTON			

BAGGAGE CHECKED UNCK'D — PCS / WT / UNCK WT

BAe 146
Hatfield

NOTICE: THIS VOUCHER IS ISSUED FREE OF CHARGE BY BRITISH AEROSPACE PLC TO APPROVED PERSONNEL AND GUESTS OF THE COMPANY ONLY.

Bill Batchelor

After the proving flights, the British Air Ferries liveried 146 toured African continent in 1983 with a modified livery. Pictured over the pyramids in Egypt.

February 1983 over the famed Victoria Falls between Zimbabwe and Zambia.

BAF aircraft on display at the Farnborough Air Show in 1982. *John Maydew*

exact position of the aircraft in space at known times. Every site had to be surveyed with reference to "start of roll" position on the runway and the furthest was 5 ½ miles away. Positional accuracy of around 10ft was usually achieved.

The trials involved flying the aircraft with the utmost precision, as defined speeds had to be maintained to a tolerance of +/-3kts. The throttles had to be at a constant setting while the aircraft passed through each checkpoint during the climb out, as well as during landing. Measurements were required for two takeoff techniques at several aircraft weights and various throttle settings; and two angles of approach to land, again at several weights. The main control site was known as "Tibbet 66", Tibbet being the Hatfield call sign and 66 the designated identification allocated to a noise station for radio transmissions. The whole test programme was controlled from this site which was under the flightpath, 3½ nautical miles from the start of takeoff roll. Wind speed and direction, along with relative humidity, was measured at 10 metres and 1.2 metres above the ground. The noise during each overflight was recorded on tape, and the aircraft was photographed when it was exactly overhead. Data captured on board the aircraft was transmitted via the radio after each test run.

Each measuring station also had a transmit/receive radio to report any doubts on a successful recording of data to "Tibbet 66". If everyone was satisfied the technical manager could put a provisional tick against that particular run. The whole exercise required a total of around 70-80 ticks. Data was recorded on magnetic tape by the ground-based measuring stations and on the aircraft, while instrumentation was checked at the end of each day's flying to confirm the ticks provisionally inserted on the master scoreboard. Camera film was developed and inspected regularly for the same reason.

While this activity was in progress, the local population became extremely inquisitive. Mark Sansome of BAe was at the site furthest from the runway. After a short period, many of the local village people including elders, workers, children, and a donkey set up their own encampment nearby, complete with a coffee pot. Sansome was kept well supplied with refreshments, and the children would help by pointing out the 146. Everyone knew to keep extremely quiet, allowing the test to take place each time the aircraft passed overhead.

BAF operated a series of route proving flights using a 146-100 in their own livery, carrying more than 6,600 BAe employees from a variety of locations. This aircraft was also used during the December holiday season, the flights heading overseas and back accumulating a total of 200 hours of intensive operations. Trips operated every weekday between early December 1982 and concluded on January 16, 1983, departing Luton to destinations including Toulouse, Jersey, Munich, and Beauvais. A light snack was served on the outbound services and a more substantial cold meal on the return. The bar was always open and a limited amount of Duty Free sold onboard. During one route proving flight a problem was encountered with the APU, which filled the cabin with smoke. The aircraft made an emergency landing in Munich on December 8, 1982 along with evacuation of the aircraft, and passengers were returned to Hatfield in a BAF Viscount.

At the conclusion of flight testing, which involved six aircraft, 2,000 total test hours had been flown in accordance with CAA rules (of a required 1,000 test hours). Five series -100 aircraft and one series -200 aircraft were employed with aircraft E1001 garnering the most test time – a total of 640 hours. The Series -200 had the least amount of time at 145 hours.

On February 4, 1983, the CAA issued a type certificate for the BAe 146-100 and series -200 to BAe, certifying the aircraft could be operated by airlines in the United Kingdom. It was the first time such certification had been awarded on the basis of European Joint Airworthiness Requirements, a new international standard modeled closely on the U.S. Federal Airworthiness Regulations (FAR). A total of 20 specialists from the CAA had worked closely with BAe during 65 hours of flights performed by BAe's Mike Goodfellow and his team of test pilots, during which the test data was verified.

The initial type certificate covered the aircraft as a whole, but each individual airframe produced would be tested to ensure it had been built to the standard defined in the type certificate. A secondary certificate of airworthiness was issued for each individual aircraft. Crucially, the certificate was only awarded once the aircraft manufacturer demonstrated the aircraft could be evacuated in an emergency in less than 90 seconds, and in total darkness, which BAe completed in September 1982 at Woodford.

Having completed CAA certification, the 146 was also subject to FAA scrutiny. One interesting note is that BAe 146 aircraft destined for U.S. customers had an 'A' added to their designation when they met FAA requirements. Air Wisconsin's BAe 146 would therefore be designated "BAe 146-200A" models.

The BAe 146 was now a certified aircraft ready to enter service with the two airlines that had ordered the type. But BAe was still struggling to gain additional firm orders. The firm knew it was important to get the few aircraft that had been sold into service, and to prove to those airlines on the sidelines hesitating over making a commitment that the 146 was viable airliner.

Water ingestion trials, with the engine nacelles coated in black water soluble paint to show the path of any water kicked up by the landing gear.

Series 200 prototype in Air Wisconsin colors was used for noise testing in Casablanca.

BAe's testing team setup in the old Casablanca air terminal.

Photographing the spray patterns from the water ingestion test.

The first BAe 146-100 delivered enters service with Dan-Air.

CHAPTER 04

May the Fours Be With You

Dan Air

Dan Air made 146 history when it took delivery of the very first production aircraft, a series -100 (E1006/G-BKMN), during a lavish ceremony on May 23, 1983. Its cabin was configured with 88 seats, 6-abreast, set at a 31-inch pitch featuring alternating orange and brown patterned covers which was the default fabric offered by British Aerospace. The aircraft also incorporated an extra galley up front. The airline wasted no time putting the BAe 146 to work and the first revenue service occurred just four days later on May 27. Based at Gatwick airport in the United Kingdom, the aircraft began flying daily scheduled service between cities on the Dan Air network including Dublin and Toulouse, as well as Berne where it became the only commercial jet airliner capable of operating from the 1,300m (4,265ft) runway. Later during each day, after its regular runs had been completed, the aircraft flew charters that carried tour passengers to destinations such as Palma, Venice, and Gerona. Over the first two months of revenue service the 146 proved to be a valuable resource and maintained utilization rates of greater than eight hours per day, with the average flight time clocking in at 90 minutes.

The second aircraft (#1007/G-BKHT) was delivered on June 19 and also went to Dan Air, which deployed it on routes between London (Gatwick), Stavanger, and Bergen from a base at Newcastle in northern England. Like the first aircraft, it had an equally busy schedule with utilization typically reaching nearly 13 hours a day in normal operations. Similarly to BKMN, it was utilized for charter and holiday flights from Tees-side airport during the evening and weekends, visiting a variety of destinations with the farthest being Alicante on the Mediterranean coast of Spain. Dan Air focused on scheduled operations Monday through Friday, with charters dominating the weekends. The airline was able to report even better fuel burn than BAe promised (by about 8%), using its own flight profile and cruising at Mach 0.67. Publicly Dan Air reported no major issues with the reliability of the aircraft with regards to the engines, but behind the scenes there were issues with the Thrust Management System (TMS). Nevertheless, the initial dispatch rate was 95.5%, improving to 97.8% after six months of service – very good figures for an all new aircraft type.

The bulkheads at the front and rear of the aircraft featured monochromatic drawings of locations around England.

Interior of Dan-Air BAe 146 aircraft, utilizing brown and orange alternating patterned seats.

Dan Air had previously grown and prospered through the acquisition of second-hand aircraft which were generally mature, reliable and well proven. The BAe 146 was the first brand new aircraft Dan Air had purchased. The airline was not really equipped to cope with the challenges that would inevitably arise while operating a brand-new aircraft type, or the teething troubles that plagued the 146 in the early days. Over time the aircraft began to exhibit reliability issues, and the Dan Air mechanics nicknamed G-BKMN (MN) "Mechanical Nightmare" while G-BKHT (HT) became known as "Hatfield Trash". The airline's main operational base was at Gatwick but its engineering support centre was 60 miles away at Lasham airfield, which made immediate servicing challenging. The arrival of E1005 (G-SCHH) on lease in June 1984 eased the situation, and following modifications that increased reliability, fleet performance soon improved.

RAF – Royal Air Force (written by former Sqn Ldr David Gale)

During March 1983, the Ministry of Defense ordered the purchase of two BAe146-100s for a two-year evaluation aimed at assessing whether the type would be suitable to replace The Queen's Flight's aging Hawker Siddeley Andover turboprop aeroplanes. E1004 and E1005 were funded from the Falklands War reparations budget (along with six BAe 125-700 executive jets), which was going to be closed by the Treasury if it was not spent. Squadron Leader David Gale was appointed Flight Commander of the BAe146 Evaluation Flight which was formed on May 9, 1983. The flight was attached to 241 Operational Conversion Unit (OCU) based at RAF Brize Norton, where RAF VC10 crews were taught and evaluated. Both 146s were pre-production series -100 aircraft that had been used by BAe during initial flight-testing and were designated BAe146 C Mark 1 by the RAF.

British Aerospace** invite you to watch the first flight over **London** of the **world's quietest jetliner, the BAe 146

on Monday 23rd May

at 14:30 hours.

The new 146 will fly across the City of London and River Thames at 1,000 ft. (weather permitting) on its delivery flight to its first UK customer, Dan-Air. It is Britain's newest jetliner and it is so quiet that unless you look you will not notice it's there. The 146 will bring new standards of quiet comfort to the world's short-haul air routes.

The flight path is shown on the reverse.

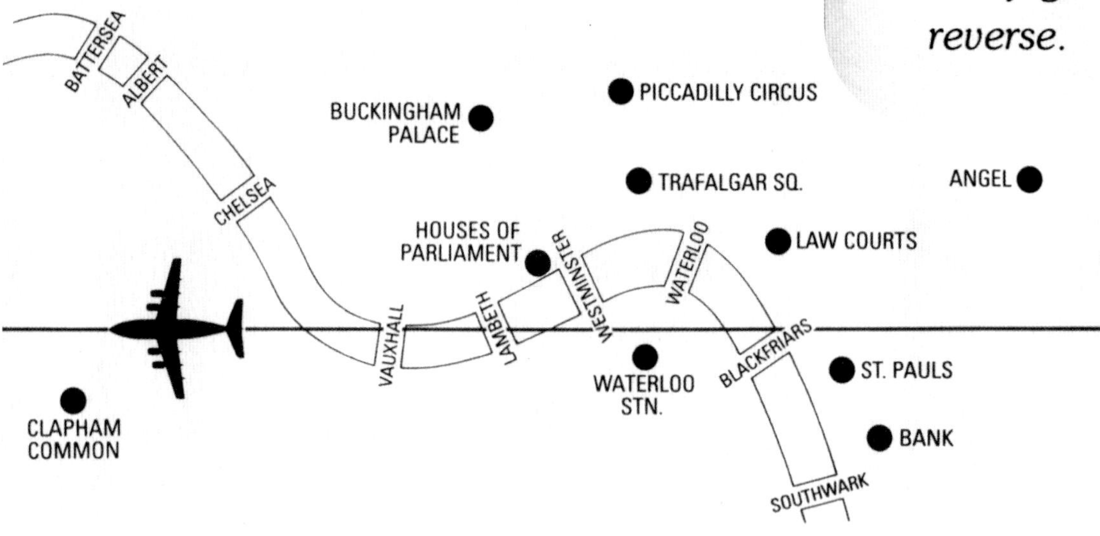

Sqn Ldr Gale initially spent a month at RAF Upavon, the headquarters of the RAF's transport fleet, writing the Standard Operational Procedures (SOPs) for the 146 using the existing SOPs for the Lockheed C-130 Hercules and the Vickers VC10 as a reference. The C-130's configuration (high wing, 4 engines) and the VC10's jet performance provided invaluable sources of information. Military flying placed some unique demands on the aircraft though, and unlike its civilian counterparts, the RAF's BAe 146s would be authorized to take off on three engines for ferrying purposes.

Crew training commenced at BAe Hatfield in early June 1983 using one of BAe's pre-production airplanes, G-SSCH, which was to be re-registered as ZD696. On June 13, Sqn Ldr Gale performed a full air test on the RAF's first BAe146 C Mk 1, ZD696, followed by an acceptance check flight. The Royal Air Force took delivery of the aircraft during a large ceremony on June 14, 1983, and it went into service immediately after it was handed over. The following day, Sqn Ldr Gale flew the 146 to RAF Northolt and then on to RAF Benson where members of The Queen's Flight had their first opportunity to see up close what was, potentially, to become their new aircraft type. Later that afternoon, ZD696 continued to its new home at RAF Brize Norton.

The BAe146 Evaluation Flight airplanes were almost identical to their civilian counterparts although they had some minor equipment additions. HF radios were installed along with Tacan and a military transponder

(IFF). The cabin had a standard galley and a full complement of airline seats. The goal of the evaluation was to assess the BAe146, to see if it performed well and was suitable for the type of flying undertaken for the Royal Family. That necessitated evaluating its general reliability as well as its performance at hot and high airports and in extremely cold weather.

Flight and ground crews were trained on the airplane for several few weeks, interspersed with it undertaking a few VIP flights - senior officers just had to come and see it! In mid-July 1983, a request was made to fly the Chief of the Air Staff from Grosseto, Italy to RAF Northolt. Unfortunately, during the approach into Grosseto the landing gear failed to lower so the emergency gear lowering procedure was accomplished. As the airplane was still very new, a British Aerospace pilot and engineer were on board while the RAF crew was completing their training. The engineer tried to fix the problem and a quick air test was flown. The gear failed to retract after takeoff and the aircraft landed again at Grosseto. It was then decided to make a gear-down ferry back to Northolt with the Chief of the Air Staff on board. There was no performance, range or fuel burn data available for gear-down operations in the flight manual so it was unclear whether the aircraft would be able to make it all the way to Northolt without refueling. It did not. A diversion had to be made to Lyon, France while en route to refuel before the aircraft continued onwards to Northolt. Needless to say, the BAe 146 did not make a great first impression with the most senior officer in the RAF. However, the fault was diagnosed as a one-off mechanical issue and did not reoccur during the rest of the two-year evaluation.

The BAe146 was a new airplane and there was a great deal of public interest in it. During the summer of 1983 it appeared at numerous airshows in static displays as well as participating in the flying displays during the International Air Tattoo at RAF Greenham Common and the Battle of Britain airshows at RAF Akrotiri (Cyprus), RAF St Athan, and RAF Abingdon.

In late August, the BAe146 Evaluation Flight was called on to support the rotation of troops between RAF Gutersloh, Germany and RAF Aldergrove, Northern Ireland. On the very first flight, a starter motor drive shaft failed during engine start at Gutersloh. The SOPs which had been written by Sqn Ldr Gale in May had just been printed and distributed to all relevant departments, so he was able to point them immediately to the page and paragraph number for the authorization to perform a three-engine ferry. The troops and baggage were offloaded and the aircraft performed the first three-engined takeoff from RAF Gutersloh to RAF Brize Norton on August 29, 1983.

The BAe146 Standard Operating Procedures for a three-engine takeoff:

1. Set takeoff power on the two symmetrical engines with brakes applied.

2. Release the brakes and slowly apply power on the asymmetrical good engine in order to achieve full power by rotation speed.

3. If rotation speed hasn't been achieved by the 3,000ft remaining point on the runway, abort the takeoff.

4. Achieve Vmca2 as soon as possible after takeoff. (Vmca2 is the minimum speed at which directional or lateral control can be maintained on 2 engines – i.e. worst case of losing 2 engines on the same side).

Handover ceremony for the first RAF BAe 146 aircraft.

ZD696 departs Hatfield after the handover ceremony.

The first BAe 146 for RAF evaluation arrived at Brize Norton on June 14 after the delivery ceremony. Sqn Ldr David Gale is front and center.

The only time both RAF BAe 146 aircraft were flown together.

The radome nose had to be removed in order to close the hangar door.

Starter motor drive failures became the only recurring mechanical issue during the initial months of the RAF evaluation. They usually occurred during a quick turn-around while the engine was still hot so the Evaluation Flight started carrying spare shafts on all trips. They could be changed in about an hour so delays were relatively minor. The problem was solved a few months later when modified drive shafts were made available.

The RAF likes to have defined operational limits but the BAe 146 had no published crosswind landing limit. In the Autumn of 1983 the wind was forecast to increase from 20kts out of the north to over 40kts. As RAF Brize Norton had an east-west runway, it was the perfect opportunity to assess what the airplane could do. Sqn Ldr Gale flew numerous landings as the wind increased and eventually called it quits when aileron controls reached its limit in the 40kts crosswind. So, he set the RAF limit at 35kts.

A second BAe146, registration ZD695, was delivered to the Evaluation Flight on September 16, 1983 enabling the evaluation to really commence in earnest. Flights between Northern Ireland and Germany were scheduled a couple of times each week and the aeroplanes also visited Norway, Jordan, India, West Africa, the USA and Canada. During this period time the 146s were assessed for their suitability to perform hot and cold weather operations. While visiting Goose Bay, Canada in the winter, temperatures that plummeted to -48C (-54F) were encountered and it became necessary to put the airplane in a hangar for the night. The only facility available was too small for the BAe146 until the crew became very resourceful and removed the radome from the nose. By maneuvering the airplane cautiously into the hangar and with the roof supports between the engines, the crew managed to close the doors with inches to spare.

One of the issues that the Evaluation Flight had to overcome was that both aircraft had been used in British Aerospace's testing and certification process, which meant that numerous modifications had been made to various systems. As a result, the technical manuals did not always correspond with what was actually installed in the airplanes and sometimes that meant extra time was required to find and fix technical issues.

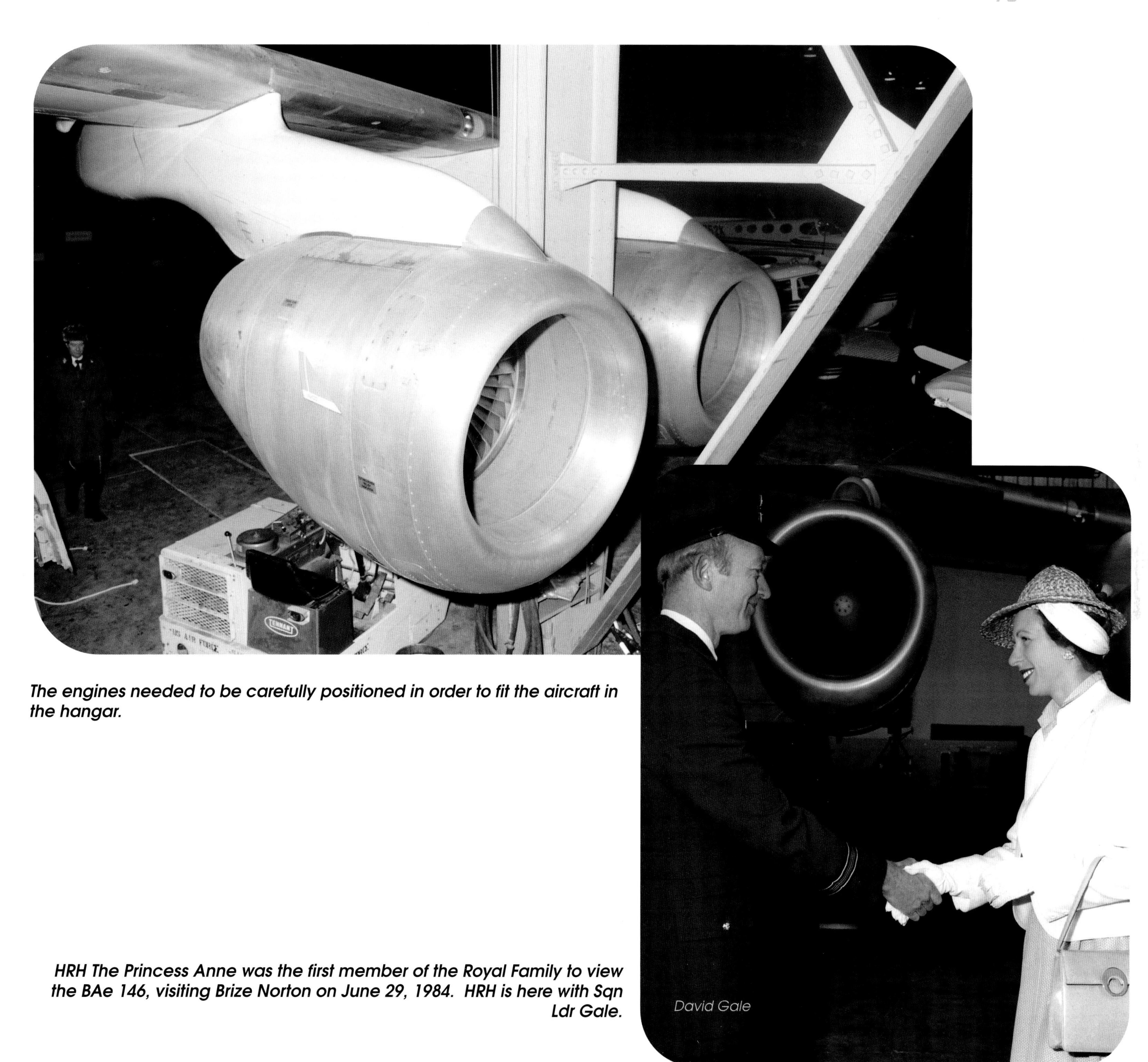

The engines needed to be carefully positioned in order to fit the aircraft in the hangar.

HRH The Princess Anne was the first member of the Royal Family to view the BAe 146, visiting Brize Norton on June 29, 1984. HRH is here with Sqn Ldr Gale.

The BAe146 was to become a very reliable airplane. The Evaluation Flight operated for a seven-month period with no mechanical delays whatsoever, which was quite an accomplishment for a new airplane type. During the entire two-year evaluation, the aircraft suffered a total of nineteen technical delays on a total of 1,025 departures, representing a dispatch reliability rate of 98.15%. It was also noteworthy that many of the delays were caused by a lack of spare parts for easily fixable faults.

The Evaluation Flight BAe146s carried a total of 25,181 passengers together with 1,030,053 lbs of baggage plus 110,020 lbs of freight. Passengers liked the aircraft and the crews loved it. It was easy to handle, very stable and extremely maneuverable. The ground effect from the large flaps and the trailing link landing gear helped to soften landings and the four-engined performance made landings and takeoffs from short runways very safe. It was quiet outside and inside and very fuel efficient.

On June 7, 1984 ZD696 was sent back to British Aerospace at Hatfield thus reducing the BAe146 Evaluation Flight to just one aircraft for the remainder of the evaluation. In reality, by that time the programme had proved that the 146 would be a very suitable replacement for the Andovers of The Queen's Flight. The vast majority of The Queen's Flight trips are short range around the UK and Europe and some require short field capability. The BAe146 was ideally suited to that kind of flying and could safely land and depart from a 3,000ft long runway.

The remaining BAe146, ZD695, continued flying twice-weekly shuttles between Northern Ireland and Germany until March 29, 1985 after which it too was handed back to British Aerospace at Hatfield. The BAe 146 Evaluation Flight was disbanded but as a result of the successful evaluation, two new BAe146s were ordered in VIP configuration and fitted with extended range fuel tanks for The Queen's Flight.

The BAe 146s of the RAF were one of the only transport aircraft in the RAF without a refueling probe. Near the end of the evaluation, a fake probe was constructed and fitted on the aircraft. This image was published in the RAF's Flight Safety Magazine 'Air Clues' which was widely distibuted. Sqn Ldr Gale began receiving phone calls from the Ministry of Defence, British Aerospace, and elsewhere asking how the evaluation was going including the refueling probe. In the end, most saw the humor in the situation and no one was reprimanded.

RAF BAe 146 Evaluation Team in front of ZD695 just before the final flight back to BAe Hatfield (to return the aircraft). The refueling probe was removed prior to flight.

Air Wisconsin

On the opposite side of the pond, regional airline Air Wisconsin took delivery of the very first BAe 146-200 (E2012/N601AW) just two days after the RAF took on its first BAe 146-100. E2012 left Hatfield for a long distance ferry flight on June 16, taking E2012 through Prestwick, Keflavik, Sondre Stromfjord, and Sept Iles before arriving at Appleton just after 8pm and 16 hours of flight time. Local residents and television crews came out to see the new aircraft, while U.S. Customs formalities were completed just before midnight. Plans envisaged the aircraft would enter service on June 27, after route proving flights for the FAA had been completed.

The 146 became the biggest aircraft in Air Wisconsin's fleet and started carrying the largest passengers loads in the airline's history. The carrier's inaugural flight involved a full aircraft flying from Fort Wayne, Indiana to Chicago, Illinois including 79 fare paying passengers, with the remainder being VIPs and press. This marked the first 146 service in the United States. Air Wisconsin's President, Preston Wilbourne also accompanied the flight and sat in an open jump seat at the rear of the cabin. The day started with a mere 32 minutes of flying time at the beginning of a 13-sector day, with very tight turnarounds scheduled including stops at Chicago O'Hare airport. Air Wisconsin's network featured shorter flights than Dan Air's, some as brief as 30 minutes, and with congestion at O'Hare it was vital for the 146 to be punctual and reliable or the days' schedule would fall apart. On a typical day the aircraft left Fort Wayne, Indiana (the airline's maintenance base) at 7am and returned there at 9:10pm for overnight servicing and maintenance. O'Hare is the primary hub for its busy routing, but the 146 visited other destinations on the network included Appleton, Wisconsin; Toledo and Akron/Canton, Ohio; as well as the two stops at Fort Wayne. Air Wisconsin put the second aircraft to arrive into service immediately, introducing it to new destinations including Milwaukee and Green Bay via Chicago.

The history of Air Wisconsin was tied to Hatfield when the airline began operations with two de Havilland Dove aircraft in 1965. Formed with only 20 employees, three of which were pilots, no one expected it would grow into a large regional airline 17 years later, or become a launch carrier for a new aircraft type, the 146-200. But the 146 fit right in, providing jet services with excellent economics on smaller point-to-point routes. The average stage length for Air Wisconsin was a mere 145 miles, reinforcing the BAe's promotion of the 146 as a flying bus.

In comparison to other aircraft in the Air Wisconsin fleet, the 146's economics and fuel burn were unmatched. The 16-passenger Swearingen Metros burned 100 gallons per hour at 260MPH and the 50-passenger DHC Dash 7s burned 290 gallons per hour at 260MPH. Meanwhile, the 100-passenger BAe 146-200 burned about 600 gallons per hour at 460MPH, which represented a fuel saving for every passenger/mile flown. Initial calculations showed seat mile costs and capacity for a single 146 were better than operating two Dash 7 aircraft. After the 146 entered service, it turned out those initial calculations were to conservative. The actual cost of operating a single BAe 146 was equivalent to operating three Dash 7 aircraft. Breakeven load factor on the 146 was approximately 36% and the type eventually went on to replace the turboprops

The second BAe 146-200 to be delivered to Air Wisconsin.

1.
AIR WISCONSIN BRITISH AEROSPACE 146 JET
WEMI
FM 100
LONDON TRANSPORT
2.
air wisconsin
BAe 146
601
4.
5.
6.
air wisconsin

Courtesy of Air Wisconsin

in Air Wisconsin's fleet. Conversion training for turboprop pilots was 25 hours of training, but pilots typically grasped the 146 and its flight characteristics in less than half the time. About the only complaint that cropped up was that the APU was louder than all four engines.

The BAe 146 was Air Wisconsin's first jet aircraft, and it proved popular with Air Wisconsin's passengers who preferred it over the turboprops that they had previously travelled. As a result, the airline's passenger traffic increased considerably. According to Tom Helm, manager of crew training, customers especially liked the landings. President Preston Wilbourne was so smitten with the type that he acquired the license plate "BAE146" for his personal car, and one of the aircraft (N606AW) was named "Kitty" after his wife Catherine. The airline took delivery of two 146s in 1983, and ultimately went on to order substantially more aircraft and launch a new variant in 1987. In 1986, Air Wisconsin became a branded carrier for United Airlines (adopting the United Express livery), and the original colour scheme began to disappear from the fleet. The BAe 146 series operating with Air Wisconsin and the partnership with United Airlines meant the 146 would truly live up to the "feeder liner" designation British Aerospace used in its marketing.

1. Air Wisconsin went to great lengths to promote the arrival of the British jet, chartering an AEC Double Decker bus to shuttle members of the press.

2. The inaugural flight is preparing to get underway at the break of dawn.

3. Preston Wilbourne accepted the logbook for the very first jet aircraft to join the Air Wisconsin fleet at Hatfield.

4. The inaugural flight was completely full with paying passengers and press.

5. As a result of the full flight, founder Preston Wilbourne ended up sitting in the very last seat in the back of the plane.

6. The mechanics at Air Wisconsin built a raft to compete in an annual rafting race put on by a local radio station (they came in 2nd place).

Republique du Mali

May 1983, BAe took a BAe 146-100 to the Paris Air Show, where the type was put into the flying display. Air shows are very expensive operations for any manufacturer to participate in. Setting up a chalet is not an trivial endeavor, and participation requires staff to run it, reserving space to park the aircraft in the static display, the air crew, and performing maintenance between the flying displays, not to mention the receptions and dinners day after day, can be incredibly costly. G-SSHH appeared carrying TABA titles and a Brazilian flag and took part in the flying display. G-SCHH which had been repainted in RAF colors, visited the show site at Le Bourget airport briefly on June 1 to pick up media guests and return them to Chester for the BAe 125-800 roll-out ceremony. Managing the resources involved was incredibly taxing for the company and the employees, and often at the conclusion of such events there was not much evidence that anything worthwhile had been achieved.

The show did however result in an order. The Republic of Mali made a commitment to take a single -100 aircraft (E1009) which was delivered October 15, 1983. It was the first aircraft sold under a program of loan guarantees from Britain's Export Credits Guarantee Department (ECGD). Initially the BAe 146 wasn't considered eligible for the program given its high American-produced content. After a protracted battle BAe won the argument and ECGD agreed to consider the wings as UK content, making future sales of the 146 eligible for ECGD financing. The Republic of Mali deal consisted of the Government of Mali guaranteeing 15% of the aircraft purchase price and the ECGD the other 85%, with Barclays Bank underwriting the deal. The aircraft would have dual use, serving as a commercial airliner but capable of being used as a VIP transport when needed. The Republique du Mali aircraft featured a quick-change-like interior enabling it to operate in a partial VIP role, or to carry 87 economy passengers. On occasion it was used for Presidential charters, necessitating the interior being split into two. A coach section (outfitted with the standard brown BAe 'camouflage' interior) with six abreast seating was at the rear, while up front there were 12 first class seats, along with two executive tables that lifted and folded into storage against the bulkhead.

The cabin featured brown carpeting, and the sidewalls and bulkheads had extensive native artwork stenciled throughout. The front bulkheads also featured the coat of arms of Mali. TZ-ADT "Nioro de Sahel" was delivered October 15, 1983. After arriving in Bamako, it operated a VIP flight to the airport and town after which the aircraft was named, Nioro, where the runway was a gravel strip. No jet had ever landed there previously, and the local mayor along with a large turnout of locals came to see the 146 arrive. The Mayor was presented with a model of the aircraft in the airline's livery.

The aircraft also visited neighboring countries such as Niger and Senegal. Subsequently it was converted to all economy high density seating for scheduled flights to Lagos, Abidjan, and Brazzaville. It also operated weekly scheduled service to Paris, France, and additional destinations including cities in Ivory Coast, Togo, Nigeria and Algeria were added. It was possible for the 146 to fly these routes as it was

the first of the type to be equipped with wing root fillet extended range fuel tanks.

The presence of the 146 at the Paris Air Show also attracted the attention of a very large low-cost carrier (LCC) from the United States, and this would bear fruit for BAe before the end of the year.

BRAZIL
Santarem
Belém
Altamira
Itaituba
Porto Velho
Ji-Paraná
Alta Floresta
Vilhena
Cuiabá
Operations from 2 Jan 84
Demonstrations
TABA

TABA

TABA Air (Amazon Basin Air Transport Company) handled air traffic to many destinations in the Amazon basin and northern Brazil. It was originally an air-taxi operator, and with many its destinations being remote, its customers had a choice of travel: air or boat. Air travel sometimes involved trips of four and a half hours (with stops), but some boats took up to 38 days to get to their destination. Four hours in an airplane represented lightning speed in comparison to surface transport, and was quite a compelling alternative.

Bruno Gibson, son of TABA's founder (and commercial director of the airline) had pitted the Boeing 737-100, McDonnell Douglas DC-9-40 and the Fokker F.28 against the BAe 146. With the airline's average stage length being just 342 miles, the Boeing and McDonnell Douglas aircraft were soon eliminated from consideration. TABA's routes could take the 146 up to 1,550 miles from its main base, and the four engines meant the aircraft could be ferried back to base with an engine out if necessary, while the F.28 would have to remain grounded until parts and the personnel were brought in. The point made by BAe that the engines could be worked on while still attached to the aircraft became another plus. A further advantage Bruno cited was that a spare engine could be carried in the hold of the 146, something that the F.28 could not do. Finally, many destinations were not only remote but had unpaved runways, which the 146 was designed to handle. The perception was that the F.28 was at the end of its development life while the BAe 146 had just begun, making the choice easy.

TABA leased two -100s carrying a colour scheme based on the blue house livery of the British Aerospace 146 demonstrator E1002, merely swapping the titles from BAe to TABA. The airline planned to purchase more BAe 146s in 1985 if the aircraft performed as expected. The first aircraft arrived in Brazil on December 8, 1983, configured with 80 seats, after being used as a demonstrator at the Paris Air Show. The ferry flight from Hatfield to Brazil was the longest sector recorded by a 146. Instead of taking a route over the North Atlantic and then through North, Central and finally South America, the aircraft headed to the West Coast of Africa via Liberia, then over the Mid-Atlantic at its shortest point of 1,880 nautical miles to the Brazilian town of Natal. The flying times were more than four hours but the aircraft still had ample fuel reserves for loading and possible diversion. TABA was based out of Belem on the Equator near the mouth of the Amazon, and put the aircraft to work on routes between Manaus, La Cuiaba, and Ji Parana. It also became the first quad jet to operate from Rio De Janeiro's Santos Dumont airport, where the short 4,200-foot runway had forced airlines to use turboprop aircraft. The aircraft was regarded as being perfect for the Brazilian Air Bridge out of the Rio airport to downtown Sao Paulo. TABA planned to fly the 146 on up to 15 sectors per day, with one aircraft on the Belem-Manaus-Tabatinga route and the second operating Belem-Cuiba-Porto Velho.

The 146 was finally beginning to pick up a few sales, with deals having completed in both North and South America, as well as Africa and its home country the United Kingdom. Additional orders were still trickling in but the 146 needed a commitment from a large carrier to establish it as a serious competitor to other regional aircraft, as well as moving BAe into Boeing and McDonnell Douglas jet territory. The 146-200 was in a good position to begin replacing large turboprops and small jets like the DC-9 aircraft on certain routes and after some benign inquiries, British Aerospace received its largest order yet from a LCC (low cost carrier) based in sunny California. This commitment was to finally put the 146 on the map as an aircraft to be taken seriously.

Demonstration aircraft at the Paris Air Show.

Cabin of TABA during inaugural service.

TABA
BELÉM
PT-LEP

CHAPTER 05

Catch Our Smile, British Aerospace

San Diego, California: May 6, 1949 marked the inaugural service of what would become the largest low-cost carrier (LCC) in the United States, Pacific Southwest Airlines (PSA). Before deregulation in 1978 the federal government of the United States, through the Civil Aeronautics Board (CAB), controlled everything relating to interstate airlines from route awards to the fares that could be charged. However, PSA was not an interstate carrier, but an intrastate carrier and thus not subject to CAB route and fare rules. Instead the state of California, where PSA was based, and the respective Public Utilities Commission had oversight of PSA's routes and pricing. Only when PSA elected to fly beyond the State of California would it become subject to CAB airline route awards and pricing.

PSA began with a leased Douglas DC-3 aircraft, flying from San Diego to Oakland with a stop in Burbank. Compared to its competitors at the time, it charged almost 30% less for the same seats. The airline grew over the years, thanks in part to the large naval base near San Diego airport. PSA even earned a nickname as a result of the cash strapped navy servicemen who often chose the carrier to fly to other parts of California: "Poor Sailors

Airline". PSA also grew in popularity with the public, not just because it made flying more affordable, but because of the friendly nature of the entire operation. PSA soon adopted the slogan "Worlds Friendliest Airline", and its founder Kenny Friedkin encouraged the pilots and flight attendants to joke with passengers. This mantra became popular, and advertising material was introduced featuring PSA aircraft with a smile. That led to the infamous 'smile' on the nose of the aircraft. There are several stories explaining how it came to be, ranging from a painted line to help align radome installation on the nose to a marketing stunt to emphasise the "Worlds Friendliest Airline" slogan. Either way, the smile became standard after it received an enthusiastic reception from the public.

PSA continued to grow during the late 1960s, adding more flights and larger aircraft to accommodate the increasing passenger traffic. It introduced brand new McDonnell Douglas DC-9s, Boeing 737s and Boeing 727s to its fleet to supplement the 92-seat Lockheed Electras that had been the primary fleet. The early 1970s brought even greater changes: a new livery (pink stripes), new flight attendant uniforms (the hot

Sales presentation model given to PSA during initial discussions.

Presentation folder for PSA and the BAe 146.

pants era), and widebody aircraft. One of those did not last long. Manufacturers had been marketing larger aircraft to PSA for its popular and often sold out services up and down the California coast. McDonnell Douglas pushed the DC-10, Airbus the A300, and Lockheed the L-1011 TriStar. PSA, convinced it could fill the aircraft with what was then very cheap fuel, ordered five L-1011s with delivery of the first two set for 1974, and one a year until the order was fulfilled in 1977.

The L-1011 turned out to be a disaster for PSA. The oil crisis hit in 1973, resulting in fuel shortages and price increases from an average of $0.10 per gallon to more than $0.33 a gallon. The TriStar might have been profitable if it had been filled with passengers and cheap fuel, but almost overnight one of its key selling points was wiped out. To add insult to injury, PSA could not turnaround the L-1011 at airports as quickly as it could its narrow body aircraft, and as a result utilization was low. The aircraft's capacity and size also meant PSA could not offer the frequency it was able to provide flying smaller aircraft. PSA was the only airline to take the option of a 'lower lounge' for passengers (and integral air stairs that enabled it to serve airports that could not support larger aircraft), eliminating the lower galley and cargo space that made the type suitable for longer-haul operations. Burdened with aircraft that were not making money and configured in a way that made them unappealing to other airlines, PSA struggled to dispose of the two L-1011s that had already been delivered. After eight months, the airline discontinued L-1011 operations and parked them in the desert while buyers were sought. The remaining orders were cancelled, and PSA never ventured into widebody territory again. In fact, it's longer term survival and future would hinge on smaller aircraft.

With exception of a couple bumps in the road, PSA continued to grow and remain profitable until post 1978, when deregulation of the airline market occurred. From 1980 onward, PSA faced a massive increase in competition from existing airlines as well as new entrants. The restrictions imposed by the CAB were dropped – any airline could fly anywhere it wanted and charge whatever price it wished. PSA's network was expanded to include destinations in Washington and Oregon, along with Nevada, Idaho, and eventually international routes were opened starting with Mexico. This was also the period when PSA began to adopt what became the gospel of today's low-cost airlines: a simplified fleet.

PSA became the launch customer for the new fuel-efficient McDonnell Douglas MD-80, placing an initial order for 10 and ultimately taking 35 MD-80 series aircraft. Plans envisaged they would replace the existing Boeing 737s and some Boeing 727s. During the early 1980s (1980-1983), the airline began looking for another aircraft type to supplement the MD-80 fleet. Airbus pitched the A310 and Boeing the 757, the latter garnering some serious consideration within PSA. The Boeing 757 would have given the carrier a capacity increase over the MD-80 series and would have enabled the opening of additional routes that the MD-80 was unable to serve.

With the PSA order for 20 aircraft, the production line consisted of mostly aircraft for PSA. *San Diego Air and Space Museum*

Paul Barkley, President of PSA (center), with Byron Miller, Vice President of PSA on the right.

Delivery ceremony for the first two PSA BAe 146-200A aircraft.

On its home turf, PSA was starting to see the results of increased competition: financial losses. Additionally, airports in noise sensitive areas were limiting daily aircraft movements, and the situation was expected to get worse as neighbours increased their complaints. Byron Miller, Vice President of Market and Fleet Planning watched the BAe 146 during a flying display at the 1983 Paris Air Show, noting how quiet it was. Byron said "we operate in noise sensitive environment and we needed the quietest airplane available." PSA sent Doug Gentzkow, Director of Pilot Training and Don Lewis, Director of Flight Operations to England to evaluate the 146 in September 1983. After undertaking some flight checks they reported back: "The 146 is one of the best planes in terms of handling and quietness."

President and Chairman of PSA Paul Barkley accepting the log book for the first aircraft.
Colorized retouching by Benoit Vne

On November 16, 1983, PSA purchased 20 BAe 146-200 aircraft, with options for an additional 25. It was the largest order the manufacturer would receive from a passenger airline for a decade, and it was exactly what British Aerospace needed: an established and well-known airline buying a large number of 146s, demonstrating it was a formidable competitor in the marketplace. One airline's decision to actually downsize it's aircraft fleet would become an extremely positive marketing point that British Aerospace would use to sell the effectiveness of the 146 to other carriers. PSA bought the BAe 146 series to replace its Boeing 727 and Lockheed Electra fleets, and the BAe 146-200 would burn 50% less fuel than the Boeing 727s that PSA was already operating. Deliveries to PSA were scheduled for eight aircraft in 1984, and the remaining 12 in 1985. Because the backlog of deliveries to other carriers was rather small, PSA would receive nearly all of the next 20 aircraft to come off the production line. Starting with line numbers E2022 through E2025 there were very few breaks all the way up to E2048, although PSA subsequently took up four options that were airframes E2072 to E2075.

The order valued at approximately $300 million USD encompassed aircraft, including engines and spare parts. Its first ten aircraft were acquired on leveraged leases, and the remaining ten were purchased outright. But the order wasn't a slam dunk for BAe. Byron Miller, Vice President of Market Planning for PSA said that "there was some (internal) resistance" to buying the 146. Only a couple of years earlier, PSA had been looking at larger aircraft such as the Boeing 757 and 767, along with the Airbus A310. Miller had said that the large order was "a hell of a gamble" and that PSA "had our eyes open – we knew that (if the strategy worked), we'd end up being the big frog in the pond."

Paul C. Barkley, PSA's President and COO said "We chose the 146 after careful evaluation of all available aircraft because it is the only one to satisfy our requirements for fuel efficiency, low operating costs and the lowest possible noise levels." At the time of the order, PSA was operating a fleet of 24 McDonnell Douglas MD-80 (Super 80) aircraft, along with eight Boeing 727-200s and four McDonnell Douglas DC-9-30s, the latter two types being phased out. British Aerospace Chairman Sir Austin Pearce called the PSA order "probably the most significant development in British Civil aviation in the last 20 years." An accurate and prophetic statement, and once the order was placed, McDonnell Douglas dropped its projected short body challenger, the Super 90.

Delivery ceremony for the first two PSA BAe 146-200A aircraft.

Colorized retouching by Benoit Vne

The order increased British Aerospace's 146 production backlog significantly, increasing the number of firm commitments to 38 including the PSA order, along with another 40 options. At the time PSA represented more than 50% of the total BAe 146 orders and 50% of the options. Byron Miller went on to say "We bought the airplane because we felt it was the right size for our uses, and because of what was happening in the California marketplace."

PSA was not only not buying larger aircraft but was adding smaller aircraft that would enable it to offer increased frequency to its business travelers. The 146 would allow the airline to carry fewer passengers in off-peak hours and still remain profitable. The MD-80s would be used during peak hours where capacity was needed, while the 146 could fill in at other times. But that wasn't the only reason why PSA purchased the 146. It could be used to open up smaller markets that linked into the airline's wider route structure, demonstrating another characteristic of the aircraft that British Aerospace promoted heavily: feeder liner. At the time it placed the order, PSA was using San Francisco (SFO), Los Angeles (LAX), and San Diego (SAN) as its major hubs. The 146s would provide additional traffic from locations that could not support the MD-80 due to its physical size or lacked sufficient traffic to fill the larger aircraft. The 146s would bring passengers from smaller cities (e.g. Eureka, California) into a PSA hub such as San Francisco, where they could then continue onto another PSA or a partner's flight (Northwest Orient or Air Canada at the time). Unlike most LCCs today, PSA actually allowed passengers to connect onto other flights and also interlined their bags to partner airlines.

PSA Employee Jay Quarles named the first aircraft "The PSA Smile". Subsequent naming convention for aircraft would have "The Smile of" and the name of a city PSA served.

Photo by San Diego Air & Space Museum Colorized retouching by Benoit Vne

Prior to delivery of the first of their own aircraft PSA, working in conjunction with BAe and Air Wisconsin, brought Air Wisconsin's third aircraft N603AW to San Diego from where it flew to Burbank and San Jose for noise demonstrations. Both airports have stringent ordinances with regards to aircraft operations. An MD-80 taking off from Burbank generated a 97.6dB reading while a 727 generated a 106db reading. By comparison, the BAe 146 barely registered 82.3dB. At San Jose airport, the BAe 146 registered 72.5dB against 90.3dB from a Sabreliner business jet. During one take off test, the BAe 146 made a turn during its initial climb out and no reading at all was obtained. Even during a 500- feet flyover, the aircraft failed to register on the sound meter. Nearly a dozen television stations covered the demonstrations in San Diego and Burbank.

Name PSA's newest plane and win it for a day!

We've just committed $300,000,000 for 20 of the most advanced, most exciting jetliners ever built—the new British Aerospace BAe146. And we've taken an option on 25 more!

Listen To The Quiet

How quiet is it? Unless you see PSA's new plane flying by, you probably won't know we're around. Its four engines produce less noise than any other commercial jet aircraft in the world.

Enjoy The Benefits of More Frequent Flights

PSA leads the airline industry with a high performance fleet of DC-9 Super 80's, famous low fares and a convenient schedule of frequent flights. With the addition of 45 new BAe146's, PSA can increase the number of flights available to you all over the West by up to 100%.

4 dependable engines and innovative high wing placement give PSA's BAe146 unsurpassed lift, resulting in the fastest climb rate and the shortest landing distance of *any* commercial jetliner. Now PSA will be able to fly you to more cities more often.

Experience The Ergonomic Appeal

"Ergonomically perfect" seating design means that our seats have been designed for ideal human comfort. Lavish legroom, fold-down middle seats (when available) for first class elbow/desk space, and splendid viewing from every window are intelligently coordinated for added passenger appeal. Easy-access overhead storage and more underseat storage than on any other airline add to the inner space. All this in a spacious, wide cabin and, of course, the BAe146's quiet interior gives you an exceptional ride.

What Would You Call This Perfect Plane?

PSA's new jetliner is so advanced only one small detail is missing—its name. We call it "the most terrific concept in flying convenience ever developed." But our plane painter tells us that won't fit. So please help us. Suggest a name. If it's selected, we'll call PSA's new plane *yours* for a whole day!

Win PSA's New Plane For The Most Exciting One-Day Party Of Your Lifetime!

Win a chance to do something very few people will ever have the opportunity to do. Win the world's most perfect jetliner for your personal use for an entire day. A mile-high, jet-setting, party-in-the-sky day! As our Grand Prize winner, you'll receive the new PSA plane complete with pilot, flight crew, flight attendants and seats for 99 of your closest friends! We'll fly you between any two PSA cities. We'll even provide up to $5,000 worth of food, refreshments and ground transportation at your destination!

25 Second Prizes: PSA round-trip tickets for 2 between any 2 PSA cities.

1000 Third Prizes: PSA T-Shirts emblazoned with the perfect plane's new name!

The 146s could not have come at a better time for PSA, which was struggling to fill its MD-80 aircraft. It turned out smaller was better for PSA. The airline wanted to come up with a name for its new aircraft type, and ran a contest seeking suggestions. A brochure, which included an entry form featured a colour caricature of the front of the 146 in PSA livery (smiling, of course) with the caption "What would you call the worlds quietest jetliner?" Dubiously, the rules mentioned "the winning person who submits the chosen name – matching the "official" name designated by Lippincott & Margulies, Inc.", suggesting that a choice had already been made and the contest merely required entrants to submit the name that the agency had already selected. The winner, Dr. Hugh Jordan of Whittier, California, got to take 99 of his friends on a round-trip flight to a destination PSA served, as well as enough money to cover food for his entire entourage. And the name that was chosen? Smiliner.

The first semblance of a PSA Smiliner was to arrive in the form of G-OPSA. The series -100 aircraft carried British Aerospace's demonstrator livery, only this time it had a smile painted under the nose and eyebrows above the cockpit windows. The test aircraft, formally G-SSHH, was refurbished inside with a passenger interior that could accommodate 74 passengers, a large forward galley for hot meals, and two bathrooms at the rear along with wardrobe space.

On board the aircraft was a substantial amount of sales material that was passed out during the visit. It was flown into the United States for a 16-day tour by Captain D. Gurney and Mike Goodfellow, arriving on March 17, 1984. While in the USA, G-OPSA was flown to Stratford, Connecticut; New York, Washington, Miami, Dulles, Denver, and Phoenix. Over the following next six weeks in April and May of 1984, G-OPSA was used to train and provide a type rating for more than 60 PSA pilots, with the airframe accumulating more than 600 hours and nearly 1,400 cycles. The aircraft was flown from San Diego to Blythe, California each day where a series of landings, take-offs, and touch-and-gos were completed by each pilot undergoing training.

BAe rep Clive Nicholson (who was stationed at PSA in San Diego, along with fellow BAe reps Roger Pascoe at Burbank, and Derek Taylor at LAX) claimed that they were the only persons permitted to work on G-OPSA, because it carried a foreign registration. Clive noted that the PSA pilots were going through tires every two days. At one point, some of the pilots got annoyed after a day of practicing touch-and-go landings. Once the aircraft stopped and even before it had been powered down, the pilots noticed the aircraft tilting to the left side. The BAe reps had begun jacking up the aircraft to change the tires before the pilots had fully shut the aircraft down.

PSA's Chairman and CEO Paul Barkley commented "we have seen the aircraft averaging nine hours of flying and 20 departures every day. We are very encouraged to find the 146 lives up to British Aerospace's claims for day-to-day regularity in this manner. The whole airline is enthusiastic about the 146 and is looking forward to seeing it in service during June." The aircraft was shown to the public during PSA's 35th anniversary celebrations and at an employee event before being reregistered as N5828B, after which it was used in passenger service for five months.

The first two BAe 146-200 aircraft were handed over to Paul Barkley on May 30, 1984 at 11:45am, in a spectacle of a delivery ceremony. Being the local and official airline of Disneyland, it was no surprise to find that Mickey Mouse and Donald Duck were present during the delivery of the two aircraft. How do you get an aircraft with only about 1,600 nautical miles range to its new home over 5,000 miles away? Both engaged in long multi-stop ferry flights which became the norm for PSA's 146s. The aircraft would make a hop to Reykjavik, Iceland where the crew would overnight. From there, the longest leg of the journey continued with stops in Sondre Stromfjord, Greenland; Goose Bay, Labrador (both of these were the most desolate airports on the trip) before an overnight stop in Montréal, Canada. The final part of the trip had the aircraft landing in Green Bay, Wisconsin (and clearing customs), followed by Kansas City, Salt Lake City, and after 19 hours of flying, an arrival in San Diego, California. Further BAe 146 deliveries involved aircraft painted in green primer and without seats, which were delivered to and installed by Standard Aero (a subcontractor to British Aerospace) in Phoenix, Arizona. The same firm also applied PSA's colourful livery.

On June 14, 1984 the first two BAe 146-200s painted in full PSA colors arrived in San Diego to a full celebration with employees and management. As the aircraft descended towards the airport, with approval of air traffic control, both performed a formation fly-by before landing. The ceremony was attended by the press and PSA employees. Maintenance Inspector Jay Quarles, the oldest PSA employee in terms seniority, was given the honor of christening and naming the first aircraft "The PSA Smile". Naming would become a trend with the 146 aircraft in PSA's fleet. But there were other changes too. Not only did PSA take delivery of a brand-new aircraft type, not only would each carry a unique name, but PSA introduced new uniforms for flight attendants and station personnel. The hot pants that made the airline popular in the 60s and 70s were out, and more professional and traditional uniforms were introduced.

PSA was the official carrier for Disneyland, and they brought out park mascots Mickey Mouse and Donald Duck to the delivery ceremony. With Paul Barkley.

The interior that PSA chose for the 146 was a complete opposite that in the MD-80 and Boeing 727 series that the airline was already flying. Alternating orange and purple flower-patterned seats that adorned those earlier aircraft would be retired, the 146 featured tan colored seating with multi-coloured square pattern that created a light and airy interior. The layout was a 3-3 configuration, with a 34-inch pitch and a seat width at 16.5 inches, which allowed the aircraft to accommodate up to 100 passengers. There was a smoking section in the rear of the cabin. PSA's 146s also incorporated a revised high capacity overhead locker to cater to high frequency operations involving carry-on baggage. When the aircraft wasn't full for a flight, there were restrictions prohibiting seating in last two rows as well as the first row for weight and balance purposes.

146 in Operation for PSA

PSA initially inaugurated 146 service from Burbank, though the type was rapidly rolled out to other markets. Reaction to the new aircraft was initially generally good, but for those seated next to the windows it was a bit of a tight fit, especially for taller or larger passengers due to the curvature of the fuselage. Some customers said that they preferred the MD-80, which featured 2-3 seat layout. The 3-3 seating layout in the 146 was tighter than expected and the curvature of the fuselage impinged on personal space, leading to passengers 'rubbing shoulders' with each other. Investigations revealed that PSA's purchasing department had selected an inappropriate seat height, and passengers were becoming quite vocal about it. The answer involved the manufacturer, Fairchild Burns, lowering the height of the seat rails by 2.5 inches which increased headroom and eliminated the tight fit for those sitting next to the window.

PSA
N346PS
PSA

Jeff Zaruba

PSA also changed the layout to five-abreast mimicking the MD-80 series. This reduced accommodation from 100 to 85 passengers while increasing the width of each seat to 18.5 inches, which resolved most of the complaints. The changes to the interior layout took place in early 1985, starting with N353PS which arrived from Hatfield on its ferry flight to Phoenix. All aircraft already in service prior to N353PS were retrofitted by May 1985, and further new aircraft delivered with the revised configuration.

Passengers appreciated the light-colored interior, indirect lighting, and reported the uncluttered lines as "soothing", with 92% giving the PSA's configuration a favorable rating. One area where male passengers weren't a fan of the 146 was the bathroom. In a complaint sent to PSA (and forwarded to BAe), a customer suggested British designers either did not know or did not care about the bathroom, with the design suitable more for 'little people'. The severe curvature of the sidewall and the small space ensured anyone with any sense of height would have 'trouble' going to the toilet.

Another early operational issue that gave rise to complaints was oil fumes from the APU leaking into the cabin air filtration system. This would plague the 146 throughout its life, and a documentary claiming it created health issues was released in 2011. Reportedly, PSA only learned about the problem from crews who

were exposed to it over long periods every day. British Aerospace had a resolution within three weeks that supposedly eliminated the problem, but the APU along with the engines would continue to plague the 146 throughout its service life. Other items needing attention included the door seals and baffle system, the inadequacies resulted in higher than normal interior noise during flight. The inflight PA system also initially suffered from static interference, making it difficult for passengers to hear announcements.

Unfortunately, the AVCO-Lycoming engines also proved to be problematic. Dispatch reliability was 98.2% initially, but the airline suffered many operational issues relating to the powerplants. It was suggested that PSA staff coined the phrase Bring Another Engine as an acronym for BAe. As the largest operator of the type, and with the spares available (because of the large fleet), when there was a mechanical issue PSA could remove the problematic engine and install a fresh one in its place. But it came as no surprise that the teething troubles and maintenance issues did not create a favorable impression of the aircraft. Despite the problems it was having with the new aircraft, in a letter to shareholders PSA noted that the 146 had a 31% lower fuel consumption per flight hour vs. the MD-80 (and 49% less than the 727). It also required fewer crew members per flight which results in lower per trip costs when compared to other aircraft. But the 146's maintenance costs per flight hour were higher than expected (and compared to the MD-80s that PSA also flew), and the engines had to be put through more frequent overhauls to keep them running.

Over time, the AVCO-Lycoming engines would become a source of frustration for many airlines. Given the number of BAe 146-200s PSA ordered (not to mention delayed deliveries due to a labour strike at BAe), BAe had to loan the airline two BAe 146-100 aircraft to cover flight cancellations arising from mechanical issues at Los Angeles and San Francisco. The two -100s involved were the demonstrators with the original interiors, and they were not reconfigured to PSA standards. Therefore, they had the brown camouflage-like interior, and six-abreast seating. The Spare Aircraft Reserve Crew (SPARC) was kept on standby at a local hotel and were called out if one of PSA's 146 became inoperative and a replacement had to be deployed. There were no cellular phones in the mid-1980s, and pagers were very pricey to own and maintain service. The result was the SPARC crews were generally confined to their hotel rooms in order to be reachable should a call come in.

Clive believed that the ALF-502 was not a mature power plant for airline operations at the time. During descent at low or even idle power, the air pressure put pressure on the fan which subsequently put pressure on the turbine, resulting in wear and ultimately leading to the need for an engine change. During the first four months of operation, between all the PSA bases where maintenance was performed, 48 engine changes were undertaken. This was far in excess of what a typical airline would expect operating any other aircraft. While the BAe 146 engine was designed to ease maintenance, it turned out changing modules was not particularly practical and customers found it quicker to just swap powerplants. Initially it took six hours to replace an engine, but PSA developed its own change kit and cut the time down to four hours. Every engine change was estimated to cost around $250,000 USD in parts and time to accomplish. While the procedure for changing an engine was refined, it still took eight hours to change an APU and mechanics found accomplishing the task was challenging. A lot of that went back to the designers of the aircraft and its systems not spending time in the assembly hall, assessing the effort required to access and replace defective components.

Despite the teething problems, PSA used the 146 in conjunction with MD-80s to create the "PSA Expressway", which featured flights between Los Angeles (LAX) and San Francisco (SFO) every 30 minutes. It was a compelling proposition for business passengers who wanted the convenience of frequency. PSA also continued to use the 146s for non-peak travel and to connect smaller airports and destinations into the PSA Expressway, making for efficient hub-and-spoke travel. Fred Lind, Vice President of Engineering and Maintenance said "Our clientele are schedule seekers, the main thing is convenience and reliability. The 146 has allowed PSA to run to the limits of noise curfews at airports. This allows the airline to more efficiently use its airplanes to schedule for a passengers convenience."

The 146 Competitive Advantage: Noise Sensitive Airports

PSA was about to learn just how important the "Quietliner" was in California markets, specifically at Orange County Airport (also known as John Wayne Airport/Santa Ana-SNA).

John Wayne Airport lies approximately 40 miles south of Los Angeles International, where PSA had a hub. Technically in the city of Santa Ana (and on the fringes of Santa Ana, Costa Mesa, Irvine, and Newport Beach), the regional facility caters to business travelers as well as recreational flyers. The neighbors to the south, in particular Newport Beach, have always been very vocal in complaining to the county about aircraft noise and pollution, and as a result, have generally stifled growth at SNA for much of its life. The noise abatement rules at SNA are the strictest in the country, with ten sound meters taking measurements along

Center
6 across vs. the later 5 across seating.

Below:
PSA would assign seats at the gate, and put a sticker of the seat assigned on a passengers ticket. This is an unused sticker sheet.

Roger Pascoe

the take-off routes south of the airport. Any airline that is caught exceeding the noise limits are penalized financially and risks losing the limited number of aircraft slots it has been allocated. To make operations even more difficult, the runway is one of the shortest in the US that is used regularly by commercial airlines, measuring a mere 5,700 feet. The runway length and noise abatement procedures often mean commercial jetliners cannot not take off fully loaded, with either passenger or cargo capacity having to be reduced as a result. This makes it more difficult for airlines to make money, and flights out of SNA are typically priced higher than those from other local airports.

The runway length and noise rules at SNA were an issue for operators of Boeing and McDonnell Douglas aircraft. But the challenges posed by operating through SNA were not a problem for PSA's BAe 146s, which could take off fully loaded and without any range restrictions. PSA pilot Chuck Ross said that even though the 146 was a quiet aircraft, the airline's procedures dictated a power reduction after takeoff to avoid triggering the sound measuring equipment in any way. This contributed to the BAe 146 being allowed (technically) unlimited operations in and out of SNA.

PSA was the first airline to receive BAe 146 simulators.

The authorities at SNA originally set a noise limit of 98dB, and any aircraft that exceeded that threshold were banned from operations. Noisy aircraft were allocated a limited number of operational movements ('slots') per day, which was capped at 41 shared between all airlines pre-1985 (in 2019, it's 89). For airlines using turboprops this wasn't an issue as they all fell below the noise threshold. But the constraints meant that airlines using older DC-9 and 737 series aircraft soon had to cease using them. The 'home' airline at SNA was none other than PSA competitor AirCal (formerly Air California). AirCal had recently rebranded, changed its livery and purchased new fuel-efficient aircraft to operate out of SNA. The airline took delivery of a brand-new Boeing 737-300 the week of February 10, 1985, and used it for noise tests with county and airport officials. With a full load of employees who volunteered to travel on the test flights, fuel, and ballast to simulate a load of 113,700 pounds, the aircraft registered between 90.6dB and 91.7dB noise. In comparison, an American Airlines MD-80 that departed previously registered 96dB. If AirCal operated their 737-300 aircraft with a weight penalty limiting fuel, range and capacity to just 100 passengers (somewhat below the normal 140 passenger capacity), the aircraft recorded just below the 89.5dB limit at which SNA offered an incentive.

PSA brought the 146 to SNA for testing on February 15, 1985. Upon departure, with 80,000 pounds capacity the aircraft registered less than 86dB (82.5-84.9dB). It ushered in a new era of quiet aircraft to the boisterous local community who wouldn't put up with the slightest increase in noise. Thomas F. Riley, County Supervisor for Orange County, was at the airport for tests. Thomas called his friend, Roger Luby, who lived under the flight path in the upper Newport Bay for his reaction. Roger responded "we were standing out in the backyard, and we applauded, the dog applauded, the gardener and even the maid applauded. I guess if you're going to have problems, it's best to have these kinds." The county wanted a few more weeks to devise a plan as the BAe 146 registered noise levels so low (86dB) that it would have been exempted from the county's cap of 41 flights per day (which was scheduled to increase to 55 flights on April 1, 1985). The airport could not consider revising the noise profile to below 86dB as it would prohibit some commercial and even some general aviation planes. Airport Manager Murry Cable said "we'd have everybody against us: the FAA, the airlines, we'd be out there by ourselves in never-never land." However, county officials were actively discussing options to prevent PSA from qualifying for an unlimited number of flights for the BAe 146.

But the noise trials did not go perfectly. One of the sensors kept giving an alarm that the noise limit had been exceeded. After a number of incidents one of the pilots drove over to the sensor reporting high noise while the test aircraft taxied to take off. The pilot and BAe's customer service rep parked near the sensor

How PSA Plans to Save Money With the BAe 146
Operating costs per flight hour*

	BAe 146	Super 80	DC-9-30	B727-200†
Labor	$840	$1,050	$1,006	$1,252
Fuel	697	1,024	956	1,430
Other (Landing fees, parts, insurance)	154	197	356	261
Total	$1,691	$2,271	$2,318	$2,943

* Does not reflect depreciation, interest expense or new wage-concession package.
† Company sold off the last of its 35 B727-200s last month.
Source: PSA Inc.

PSA branded promotional foam glider. This could be assembled and flown.

Shown is the noise profile out of Burbank airport for the PSA BAe 146 in comparison to the MD-80 and Boeing 727-200.

which was in a homeowner's backyard surrounded by a high wooden fence, and waited for the aircraft to fly overhead. As the aircraft took off and began its steep ascent, they saw a man come down the street with a stick in his hand. The man ran the stick across the top slats of the fence, irritating a large dog in the yard which started barking. It was this event that caused the noise sensor to give erroneous readings, and was clearly premeditated. The PSA staff approached the man who immediately took off running around down the street and disappeared around the corner. The behavior and the event that had been witnessed provided a clear explanation of the erroneous data to airport officials, and there wasn't a repeat event. Turning off the APU on departure also further reduced the noise footprint of the aircraft on departure.

PSA was so confident that it could acquire additional takeoff slots at SNA that it began taking reservations several months before increasing its scheduled operations, even though it did not have formal approval for the extra services. PSA representative Margery Craig acknowledged that the airline began selling tickets on unapproved flights out of Orange County because of the deadline for filing schedule changes with the Official Airline Guide (OAG), although PSA had been given two additional daily flights during the 90-day test period. PSA's BAe 146 aircraft equipped to carry 89,500 pounds which would give it a range to fly to Kansas City and still registe below the 89.5dB limit (it had registered as low as 87.4dB).

John Wayne Airport, on the other hand was pushing back, claiming that an environmental study could take well in excess of six months, after which a report and decision would be be made. In the end, after a short time in service, John Wayne Airport began offering any carrier operating an aircraft that registered below 89.5dB a two-for-one swap: remove one aircraft/flight that registered above 89.5dB with a quieter aircraft, and two slots would be allocated. That meant PSA's 20 slots at SNA could be doubled to 40 aircraft movements per day. That put PSA in the drivers seat in what was becoming a very crowded LCC market with AirCal, Royal West, Western Airlines, and United Airlines all trying to enhance their operations in the California corridor which PSA typically had 50% of. Airlines who did not operate or plan to operate the BAe 146 – Continental Airlines for example which was using Boeing 737-300s out of SNA – felt PSA should be confined to a fixed number of slots like the rest of the competition.

AirCal found itself in a challenging position with respect to its competitor, PSA, that was moving to dominate SNA. However, the airport gave preferential treatment to the local carrier when it awarded more take-off slots to AirCal than to PSA. Even though the BAe 146 was technically exempt from departures due to its below average noise footprint, the county imposed a limitation of no more than 4.75 million passengers per year through the airport. PSA won a judgement in Federal Court from U.S. District Court Judge Terry J. Hatter Jr., who ruled that the Board of Supervisors showed favoritism to AirCal and required SNA to evenly distribute any additional departure slots. As a result, AirCal ordered six BAe 146 aircraft to replace its MD-80s and supplement the Boeing 737-300s it was using to counter PSA's encroachment into its home base. At the very least, AirCal's 146s would give it 12 additional flights per day to counter PSA.

PSA began to use its BAe 146 aircraft to enter airports that either could not support jet aircraft due to runway length, or were on the noise sensitive list. After LGB (Long Beach), BUR (Burbank) and SNA, PSA focused on Concord, California where the 146 passed noise tests with flying colors and the short runway was not a concern. The FAA engaged in further tests after a small light plane crashed into a local mall which delayed the start of PSA services and added to local concerns about a larger commercial plane doing the same. Not long after, however, the FAA cleared the BAe 146 and PSA to use Concord and they became the first scheduled jet airline flights into the airport.

It was beginning to become clear that the BAe 146 was giving PSA a competitive advantage over its competitors in California due to the aircraft's ability to operate into airports with short runways, and by keeping residents happy with a negligible and unmatched noise footprint. It is the authors belief that had PSA not selected the BAe 146, it would have struggled to survive like other airlines in the California corridor and likely would have shrank, not grown, at an accelerated rate. The BAe 146 gave the airline flexibility when demand wasn't there for larger aircraft; it provided PSA with increased departures out of SNA; it

British Aerospace took the results of the noise test at John Wayne Airport and began advertising the BAe 146 noise benefits.

John Wayne Airport - British Aerospace (BAe) 146 Operational History	1985	1986	1987	1988	1989	1990	1991	1992	1993
Pacific Soutwest Airlines (PSA)[1]									
Total Operations*	6,390	14,156	14,268	12,484	12,100	11,788	3,302	-	-
Avg. Daily Departures	8.75	19.39	19.55	17.05	16.57	16.14	4.52	-	-
Air California (AirCal)[2]									
Total Operations*	-	7,750	9,850	11,590	8,788	3,222	-	-	-
Avg. Daily Departures	-	10.61	13.49	15.83	12.03	4.41	-	-	-
United Airlines[3]									
Total Operations*	-	-	-	-	26	2,082	4,226	4,246	1,894
Avg. Daily Departures	-	-	-	-	0.03	2.85	5.78	5.82	3.47

*Arrivals and Departures

Total British Aerospace 146 scheduled operations out of John Wayne Airport (SNA) for all operators.

opened up new markets and airports that couldn't support larger or jet aircraft; and it offered lower operating costs versus larger aircraft. PSA prospered as a result of the 146.

Boeing recognized the BAe 146 as a threat at noise-controlled airports, particularly for customers flying the new Boeing 737-300. One weekend, AirCal took one of their new 737s up for a test to see if the readings from noise monitors could be improved. According to David Banmiller, former President of AirCal, he and the chief pilot tried something different on departure. They turned off the air conditioning packs, got the aircraft off the runway and retracted the gear quickly, and then immediately idled the #2 engine and reduced power on the #1 engine and 'floated' the plane to the coastline, keeping it around 600ft for the entire time. They repeated these tests on a number of occasions and then approached Boeing and the FAA for approval to change the departure procedures. David and his pilots met with Boeing's chief pilot and engineers, who claimed the procedure impossible until David's team performed a demonstration. They were all surprised, but the FAA said that procedure could not be approved because it felt that it offered insufficient time to spool up the idling engine in case of emergency.

Boeing however, working with the FAA, developed and received approval for a new procedure called a 'deep cutback'. This had the aircraft throttle up against the wheel brakes at the beginning of

the runway, which were then released to 'launch' the jet down the runway. As the aircraft rotated and became airborne, it climbed quickly and steeply. Approximately five seconds after becoming airborne, the pilot would initiate the 'deep cutback' procedure which resulted in N2 power (the high-pressure compressor of the engine) being reduced by up to 60%, reducing noise and avoiding triggering the noise measuring equipment, with N1 (the engine fan) generating more thrust on departure.

A passenger experienced a 'launch' down the runway followed by a quick rotation and a steep climb, then moments later the aircraft seemed to 'fall forward' as the reduction in power took place and a much slower climb was the result. This continued for approximately two minutes until the aircraft cleared the coastline at Newport Beach, when normal power and climb was resumed. The procedure makes passengers nervous, and nearly all pilots notify passengers of the procedure just before take-off. The airport receives an average of approximately 400 phone calls complaining about noise every ninety days. Airlines that trigger the noise monitors receive a warning, and if the transgressions continue then a financial penalty is imposed. Further violations result in bans on aircraft and loss of departure slots. But with the 'deep cutback' procedure in place, the Boeing 737-300 now qualified for the unlimited movements rule as well.

Former PSA Pilot Chuck Ross loved to fly the BAe 146 and looked forward to each and every flight. The overhead panel was the "best he ever saw" with the rocker switches more useful than the push buttons Airbus and Boeing use. He explained that pilots can quickly glance at the switch position and instantly tell what setting or mode it is in. With a push button, it is necessary to look for the light or colour of the light, but if the bulb burns out there is no light (making it difficult to tell whether it has been pushed and the light is defective, or if it has not been pushed). The arrangement in the 146 was certainly clearer for the pilots.

But Chuck's favorite feature of the BAe 146 was the tail petal speed brake. Wing mounted speed brakes induce an aircraft shudder, whereas the tail brake did not. In fact it would result in up to a 2,000ft per minute descent with a dead-level deck angle, and passengers wouldn't feel it. Chuck recalls a story from pilot training when a British Aerospace pilot was asked by air traffic control (ATC) to make a steep descent, but the pilot said he was unable to comply. ATC responded "do you have speedbrakes?" and the BAe pilot responded "Oh, I am sorry, those are for my mistakes not yours." Chuck further recalled an event when ATC asked how soon he could drop his speed from 250kt to 180kt, and he replied "Watch this" (a phrase that should be uttered with great caution and co-pilots never permitted to use). He reduced the throttles, deployed the full speedbrake, and then almost as quickly retracted the speed brake and increased throttles to 70-percent. ATC responded "That was great! One sweep of the radar!"

Chuck noted that the BAe 146 was the only jet he ever flew that was permitted (legally) to land with the engines unspooled (at idle). In most jets an RPM must be maintained that permits rapid spooling up should a go around become necessary in the flare (just before touchdown). The 146 required full speed brake across the threshold of the runway. Additionally, unlike most aircraft, the BAe 146's fuel burn did not improve much with increased altitude. In rough turbulence at normal cruising altitude, PSA could keep the aircraft at lower altitudes, sometimes as low as 19,000 ft, to avoid the chop while not having a major impact on fuel consumption.

PSA pilots transitioning from the Boeing 727 series (which was being retired) initially had difficulties landing the BAe 146. Former rep Roger Pascoe noted that it was common practice to add about 10 knots to the target landing speed on the 727, as it was a little 'dirty' aerodynamically in the flare and could drop unexpectedly and slam down on landing. Flying a little faster decreased the chances of this happening. Unfortunately flying the BAe 146 a little bit faster produced the opposite effect when flaring on landing. Because of ground effect, which resulted in a cushion of air below the wings pushing up on the aircraft, it would float down the runway far longer. Roger recalls watching one PSA BAe 146 landing at Burbank – the approach was a bit hot (fast) and the aircraft porpoised at least three times while passing over the runway, and never once did both main wheels touch. This prevented the wheel brakes from being armed, and the ground spoilers (lift

As of result of the noise tests in favor of PSA and the BAe 146 out of SNA, AirCal was forced to buy the BAe 146 as well.

proudly introduces

SMILINER

The Shape of Flights to Come

The world's most exceptional passenger aircraft joins the PSA fleet this summer. It's the British Aerospace 146, a four-engine jet unlike any other airliner you have ever flown.

Within the next year, twenty of these smooth and nearly silent new airliners will be serving an expanded PSA network – and in coming years we plan to add twenty-five more. On the following pages you'll see why we think the BAe 146 is the plane of the future; why it is well worth our $300 million investment; and – most important of all – how it will better serve you, the PSA passenger.

We've decided to nickname this new addition the Smiliner. When you've finished reading, we think you'll understand why.

Paul C. Barkley
President and C.E.O.
PSA, Inc.

Even more frequent flights at even more convenient times – all because of PSA's new SMILINER

As we add Smiliners to our fleet you'll find our service in some cities *doubled* in frequency, with conveniently timed new departures and arrivals added to the flight schedule.

Credit Smiliner: as a 100-seat aircraft, it fits PSA's typical passenger load requirements like a glove. So eight or even ten Smiliner flights per day into many PSA cities – instead of four or five flights using larger capacity planes like the DC9 Super 80 – are as beneficial to PSA's operating efficiency as they are to your luxury of choice. Sometimes, both sides *can* win; thanks to Smiliner, this is one of them.

Space to Work...

The center seat in each row of three is designed to fold down flat. We plan to fill these center seats *last* on every flight, so when the Smiliner isn't full you can enjoy the luxurious extra work or elbow space normally provided only in First Class.

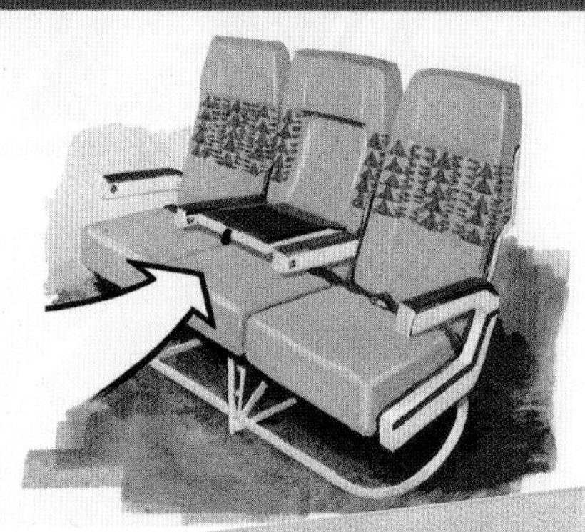

EVOLUTION vs. REVOLUTION

The plane that will evolutionize air travel in the west

PSA revolutionized air travel back in 1949 with deregulated air fares 29 years before most airlines offered them and more flights at more convenient times. Since then, we have led the evolution of convenient air travel with innovative services and a high performance fleet.

Being first in aircraft technology is a PSA habit. When the superbly quiet and efficient DC9 Super 80 first entered U.S. commercial service it flew in the colors of PSA.

Now with 26 DC9 Super 80's combined with 20 Smiliners our passengers will enjoy the benefits of the most modern fleet among all major U.S. carriers.

It just goes to show that while most airlines' aircraft move at about the same speed in the air, some airlines move faster in the area of better ideas.

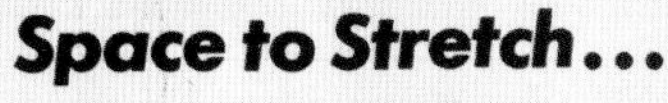

Space to Stretch...

The Smiliner's design ingenuity is all around you – even underneath you, in its remarkably comfortable, remarkably space-efficient new seats. Note that there's almost a full *yard* of space between one seat back and the next. And that you can stretch out your legs under the seat ahead, instead of jamming your knees against it.

Space to Spare...

Your Smiliner seat stands a full *12 inches* above the floor, creating more under-seat space than any other airliner can offer. Imagine the benefits in terms of extra footroom and space for your carry-on luggage. It's one of those little differences that at times can make all the difference.

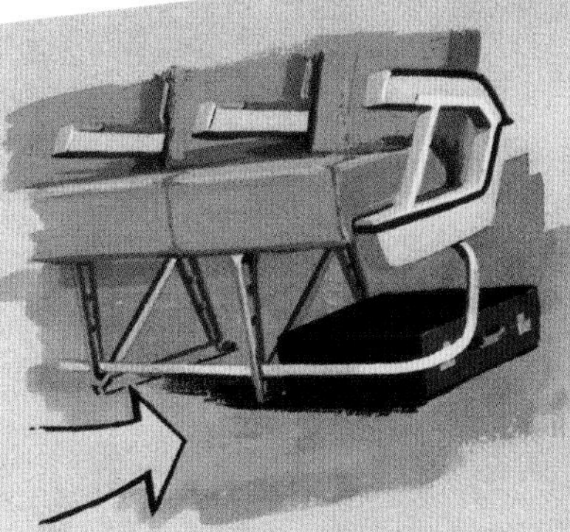

Space to Stow...

Overhead stow-away bins in the Smiliner are designed to be unusually roomy, affording each passenger plenty of space for the bags and coats and packages that so often have nowhere to go on cramped airplanes – and so often end up in your lap or stuffed beneath your seat.

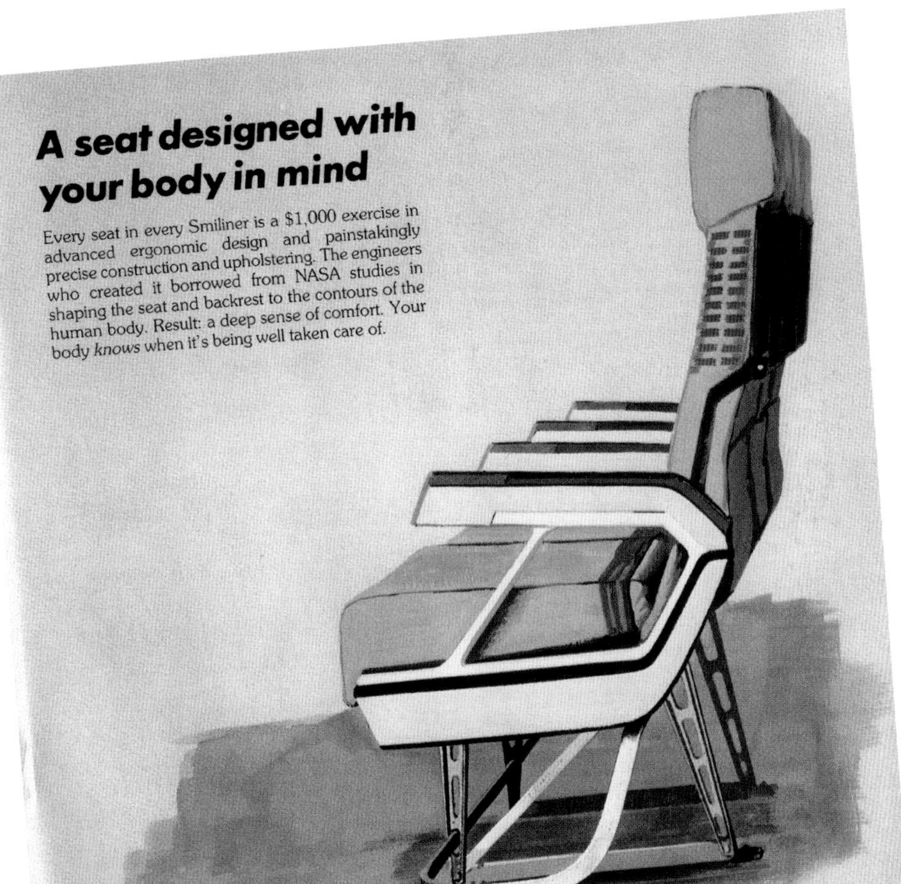

A seat designed with your body in mind

Every seat in every Smiliner is a $1,000 exercise in advanced ergonomic design and painstakingly precise construction and upholstering. The engineers who created it borrowed from NASA studies in shaping the seat and backrest to the contours of the human body. Result: a deep sense of comfort. Your body *knows* when it's being well taken care of.

Remarkably swift, sure take-offs

The Smiliner boasts one of the fastest climb rates of any commercial aircraft in the world; the combination of its four powerful engines and unique high-wing design, for instance, can lift the plane to 28,000 feet in under 17 minutes – swifter even than some much larger passenger jets.

Strong performance, and *dependable performance;* Smiliner's engines are built by Avco-Lycoming. This United States company is one of the world's most respected aerospace corporations. The engine design is further backed by millions of miles of proven flying reliability.

Height-ft
400
300 SMILINER
200 TYPICAL TWIN JETS
100
Distance from brakes off – '000 ft.
0 4 5 6 7 8 9 10

Landings you'll hardly notice

The Smiliner is designed for smooth landing approaches and soft, jolt-free touchdowns. That high wing configuration requires no leading-edge flaps – so that the plane comes in level, not tail-down. There's no screaming reverse-thrust of the engines once the Smiliner kisses the runway. And a unique trailing-arm landing gear suspension, coupled with super-flexible shock absorbers, makes it almost hard to believe you've actually touched ground.

All this comfort – and incredible efficiency, too. The startling fact is that the Smiliner not only lands smoother and softer, but in a *shorter distance* than almost any other commercial aircraft now flying.

Performance proven

The ruggedly designed plane PSA calls the Smiliner is currently in regular service on some of the most demanding routes in the world.

It's flying passengers and freight in and out of a 4,000-foot landing strip in remote Alaska – the only commercial jet that can do it.

It's serving 14 isolated communities in the steaming jungle terrain of northern Brazil.

It's a familiar sight at such unfamiliar airline destinations as Dakar and Brazzaville and even Timbuktu, in remotest Africa.

It's also the only commercial jet the tidy Swiss will allow to land at their capital city of Berne.

And soon the Smiliner will be flying the routes of PSA, and opening new routes. Helping fulfill PSA's long-standing commitment to operate the most modern possible equipment, with the most frequent and convenient possible flights, at the most affordable possible rates.

We look forward to welcoming you aboard the Smiliner soon.

from Alaska...

to the Amazon

Almost too quiet to be a jet

In terms of noise abatement, this new Smiliner is going to be the best friend ecology has ever had. Fact: it easily meets or beats all existing and planned noise-abatement regulations. Fact: it's a stunning *80 per cent* quieter than the next quietest commercial jet on take-off. Fact: it's been certified by the F.A.A. as the quietest jetliner in the air.

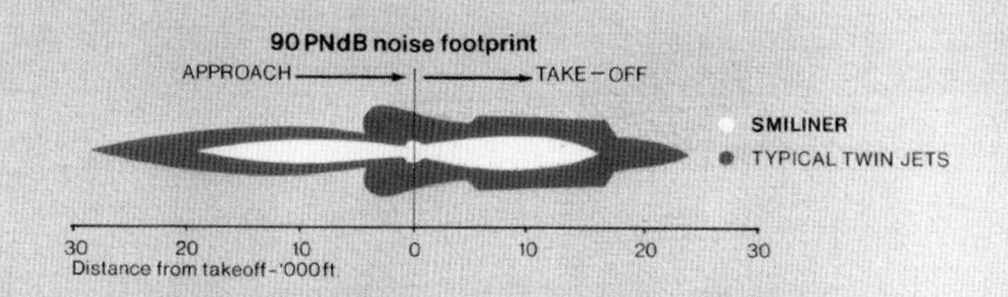

A pedigree of excellence

The Smiliner is designed and backed by British Aerospace – the company that gave the aviation world the supersonic Concorde and such earlier flying legends as the immortal Spitfire fighter of World War II.

No wonder the Smiliner bristles with high performance features – from its automatic back-up systems, to wing skins that are milled from a single piece of metal then *bonded* together for astonishing structural strength.

dumper) from deploying, resulting in the aircraft not being able to slow down. Because it was traveling downhill on the runway, it passed the intersections and the crew ended up having to make a hard-left turn to avoid the fence and railroad tracks at the far end of the runway. The inboard wheels on the left-hand side lifted off the ground approximately 18 inches as the aircraft took the last exit. It came to a stop on the taxiway and, instead of being towed, the 146 taxied to the terminal as if nothing happened. The passengers disembarked none the wiser, oblivious to the event Roger had just witnessed.

The pilots disembarked and reported the brakes weren't working. Roger explained that the brakes would not arm until the aircraft thought it was on the ground (they arm when both main gears are on the ground simultaneously). The captain shoved Roger against a wall and threatened to kill him before the aircraft had a chance to kill anyone else. The captain never flew on the 146 again, but Roger went out to the runway to investigate the tire marks and was impressed by how much rubber was left behind from the turn. The amount of roll required to lift the inboard wheels clear of the ground was remarkable. PSA, as far as Roger knows, never received a complaint about that particular landing.

By early 1986, PSA exercised options on four additional aircraft, bringing its BAe 146-200 fleet to 24 aircraft. These aircraft, when delivered in 1987, were registered to USAir Leasing and Services company.

On November 18, 1986, American Airlines agreed to acquire PSA's competitor AirCal. The motivating factor was that it would give American an immediate presence on the US West Coast, and access to two major airports it did not serve (Burbank and Orange County). Less than a month later, on December 8, 1986, USAir announced it was buying PSA for $400m, it too wanting a strong West Coast presence. Over the next 14 months, PSA began transitioning to operations under USAir. It operated its last flight on April 8, 1988 from Los Angeles to San Francisco, departing at 11:15pm and arriving just after midnight on Saturday April 9, 1988, the transition to USAir now complete. When USAir took over, fares more than doubled nearly overnight. At the time PSA was charging $140 for a full-fare round-trip between San Francisco and Los Angeles. The fare became $296, and it was not just USAir charging so much; American Airlines, United, and Delta Air Lines were also much higher too.

Only six months earlier, USAir starting having PSA begin removing the smiles from its aircraft and repainting them with the USAir maroon stripes on a bare metal fuselage (a livery Allegheny applied before it was acquired by USAir). The smiles disappeared from the aircraft, and soon afterwards from the faces of the employees of PSA as they became USAir. USAir began to impose its corporate culture, which was the opposite of PSA's. USAir issued a notice to flight crews instructing them to discontinue the 'witty banter' and 'humorous safety announcements' that had become a PSA trademark that was enjoyed by its customers. The change was leaked to the press. The self-proclaimed "world's friendliest airline" was having the life sucked right out of it. Everything that made PSA successful and unique had been stripped away by 1991, including the paint from the fuselages of the BAe 146.

Against British Aerospace directives, USAir stripped the fuselages down to the bare metal skins as part of the rebranding, so that they would match the rest of the aircraft in the fleet. British Aerospace advised against it because polishing could thin the fuselage panel skins and possibly introduce cracks. British Aerospace issued a Service Bulletin, #53-87 on January 19, 1990 regarding the polished fuselage skins. The FAA became involved, and after inspecting the aircraft issued an Airworthiness Directive (AD) # 91-08-11 noting the affected aircraft. Working off another British Aerospace service bulletin calling out specific serial numbers (#53-88, and #53-98), aircraft serial numbers E2022-2025, E2028, E2030-2031, E2034, E2036, and E2039-E2048 were affected. The AD required detailed visual inspection of the rivets (including use of dial test indicator and 10x magnifying glass), requiring the noting of loose or missing rivets, as well as those that were damaged or had significant abrasion. It went on to give even more detailed instructions for inspection and maintenance including guidance should the aircraft be painted again, aimed at ensuring no cracking or further damage occurred. Inspections were required every 1,500 flights. This added to the cost of maintaining the 146s, albeit not resulting from actions or shortcomings from BAe, but from USAir's defiance of British Aerospace's instructions.

Brian Thomas, former President and CEO of British Aerospace had previously handled support for the BAC One-Eleven, of which USAir (specifically Mohawk and Allegheny airlines) were operators. With Brian's insight and contacts within what was now USAir, he had very high hopes that after the PSA purchase, the larger carrier could be persuaded to buy even more BAe 146 aircraft. British Aerospace pitched USAir on the virtues of the BAe 146-300, but unfortunately no headway was made because USAir was fairly conservative, and the 146 was just to unconventional an aircraft for it to appreciate. In fact, quite the opposite effect was about to occur with USAir, and no one quite saw it coming. On May 2, 1991, at midnight USAir maintenance began removing all the titles on the fuselages and tails of all the BAe 146 aircraft in the fleet. No one outside the airline was aware of what was happening. USAir ceased BAe 146 operations overnight citing high operating costs, and flew all the aircraft to Mojave Airport in the hot and dry desert of Southern California where they were stored awaiting eventual disposal.

In its annual report USAir took a $21 million charge to park 18 146s in the Mojave Desert, noting the "unacceptably high maintenance and operating costs", and adding that the aircraft cost as much to keep flying as its much larger Boeing 767s. USAir also attributed the reduction of seats from 100 to 85 as one of the key reasons why it (and PSA beforehand) were unable to make money with the 146. They also noted that total aircraft maintenance costs had decreased by nearly $34 million, largely attributable to the grounding of the 146 fleet, although it faced a $44 million charge for lease obligations. USAir told its investors 'it's cheaper to park the 146 fleet than continue operating them.' This certainly did not help the aircraft's image or the marketplace for used airframes, and was a swipe at BAe that was still building new 146s. British Aerospace would have difficulty selling new aircraft when plenty of used ones were available, and having a large fleet parked in the desert does not paint a rosy picture of the type for prospective airlines.

But British Aerospace did not let USAir get away with the negative claims about the 146. It responded with a letter to the industry aviation publication Flight International, in which Managing Director Charles Masefield pointed out that most of the complaints from USAir were not BAe 146 related – namely withdrawing from most of the California market (half of which had been operated by the 146), and that the claim it was impossible to make a profit in California was also not a 146 issue. Charles told the press that the USAir 146s had been averaging a 99.1% dispatch rate, higher than the rest of the aircraft in the carrier's fleet. The 146 had also been paramount to PSA's expansion plans, whereas it was much less so under USAir.

Immediately afterwards, USAir withdrew its public criticism of the 146, noting that the closure of most of its California network was due to market conditions and not the aircraft. USAir said "It is not the aircrafts fault. It is the marketing conditions which are behind our decision. You have got low-fare carriers who are determined to beat each other's brains out." USAir did stand by PSA's decision to reduce seating capacity to 85 passengers as having a negative impact on the aircraft economics. But the damage done was two-fold: the obvious results were the parked fleet of 146s and the initial allegations that the type's operating and maintenance costs were too high. BAe claimed that the culmination of all the allegations had hurt its marketing efforts with other U.S. regional carriers, and that USAir made the 146 a scapegoat for its failure to compete in California. In a bit of irony, a wholly-owned subsidiary PSA Airmotive maintained all USAir's 737 engines (Pratt & Whitney JT8Ds) as well as the ALF-502s on the 146 – and the costs were up to four times higher than at other major U.S. carriers.

PSA surely did smile on British Aerospace, and in return it received an airplane that allowed it to open up new markets and increase frequency. It gave the carrier a competitive advantage at noise sensitive airports, and resulted in PSA becoming more relevant than had it not flown the 146. Eventually USAir and American Airlines merged in 2013, creating the world's largest airline.

USAir had polished the skin against British Aerospace recommendations, resulting in FAA directives on frequent inspections for affected aircraft.

Derek Taylor

Graham Robson

Derek Taylor

USAir parked all 18 BAe 146 aircraft in a single day, bringing 146 operations to an end overnight.

CHAPTER 06

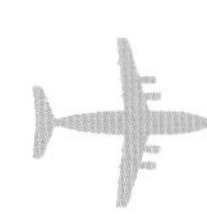

The 146 Spreads its Wings

Air Wisconsin was the first North American carrier to commit to the BAe 146, but it was PSA that kickstarted the rapid adoption of the 146 across the United States by regional carriers. Following PSA's original order, and after deliveries began, there was a notable 'take-off' in interest in the BAe 146 – initially from North America and Australia, but also from potential customers from South America, and the Caribbean. For years the project had been plagued with a lack of orders; Airlines were dubious about the economics of operating a four-engine jetliner on short routes, while commuter carriers running turboprop equipment were nervous about the step up to jets. With PSA's 146s beginning to demonstrate what the aircraft could achieve, airlines started taking notice and enquiries began to increase.

The 146 was proving to be quite the competitor in a newly developed 'regional jet' segment. Small aircraft serving smaller destinations was nothing new. The small Yakovlev Yak-40 began serving regional markets in the Soviet Union during the late 1960s, but that trend did not catch on in the West at the time. The BAC One-Eleven and Fokker F.28 were used for short hops, but neither were fuel efficient or had STOL capability and the quiet operations the BAe 146 offered. The BAe 146 offered flexibility, and most missions it performed averaged about an hour in stage length. The 146 was quickly becoming recognized as the first true "Regional Jet" that airlines could deploy profitably and productively.

G-OPSA over Washington D.C., demonstrating the 146 to airlines in a tour of the U.S. before crew training for PSA begins.

BAe took the 146 on a third tour, this time to the United States. Aircraft G-OPSA, which was assigned to crew training prior to the arrival of PSA's first 146s, toured select airports and airlines. It was first flown to Miami, demonstrating to airlines its prowess in high temperatures and strong 40kt crosswinds. It also departed behind a British Airways Concorde, which had just inaugurated service to Miami. The aircraft then ventured up to Meigs Field, a commuter airport in Chicago, Illinois, where it was the first four engine jetliner to land on the 3,900-foot- runway. Meigs Field was primarily a VIP facility and was very close to the heart of Chicago, much like London City is to downtown London. There was a planetarium at the end of the runway at Meigs, which aircraft were required to clear vertically by a minimum of 35 feet. The 146 cleared the planetarium by 850 feet, and reportedly reached the a flight level equivalent to the top of the Sears Tower (110 floors) from brake release (not rotation), faster than the building's elevator.

From Chicago the 146 headed on to Denver where snow storms were raging, but in the lull of the storm BAe arranged for flight demonstrations and an air-to-air photography session. The aircraft then continued on to Phoenix where the temperature was the opposite: hot and 100-degree heat. It was not just airlines interested in the 146, but the city too which took noise measurements to substantiate the claims of it being the "world's quietest aircraft". A mere 75dB was recorded – the lowest decibel level ever noted at the airport according to an official. After the brief tour, the aircraft moved on to California to complete its planned two months of crew training with PSA.

The 146 was beginning to flourish and show its prowess with several reputable airlines. And as a result a mixture of known carriers, new operators, and small regionals wanting to grow their markets began to provide BAe with additions to the order book. The following outlines the role the BAe 146 played at some of those customers.

AirPac - 1984

AirPac, an Alaskan based airline founded by Dick Maloney in 1977 became the second airline to put a 146 (N146AP) to work in the United States. It enabled the introduction of the first jet service between Anchorage and Dutch Harbor, one of the largest commercial fishing locations in the world. AirPac took delivery of the first series -100 to operate in the US on March 3, 1984 and had "The Alaska Connection" painted on the main landing gear doors. The aircraft was operated with 40 seats in the rear cabin, and the first 10 rows filled with collapsible containers that carried fresh seafood from Dutch Harbor (up to 1,500lbs per container). During the low season, the aircraft could be reconfigured to all passenger operations carrying 76 passengers. There were dual galleys for hot meal service, and the rear entrance had integrated air stairs to deplane passengers. These facilities meant AirPac was the first airline in Southwest Alaska to offer hot meal service during flight.

The 146 cut travel times between Dutch Harbor and Anchorage from three hours in a turboprop aircraft to just under two hours in the jet. But AirPac would be the first North American carrier (and the second in the world) to use the BAe 146 on unpaved runways, with Dutch Harbor having an unsealed 3,940-foot gravel strip right on the water and running through what used to be a giant hill (partially cut away and removed). Prior to scheduled service, AirPac had to operate a number of proving flights for the FAA. The weather at Dutch Harbor was often challenging with frequent strong crosswinds, fog, rain, sleet and snow providing operational challenges. AirPac managed just over five hours of utilization per day from its 146, nearly 40% lower than other operators. The airline operated flights as far south as Seattle (from Anchorage), served Kodiak in Alaska, and during the salmon season, Dillingham. AirPac claimed that the unreliability of the BAe 146 caused its cash drain, leading to its insolvency and shutdown in 1986. It was an early example of the perils involved when a small thinly capitalized airline attempts to run a single jet operation in a remote region.

The front of the AirPac cabin was equipped with containers to carry fresh seafood like crabs.

AirPac BAe 146 engaged in unpaved field operations testing at Dutch Harbor. The principle owners of AirPac are standing to the right of the gravel field watching the aircraft.

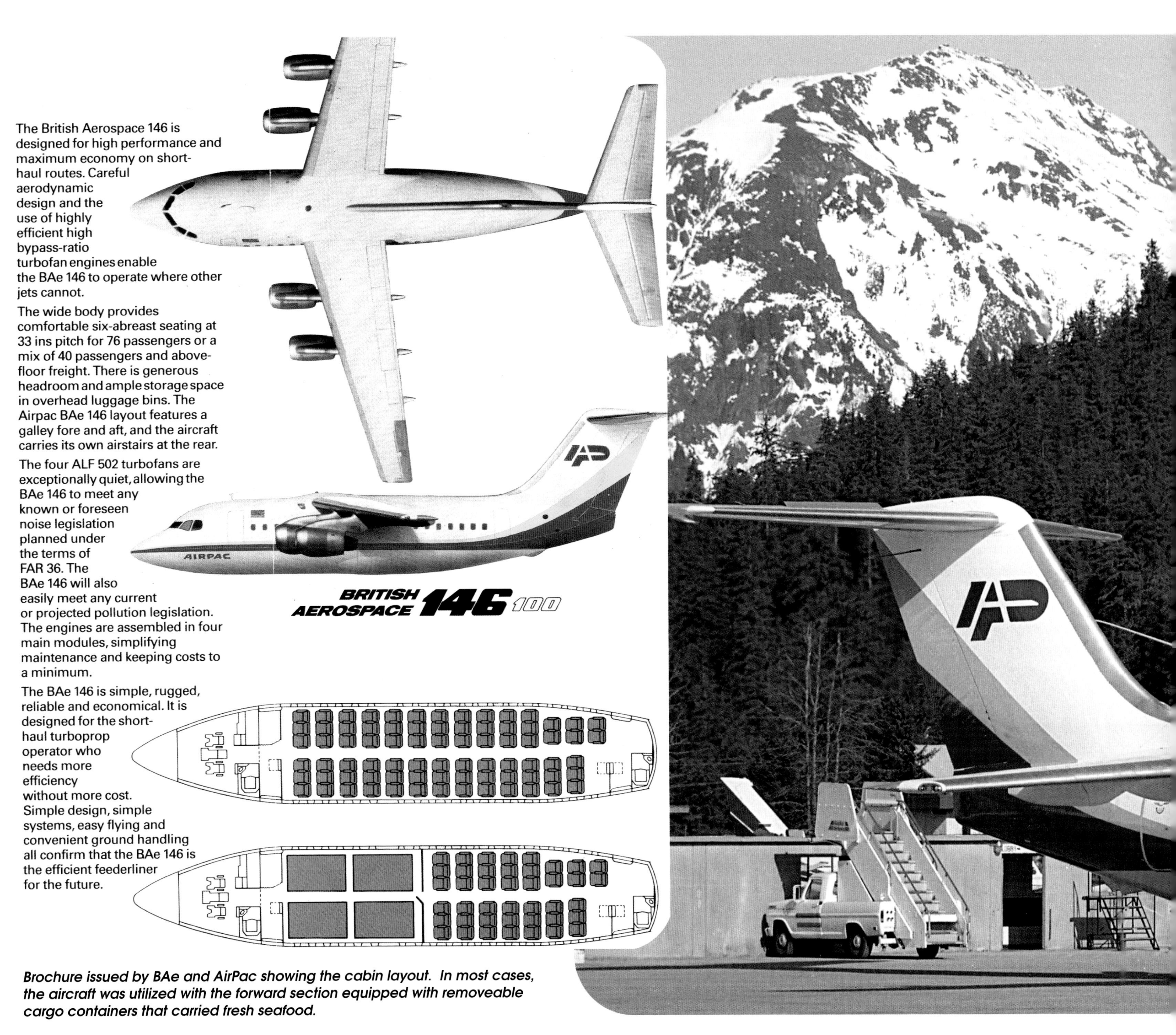

Brochure issued by BAe and AirPac showing the cabin layout. In most cases, the aircraft was utilized with the forward section equipped with removeable cargo containers that carried fresh seafood.

The Alaska Connection was painted on the landing gear doors.

Aspen Air -1984

Aspen Air became the fourth 146 operator in the United States, and the second to use the 146-100 series. Configured for 88 economy seats, its first aircraft was delivered at Hatfield on December 6, 1984 and entered service at the end of the month on the airlines route network radiating from Denver, Colorado. Aspen Air planned five daily trips between Denver and Aspen, landing on the runway 7,000 feet above sea level in the Rocky Mountains. Local air traffic controllers referred to the 146 as "the SST – Short Stubby Thing". The 146 experienced better reliability at Aspen, being utilized for eight hours a day over 12 sectors. Load factors of 75% were also higher than anticipated and the airline asked for the delivery of the second aircraft to be moved forward. Occasionally Aspen used its 146 to carry excess baggage for competitor (and former rumored 146 customer) Rocky Mountain Airways. There was an aft cabin

bin installed for this as well as excess baggage that wouldn't fit on Aspen's Convair CV580 aircraft. The bin was eventually replaced by two additional seats.

Aspen's colorful aircraft weren't completely painted. The tree that served as its logo was applied to the aircraft using hand painted, self-adhesive decals developed by FasCal. Aspen Air eventually received four BAe 146-100s which enabled it to expand service to other states: Texas, California, and Illinois providing seasonal service. Aspen Air contracted as a feeder carrier for United Express in 1986, but in 1989 its routes and Denver hub were taken over by Mesa Air Group. The aircraft, routes to Aspen, and equipment were acquired by BAe 146 operator Air Wisconsin. When Air Wisconsin retired the BAe 146, it had no replacement aircraft that could fly a full load out of Aspen. The carrier's current CRJ operations are weight restricted and often run with just 54 seats filled out of the maximum of 70, whereas the BAe 146-300 could fly in with a full load of 100 passengers.

The BAe 146 was the only commercial jet certified for operations into Aspen airport for years.

aspen
N461AP
aspen
BAe 146
aspen airways

Ansett W.A / Ansett Australia / Ansett New Zealand - 1985

On the other side of the planet, Australian carrier Ansett Western Australia (Ansett W.A., previously MacRobertson Miller Airlines) purchased two 146-200 aircraft with options for six and took delivery of its first aircraft in on May 10, 1985. When Chief Executive, Sir Peter Abeles was asked why Ansett had bought the 146, he responded "Because the first time I saw it I fell in love with it." The carrier's 146-200s were configured with 75 seats and serving west and northern Australia. Ansett reviewed the 146's performance in service, not just for a potential larger follow-on order, but also for use with other affiliates in the Ansett group. The aircraft were based in Perth, and began to serve Derby and Broome on the west coast of Australia on May 13, 1985.

Ansett's 146s were heavily modified from BAe's standard specification, reflecting the operating and competitive environment prevailing in the Australian market. The aircraft were configured with triple JAMCO galley units to enable the provision of hot meals over several consecutive sectors, and HF Radio for the long outback routes. It was the first time that BAe had to modify the aircraft for a heavy galley installation with hot ovens. These items were also Buyer Furnished Equipment (BFE) which was something else BAe had to learn to adapt to. BAe had assumed customers would generally only require aircraft with light galleys and no hot meal service. The conversion of the first aircraft

proved extremely challenging for BAe, and subsequent deliveries involved customization of the interior to the airline's requirements after the airframes were "finished" at Hatfield. The aircraft were then flown to Marshall's at Cambridge for rework, which had to partially disassemble them to install the new galleys and additional equipment specific to Ansett. This was complicated, expensive and time consuming, resulting in late deliveries which irritated Ansett. BAe was learning the hard way that its new-found customers, especially larger carriers, wanted their aircraft customized beyond the meagre standard specification it had assumed.

Only one 146 was painted in the Air W.A. livery, before the carrier was rebranded Ansett W.A. This aircraft participated in the flying display at Farnborough in 1984. After delivery, Ansett's 146s were subject to a grueling schedule every day. Departing Perth at Midnight, the 146 would run a daily schedule of over 17 hours, firstly flying to Port Hedland. It then continued on to Broome, Derby, Kununnurra, Darwin, Gove and finally Cairns, where it would be turned around to follow the same path back to base. The total daily trip covered approximately 5,300 statute miles.

Gove
Darwin
Broome
Derby
Cairns
Karratha
Port Hedland
Ansett W.A.
Perth
Air W.A.
VH-JJT

Presidential Air -1985

A brand-new regional airline based in the United States placed an order for five BAe 146-200 aircraft and spares worth $95m plus seven options, at the Paris Air Show in June 1985. Presidential Air planned to operate the 146 out of Washington Dulles International Airport (IAD) using a traditional hub-and-spoke system. The new aircraft would feed IAD from smaller airports (similar to PSA) and replace the carrier's larger Boeing 737-200s where capacity wasn't needed. The BAe 146s would also enable frequency to be increased to New York and Boston, where it would operate flights during off-peak hours.

The 146 was chosen by Presidential because it met strict FAA stage 4 noise standards and could operate in noise sensitive airports, such as Boston, after curfews came into effect. The 146s would see six to eight cycles a day, with seat mile costs expected to be approximately 6.7 cents and a break-even load factor around 35 passengers (based on a $1,100 per block hour cost including aircraft lease). President and founder Harold J. Pareti said the seat mile costs may be higher than the 737, but block hour costs are significantly lower. Presidential received rebates from BAe for advertising the 146 service, and the manufacturer provided a full-time maintenance advisor at IAD, where Presidential would perform its own maintenance.

On August 5, 1986, N401XV "Franklin Pierce" was handed over to the airline, and began its delivery flight three days later. Scheduled services began on August 29, 1986 after the carrier's second BAe 146-200 had been delivered just the day before. One of Presidential's BAe 146 appeared in the flying display during the 1986 Farnborough Air Show.

Presidential Airways chose a five-abreast seating configuration and initially signed a marketing agreement with Pan Am, as part of their Airpartner program, to increase feeder traffic between both airlines at Dulles and Miami in late 1986. On January 12, 1987, Presidential switched to a codeshare agreement with Continental Airlines, and shortly afterwards, Presidential began flying under the Continental Express banner. Two of the 146 aircraft (N406XV and N407XV) were painted in Continental Express colours after completion at Hatfield, but the rest of the fleet was not repainted and just had the titles changed. N407XV was also displayed at the 1987 Paris Air Show in June. In keeping with the literal sense of the airline's identity, Presidential Airways BAe 146-200 aircraft carried the name of a famous U.S. President on the nose of each aircraft, like N402XV "James Buchanan" and N403XV "Abraham Lincoln."

As luck would have it, due to immense competition from both New York Air and United Airlines, Continental began pulling out of Washington in late July 1987, not long after the agreement with Presidential commenced. It started to reduce the number of seats it bought on Presidential Airways, and as a result, Presidential severed ties with Continental Express almost a year after the cooperation began, on January 5, 1988. Less than two weeks later Presidential announced it was to begin flying its aircraft in its own colours again. At least Harold Pareti could acknowledge and joke about the frequent paint jobs his aircraft had been through in such a short period of time. But then on January 20, 1988, Presidential signed up as a United Express carrier. Air Wisconsin would begin transferring additional capacity back to Chicago (ORD) as a result of the new codeshare, with Presidential picking up capacity.

With sustained heavy losses over the previous two years, it seemed unlikely that the airline would turn around and spend more money. But Presidential subsequently entered into an agreement to acquire 16 50-seat deHavilland Dash 8-300 turboprops, although it took delivery of just two aircraft. This was in addition to the BAe Jetream 31 aircraft Presidential had purchased the previous year.

1989 was the only year Presidential would turn a small second quarter profit, versus heavy losses of the previous three years. In March, Eastern Air Lines had a devastating strike that ultimately grounded the airline, which clearly benefited Presidential. But Eastern's woes were not enough to enable Presidential to recover from the heavy losses it made in previous years. By September, Presidential was short of cash and had massive debts. In October it began asking employees and executives to defer between 20 and 25% of their pay respectively for only two pay periods out of ten until March 1990. Two of its aircraft (N407XV and N408XV) were also wet-leased to provide short-term capacity to British Airways due to a delay in delivering BAe ATP turboprops to the UK carrier, while also providing Presidential with some much-needed cash.

October 27, 1989 was a dark day for the 700 employees of the airline. Presidential Airways filed for bankruptcy protection, listing $30m in assets and $76.4m in liabilities. United Airlines offered up to $1.7m to keep Presidential operating, continued to review and provide funding for additional weeks on an ad hoc basis. November 31, 1989, marked the day it was all over for Presidential Airways. BAe obtained a court order to ground most of the airline's aircraft, after it failed to make $170,000 in lease payments for two BAe 146-200 aircraft and eight Jetstream 31s. December 5, 1989 was the last day for Presidential, after which all remaining flights were cancelled and employees terminated. The remaining Presidential BAe 146-200s were repossessed and placed with new operators within a few months. In 1995, Harold Pareti tried to start up another airline (Jet Aspen) using two BAe 146-200s formerly with USAir, but it never got off the ground even though one aircraft was painted.

Replacing Continental Airlines was United Airlines who signed with Presidental Airways. Shown here is N406XV painted in United Express livery. This was the beginning of the end of Presidential Airways who was in massive debt to British Aerospace. Note the Jetstream 31 in the background that Presidential Airways agreed to take a large number of.

Washington-Dulles (IAD), the home airport for Presidential Airways.

Presidential Airways CEO Harold Pareti accepting the aircraft logbook for their first BAe 146-200.

Presidential Airways configured their aircraft for 5-abreast seating.

CONTINENTAL
EXPRESS
CONTINENTAL
JET EXPRESS
CONTINENTAL
EXPRESS

AirCal -1985

AirCal, the local Orange County based airline that was home-based at John Wayne / Santa Ana (SNA) was getting clobbered by PSA's ability to run unlimited frequency out of the airport with the 146. This allowed PSA to use its coveted and limited take-off and landing slots for MD-80 aircraft. AirCal was PSA's biggest regional competitor (outside of Western Airlines and United Airlines), and yet its arch-rival PSA was dominating its hometown airport. In early 1985, AirCal began taking delivery of Boeing 737-300s which carried more passengers than PSA's BAe 146-200 aircraft. But because of noise considerations and the short runway at SNA, the 737s initially suffered weight and range

With pressure mounting from PSA at AirCal's home base airport, AirCal placed an order for six aircraft.

penalties. AirCal's Boeing 737-300s were limited to around 70% of the maximum capacity when departing out of SNA to San Francisco (SFO), just a short 55-minute flight away, and the Boeing 737s also used a take-off slot whereas PSA's BAe 146s were given 12 additional slots. David Banmiller, President of AirCal and his team headed to the United Kingdom in December, where they were picked up by chauffeur driven Jaguars, and were taken to Nast Hyde, a private company owned house five minutes from Hatfield. BAe would often entertain potential customers at Nast Hyde, and generally final negotiations for new aircraft took place over dinner.

When PSA received its 20th 146 and while exercising options for four more aircraft, in December 1985, AirCal placed an order for six BAe 146-200s for a quick delivery in March of 1986. That would give AirCal the same advantage PSA retained, being able to run the 146 in and out of SNA without any slot constraints. With no range or weight penalties, the low noise footprint would allow unrestricted operations. On March 3, 1986 – a mere three months after placing its order – AirCal received the first of its BAe 146s, configured with 100 passengers in a 3-3 layout. The second aircraft was delivered two and a half weeks later. Due to complaints from passengers,

AirCal's first BAe 146-200 prior to delivery in March 1986.

AirCal removed 15 seats changing the interiors to a 2-3 layout and capacity for 85 passengers which brought it into line with PSA. After crew familiarization flights, AirCal put the 146 into service between SNA and SAN (San Diego), PSA's home base, on April 1, 1986. The remaining 146s would be delivered fairly quickly with the last one arriving in October 1986.

AirCal was generally satisfied with the 146, recognizing its ability to get in and out of small airports quietly. But for mainline operations (e.g. LAX-SFO), David Banmiller said "there were better aircraft out there." AirCal began to centre its fleet around the Boeing 737-300 as the workhorse, with the BAe 146-200s as the supplementary aircraft dedicated primarily to SNA operations where noise sensitivity and a lack of payload restriction was key. But AirCal soon began to experience problems with its 146s in three key areas. Engine reliability was the biggest factor, followed by faulty yaw-dampers, and brakes that were constantly being replaced. BAe asked AirCal if their last aircraft (N148AC) could be handed over at the Farnborough 1986 air show, which AirCal agreed. Before David flew to the UK, he went to AirCal's Oakland maintenance facility and asked the chief mechanic to make a list of the top 10-15 consistent issues with the

Over Los Angeles prior to entry into service.

aircraft, and provide it in "Dick and Jane language" (simple laymen's terms). Once David had that list, he flew to Farnborough.

David met with high level BAe executives upon his arrival at the show. The first words out of David's mouth to the group assembled were "Guys, we have problems and we are not happy" and he proceeded to go through the prioritized list the mechanic had provided. BAe's initial response was "Do you have an A&P license? Do you have a maintenance background?" David responded "no" but BAe thought he did because he talked so intelligently about the issues. David muses "I wasn't the smartest guy in the room, but I knew what to

Handover ceremony for AirCal's first BAe 146-200.

study. I don't care about 'that', but 'this' is important." David's words of wisdom for being a CEO: "You want to be able to manage people and events, and understand stuff in simple language. Simple 'Dick and Jane' stuff."

Halfway through his presentation on the issues AirCal was having there was a pause, because CAAC was accepting its first BAe 146, a series -100, in a delivery ceremony at the Farnborough Air Show. BAe asked David to join the Chinese Ambassador and the CEO of CAAC at the podium while the acceptance of the aircraft took place. At the conclusion of the ceremony, the Chinese Ambassador and the staff of CAAC approached David with their interpreters, followed by a full suite of news and press intent on filming and writing about the event. Behind David were the very people he was chastising about the aircraft issues and they were certainly nervous about how he would respond to CAAC's inquiry. They asked David "we understand you have 146s, tell us about the performance." Before responding, David laughed and mentally recalled that "I just left the tent where I beat the shit out of these (BAe) guys, nicely, telling them 'these planes are crap guys, what are you going to do about it?'". Ever the professional though, David responded to the entourage from CAAC by saying "well gentleman, let me assure you the fuel performance – spot on. Noise performance? Spot on. And any other technical difficulties that pop up especially with new airplanes? I guarantee you British Aerospace stands behind their product."

Log book handover for E2051.

David looked over at the BAe team as he reiterated "They stand behind their product", making a point to not disclose the issues he was having while implying they needed to get their act together with AirCal or else. When the meeting with David reconvened in the tent, he continued to unload on BAe, but did not do it in front of another customer.

N148AC was delivered on September 26, 1986, while simultaneously BAe was given The Queens Award to Industry for Export Achievement with N148AC being the 50th BAe 146 aircraft to be exported. But the celebrations did not last long. By the end of 1986, not only was AirCal compounding the pressure that PSA had exerted on BAe, but AirCal had been acquired by American Airlines which continued BAe 146 operations (with aircraft repainted in AA colours). American Airlines mainline fleet had polished metal fuselages, but BAe had advised against the practice as it would thin the already thin fuselage skins and would be subject to pitting. As a result, American Airlines painted the aircraft a light gray color. American Airlines demanded that the BAe 146 conform to the airlines operations and systems, a significant change from the process that had been in place earlier. Workarounds developed in the field were no longer permitted because they were not in the flight manual, and pilots were not allowed to perform them. And BAe wouldn't put the workaround in the manuals because it would show a much higher cockpit workload. American Airlines also

Roger Pascoe

BAe rep Roger Pascoe on a stopover in Greenland during the delivery flight for AirCal's first 146.

Interior for AirCal's BAe 146 with 5-abreast seating.

reconfigured the overhead panel to conform to how their other aircraft panels were laid out.

American reconfigured the 146 interior as well to match the rest of the fleet, which included installing a First Class section in the cabin. David Banmiller (who by then had been appointed Vice President, International at American Airlines) asked American Airlines CEO Bob Crandall "why put in a First Class section?" Bob responded, "Because we're American Airlines, and we gotta have First Class for our frequent flyers." The 146s were reregistered N694AA to N699AA.

AirCal had studied Southwest Airlines fast ground turnaround service, and by the time it was acquired by American the airline was able to turnaround its own aircraft in approximately 15-20 minutes. When American took over, it went back to 45 minutes. Former AAR mechanic Kevin Govett said that the aircraft previously operated by AirCal were achieving 99% dispatch reliability. But because of the lower pay and scope clauses applicable to the 146, the pilot's union was irritated with it and started to refuse to fly with even one minor issue listed as INOP (inoperable), even if the aircraft met the MEL (minimum equipment list – items that must be working to permit a flight to depart). The BAe 146 required daily tending, whereas a Boeing 727 or McDonnell Douglas DC-9 did not. When a BAe 146 could not depart (AOG – aircraft on ground), it could generally be fixed relatively quickly. But when a Boeing or McDonnell Douglas aircraft went down, it went down hard resulting in longer and lengthier rectification work. Unlike the Boeing or McDonnell Douglas aircraft, the 146 required constant tending to daily. As a result, American quickly retired the 146 fleet as a result of rapidly decreasing dispatch reliability. American's Chairman and President was quoted as saying "The BAe 146 is an utterly unsatisfactory aircraft." Most of American's 146 operations ended on October 30, with the last aircraft retired November 1, 1990.

Rebranded AirCal livery after American Airlines acquired the airline.

Unlike USAir, American Airlines heeded BAe's directive to not polish the fuselage like the rest of American's fleet.

CAAC (Civil Aviation Administration of China) - 1985

Following the conclusion of its use for training at PSA in the United States, series -100 aircraft G-OPSA headed to China in July 1984 for a regional tour, visiting 13 airfields including Urumchi in Northwest China, Lhasa at Tibet (11,650 above sea level), and Zhangjiang on the Southern Coast. At each stop it performed demonstration flights to Chinese airlines. The aircraft carried six CAAC personnel throughout its time in China to assist with translation and navigation, as well as to note the detail of the 146s performance. Prior to leaving the country, the aircraft visited Harbin in Manchuria where the BAe 146's landing gear doors were built under subcontract.

CAAC chose from the standard BAe fabric offerings for its interior selection.

Six months later, in January 1985, CAAC indicated that it would seriously consider purchasing the 146 series for its airline if each aircraft were approximately $2,000,000 USD cheaper. BAe replied that it could not afford to sell the 146 for less than the commercial price, even though it had been pursuing a sale with CAAC for years. Manufacturers often offer discounts below list pricing to airlines, depending on quantity of airframes likely to be ordered and other factors. The demonstrations in July 1984 made an impression though, as CAAC was reported to have been very impressed with the 146's short field performance at high elevations. The aircraft would fit right in, feeding traffic from smaller points into larger destinations and could be used on routes where excellent short field performance was needed.

After two years of courting the Chinese, the 146's regional tour to China paid off and the Chinese carrier CAAC announced an order for ten 146-100 aircraft. The $150m deal was announced at the Paris Air Show in June 1985, and deliveries were to begin in the summer of 1986. It is believed that the tour of smaller airfields may have been the deciding factor in the carrier's selection the 146.

The first aircraft was flown to the Farnborough Airshow where, on September 2, 1986, its logbook was handed over to Mr. Li Zhao, Deputy Director General of CAAC by Sir Austin Pearce. This marked the first time a 146 aircraft had been handed over to a customer at an air show. The first aircraft was initially based at Lanzhou, operating flights to Chengdu and the Tibetan capital Lhasa. The next three 146s were based in the Shanghai region, and the last four at Hohhot near Beijing When CAAC was broken up and divided into regional units, the BAe 146s were transferred to China Eastern Airlines and China Northwest.

During the sales campaign, Hawaiian Airlines was presented two different livery concepts on the BAe 146. This livery was not chosen.

Signing of the MOU for the purchase of 8 BAe 146-200 aircraft with Chairman and COO of Hawaiian Airlines John Magoon (with BAe's Brian Thomas on the right).

Hawaiian Airlines – 1985

In September 1985, Hawaiian Airlines announced it was going to replace the McDonnell Douglas MD-80s it had procured only a few years earlier with an order for eight BAe 146-200s plus an option for two additional aircraft. The MD-80s were expensive and too large to operate on inter-island services. Hawaiian Airlines' Chairman of the Board, John H. Magoon, announced the order and said "Hawaiian Air has been evaluating aircraft types specifically designed to fly the high frequency short haul sectors of Hawaiian's inter-island route systems. We selected the BAe 146 because of its state-of-the-art technology and design as well as its excellent short haul characteristics." President and Chief Executive Officer of Hawaiian Airlines Paul J. Finazzo said "The changes in the Hawaiian inter-island market place require a jet such as the 146. The BAe 146 will position Hawaiian to meet these anticipated changes."

While BAe proposed two livery variations during the design phase and retouched a photo into Hawaiian Airlines colors, the airline never converted its MOU into a firm order.

Retouched photo that accompanied the press release announcing the Hawaiian Airlines order.

British Caribbean Airways - 1986

Located on the island of Tortola, British Virgin Islands, British Caribbean Airways placed an order for two aircraft, both BAe 146-100 series. The plan was to increase tourism to the British Virgin Islands by flying customers from Miami to Tortola, and with Tortola's runway limited to 3,600 feet the BAe 146 was the only jet capable of operating the service. The flight time of just under three hours permitted non-stop services without no payload restrictions. This would avoid a further stop as most passengers connected via San Juan, Puerto Rico to surrounding islands.

Key managers were hired from British Airways Consulting, with managing director John Bull heading up British Caribbean's operations. The third production aircraft [E1003], previously used by BAe, was refurbished and delivered on March 17, 1986 and entered service one month later, registered N246SS. The airline was very short lived though, shutting down by October of 1986, having never taken delivery of its second aircraft. The aircraft that was handed over found its way back to PSA to temporarily act as a spare (painted in PSA livery), and was eventually leased to a number of other U.S. airlines during the 1980s.

The British Caribbean operation proved to be economically unviable, and furthermore it turned out that the 146 was subject to payload restrictions when the runway at Tortola was wet, not an infrequent occurrence in the summer months. The owners, a US company based in Texas, sued BAe which was blamed for these shortcomings. The matter was settled out of court, but BAe was left nursing a seven-figure cost. British Aerospace also learned another hard lesson: that dealing with startup carriers often involved too much mutual wishful thinking invariably ending badly.

Royal West - 1986

BAe did another deal with a US start up carrier, Royal West, based in Las Vegas. The carrier leased three used 146-100s from BAe in February 1986, with delivery later that same year. The aircraft were configured with a 91-seat interior at a knee-pinching 29" pitch. Royal West operated out of its Las Vegas base to Los Angeles, Burbank, and Ontario, with expansion taking flights to other cities including Reno. The aircraft was also scheduled to operate seasonally to Vail, Colorado into Eagle County Airport for the ski season. Half the capacity would be sold to tour and charter operators in advance, with the remaining seats sold conventionally to the public. Royal West's first flight from Las Vegas to Los Angeles took place on June 26, 1986, just eight days after the handover of the aircraft on June 18.

Gary Ellmer who was with Royal West initially, helping gain the flight certificates, also helped take delivery of the aircraft from BAe's Chester facility. Captain John Sloan spoke of qualifying on the BAe 146, initially doing simulator time at PSA prior to delivery of Royal West's first aircraft. John spoke about the scenarios in the simulator, including a landing with inoperative flaps and engine out. He landed the aircraft but ended up blowing all the tires and stopping off the runway. The check ride also involved steep turns and three landings including one with only three (3) engines operating. When it came time to flying the actual aircraft, John said it the 146 was easier to fly versus the simulator.

Royal West
Royal West

Royal West Airlines
SYSTEM SCHEDULE
EFFECTIVE NOVEMBER 7, 1986
For reservations and information contact your professional travel agent or Royal West Airlines:
1-800-8-4-ROYAL.
In Las Vegas/Clark County Nevada:
798-6444.

Flight attendants initially went to Hatfield for training on the 146.

Among Captain Sloan's favorite routes was Las Vegas to Reno, a short hop with beautiful scenery, and some challenging flying with the updrafts over the mountains. He spoke of one incident into Reno where an aircraft in front of him (Sunworld International DC-9-10) called in to report severe turbulence to air traffic control (ATC). The pilot spoke of a flight attendant reportedly hitting the ceiling, and the passenger cabin littered with inflight service items. Captain Sloan called for the immediate discontinuing of inflight service and for the flight attendants to be seated. From that moment onwards, Captain Sloan said the pilots never saw the instruments or attitude indicators because they were bouncing all over the place. Upon landing, Captain Sloan radioed ATC telling them they should close the airport, and while an American Airlines 727 was taxiing, they saw the numbers and reinforced Captain Sloan's message that the airport should be closed until the weather settled down.

Outside of losing an engine on takeoff at Velocity One (V1) departing from Mammoth to Los Angeles (LAX), and a single incident of a windshield cracking, the 146 was a joy to fly for the pilots at Royal West. But the airline experienced plenty of engine reliability issues, and the heat of Las Vegas certainly didn't help with the electrical system. Royal West ceased commercial operations February 27, 1987, less than six months after they began. There were unsubstantiated rumors that management was siphoning off money, but whatever the truth, the carrier was now bankrupt. The last flight out from LAX to LAS, to ferry the aircraft home, was empty except for one passenger, which was Captain Sloan who deadheaded back to Las Vegas.

Royal West actually continued to operate in Chapter 11, flying charters until April 1988, including carrying U.S. Presidential candidate Michael Dukakis across the Southern part of the country. He made a number of stops along the way, from West Palm Beach, Florida to Houston, Texas. The operations with Dukakis generally went well except for two incidents, neither related to the aircraft itself. The first incident occurred when Dukakis' wife arrived on a private jet to meet up with Michael who had

Royal West Airlines

	LEAVE	ARRIVE	FLIGHT	FREQUENCY
FROM LOS ANGELES (LAX)				
To Las Vegas	10:30a	11:25a	110	D
	12:20p	**1:20p**	120	D
	3:20p	**4:20p**	140	5,7
	6:20p	**7:20p**	160	D
	9:20p	**10:20p**	170	D
	10:50p	**11:45p**	180	5,7
To Reno	8:50a	10:05a	500	D
	1:20p	**2:35p**	510	D
	10:05p	**11:20p**	540	D
FROM BURBANK				
To Las Vegas	8:00a	8:55a	320	D
	10:50a	11:45a	330	D
	1:40p	**2:35p**	350	D
	4:50p	**5:45p**	360	D
	7:40p	**8:35p**	380	D
	9:00p	**9:55p**	390	X5,6,7
	8:00p	**8:55p**	390	5,7
To Reno	**12:15p**	**1:30p**	560	D
	4:30p	**5:45p**	570	X6
FROM ONTARIO				
To Las Vegas	9:25a	10:20a	710	D
	3:45p	**4:40p**	780	D
To Reno	**6:35p**	**7:50p**	596	D
FROM LAS VEGAS				
To Los Angeles (LAX)	7:20a	8:20a	101	D
	11:55a	**12:50p**	121	D
	1:50p	**2:50p**	131	5,7
	4:50p	**5:50p**	151	D
	7:50p	**8:50p**	171	D
	9:25p	**10:20p**	181	5,7
To Burbank	9:25a	10:20a	311	D
	10:50a	11:45a	331	D
	12:15p	**1:10p**	341	D
	3:05p	**4:00p**	361	D
	6:15p	**7:10p**	371	D
	9:00p	**9:55p**	391	D
To Ontario	8:00a	8:55a	711	X7
	5:10p	**6:05p**	781	D
FROM RENO				
To Los Angeles (LAX)	8:00a	9:15a	511	D
	10:35a	11:50a	521	D
	8:20p	**9:35p**	541	D
To Burbank	**3:05p**	**4:20p**	571	[illegible]
	6:15p	**7:30p**	[illegible]	[illegible]
To Ontario	[illegible]	[illegible]	[illegible]	[illegible]

Royal West Airlines
On behalf of the entire staff of Royal West Airlines we welcome you aboard our BAe 146 inaugural flight June 26, 1986
Grant G. Murray
President and Chief Executive Officer

Certificate given to passengers on the inaugural flight.

flown in from New Orleans. A platter of seafood was brought along and deposited at a fixed based operator (FBO), from where it was supposed to be transferred to the Royal West aircraft. Apparently, the staff and crew at the FBO did not receive the instruction and instead they ate the seafood, with Royal West taking the blame. Another incident started brewing when the crew was rapidly reaching 'time out' for rest, but Michael was running late with extended campaign stops. Captain Sloan notified Michael's assistant if he wasn't onboard the aircraft by 10pm, the aircraft would not be going anywhere for the night. About five minutes before the deadline, with a police escort, Dukakis and his staff came racing up to the aircraft just as Captain Sloan started the #4 engine. Michael Dukakis apparently loved the aircraft, and generally sat up front with the press in the back. A few rows of seats were removed to create some additional room for the governor.

Former pilots of Royal West ended up contracted to train aircrew employed by Westair, a United Express carrier that had just purchased BAe 146 aircraft and planned to operate them from Fresno, California.

Royal West

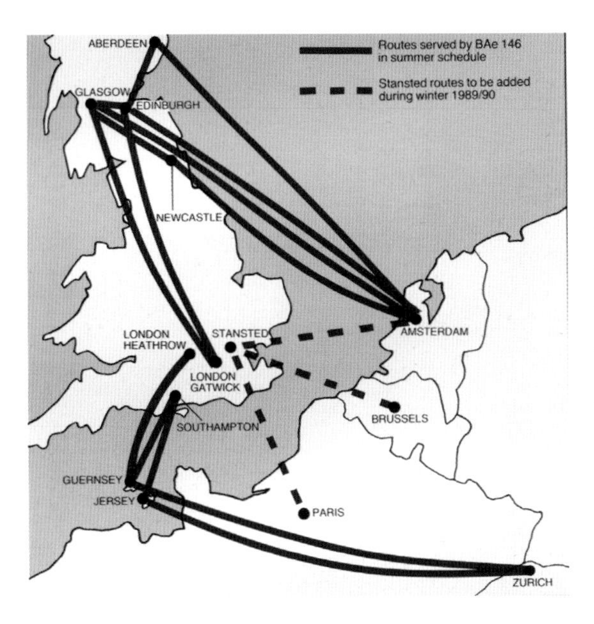

AirUK - 1987

AirUK was formed in 1980, an independent carrier from the amalgamation of four regional airlines. Unsurprisingly it had a diverse fleet of aircraft including BAC 1-11s and Fokker F28s, along with a large number of smaller turboprops. It became the third largest scheduled carrier in the United Kingdom behind British Caledonian and British Airways. AirUK went on to become a very profitable and large regional airline.

In September 1987, AirUK signed a Letter of Intent (LOI) for one BAe146 series -200, and later contracted for two aircraft. AirUK became the first British operator to order the larger 146 which was selected over the Fokker 100 because it was quieter and available immediately, despite the carrier already being a large Fokker F27 operator. Managing Director Stephen Hanscombe said, "on the basis of our arithmetic, the Fokker 100 offers no advantage in terms of economics." The first new 146 started flying from London Heathrow to Guernsey on December 1, 1987. AirUK planned to add a second 146 during the summer of 1988 to replace its only BAC One -Eleven 400 series jet which was dedicated to the Aberdeen-Edinburgh-Amsterdam route. The fleet would eventually be based at Stansted, after a new terminal and rail link were completed in 1991.

AirUK's first BAe 146-300 flying over the new Stansted Terminal under construction.

Coincidentally, AirUK renewed its colour scheme for the arrival of its first 146. The predominantly white livery included three blue stripes and a stylized British Union Jack flag on the tail (with large AirUK lettering on the fuselage). After only nine months of trouble-free operations with BAe 146-200 aircraft, AirUK placed a follow-on order for two series -300s. Scotland had become the carrier's fastest growing market, attributed to the introduction of the 146 aircraft between Glasgow, Aberdeen, Edinburgh and Amsterdam. Dutch carrier KLM had a 14.9% stake in AirUK and had recently ordered Fokker F100s, and AirUK considered ordering the Fokker because of engineering co-operation with KLM. However, in May of 1989 AirUK placed an order for more BAe 146s, this time for two used series -100 aircraft. This made AirUK the only airline in the world at the time to operate all three versions of the BAe 146. The BAe 146-100s were to come from Air Nova in Canada and were leased through BAe. Air Nova had been using the -100s until the -200s it had ordered were ready for delivery.

By July of 1990, AirUK's business was growing and so was the BAe fleet. It planned to lease an additional seven aircraft taking its fleet to 16 and making it the largest BAe 146 operator in Europe. It also took three secondhand 146s that were former PSA/USAir aircraft. In 1994, the airline rebranded again, introducing a new livery featuring a dark blue cheatline with a thin light blue line running in between. In March 1995, AirUK took delivery of its last BAe 146, a series -300. The airline was acquired by KLM in 1997, and in 1998 it was re-branded KLM UK.

WestAir - 1988

Fresno-based WestAir was a Californian commuter carrier that had a very diverse fleet. The firm operated Shorts 330s and 360s, Embraer 110 Bandeirantes and BAe Jetstream 31s. One of the founders was Maury Gallagher, who would later go on to found Allegiant Air (still in operation today). At the time, WestAir was operating as a United Express carrier, and wanting to grow its business, jumped into jets and selected the BAe 146. At the time United Airlines was having notable success working with three other commuter airlines under the United Express banner, each of which also operated the BAe 146: Air Wisconsin, Aspen Airways, and Presidential Airways.

Maury was one of the partners in Pacific Express, which ordered but never took delivery of the BAe 146. Maury mentions that the salesman at BAe were 'damn good at their jobs' when trying to convince Pacific Express to order the BAe 146, as well as later on when dealing with WestAir. Gary Ellmer, previously at Royal West, was recommended to join WestAir after Royal West folded to help with the BAe 146 fleet. WestAir initially struggled to convince the FAA that everything was in place to operate larger jet aircraft. Once Gary joined, he knew what was needed for the BAe 146 to be approved by the FAA for service with WestAir. In conjunction with BAe which enlisted the help of Presidential, the necessary operational infrastructure was quickly developed to demonstrate to the FAA that WestAir could manage the transition. After a short period the FAA granted WestAir an operating certificate and the airline's operations with the feeder liner got underway.

One of the issues Maury noticed was that even on short haul sectors, the 146's block time (the time the aircraft leaves the gate at the originating point and arrives to the gate at the destination) was extended because the aircraft flew slower than other jets (i.e. Boeing 737 or McDonnell Douglas MD/DC-9 series). As a result, its operating costs were higher simply because the BAe jet took longer to accomplish a trip. But WestAir was monetizing the aircraft by using it for its unique capabilities instead of just operating it on typical route corridors. The airline served destinations such as Telluride and Vail, both in Colorado, direct from SNA and the BAe 146 was the only jet aircraft that could operate through those remote airports.

Gary noted that while passengers and flight crew loved the aircraft, the infamous flap howl scared passengers. One incident that stood out was when a BAe 146 approaching San Francisco made a rough landing. The anti-skid system locked up the wheels (unbeknownst to the crew) and on touchdown, and all the tires were flattened as the aircraft skidded down the runway. The rubber was shredded down to the axles, disabling the aircraft and shutting down the

runway for half a day. United Airlines maintenance had to raise the aircraft using wing jacks, a makeshift cradle was cobbled together, and the stricken BAe 146 was moved to a hangar for repair. It was nearly a month later before the aircraft was back in service.

The combination of the BAe 146's higher operating costs, exacerbated by the first Iraq war which drove fuel prices 300% higher overnight, and limited liquidity led to a deterioration in WestAir's financial situation. In 1990, the airline was sold to Mesa Air Group which quickly disposed of the BAe 146s to stem losses.

Franco Mancassola, founder of both Discovery Airways and Debonair.

Prior to delivery.

Discovery Airways – 1989 (Debonair 1996)

Few will remember Discovery Airways, a short-lived startup, founded by former Continental Airlines manager Franco Mancassola that lasted five months. It was a Hawaiian inter-island carrier competing with local airlines Hawaiian and Aloha. The airline was established rapidly and an order was placed for five BAe 146-200s (along with options for seven -300s) in July of 1989, with delivery the same year.

The first two BAe 146s arrived in December 1989, with the second pair following in January 1990, and the fifth aircraft arriving in Hawaii during February. Flights didn't begin until March 25, 1990 when the five 146s began connecting the islands (Kona, Maui, Oahu, Kauai) configured with 102 seats, 8 of which were in first class. So confident was Mancassola in Discovery's growth, an order was placed for three BAe 146-300s (E3163, E3165, E3169) and the aircraft were painted in full Discovery livery.

Hawaiian and Aloha lodged legal action against Discovery alleging foreign ownership (which was against U.S. law), but were unable to stop the newcomer from launching services. Discovery was partially funded with $2 million from BAe but the U.S. Department of Transportation ruled that was not an issue because BAe had no direct control of the airline. Philip Ho, President of Nansay Hawaiian, Inc owned 75% of the stock. Ho was a U.S. Citizen who had been funded by a bonus he received from Nansay, the Japanese parent company, and by his wife who was a Japanese citizen. The two sources were considered foreign entities and as a result, the DOT revoked Discovery's operating certificate, and the airline was forced to shut down by July 13, 1990. The three BAe 146-300 were never delivered.

Franco Mancassola, undaunted, launched Debonair in the United Kingdom. The airline flew five British Aerospace five 146-200s from inception in early 1996, with two -100s added in late 1998. Each carried Debonair titles and the same livery and logo as Discovery Airways. Debonair was a low fare carrier, but provided the frills typical of an full-service carrier. The airline sold cheap tickets but provided the flexibility to change flights without penalty, as well offering connections, complimentary inflight snacks and drinks. The fleet was equipped with a small business class cabin, and economy at a 32" seat pitch. The majority of Debonair's aircraft were former PSA (USAir) aircraft that had been parked in the Mojave Desert in California. After refurbishing them, which included a refresh of the interior and reconfigured to six-abreast economy seating, the airline put them into service.

Discovery Airways shut down before it could take delivery of the first BAe 146-300s it had ordered.

Debonair featured the same exact livery as Discovery Airways.

The final livery worn by Debonair.

Paul Seymour

Based out of Luton Airport, 30 minutes north of central London, Debonair found itself in competition with other low fare startups such as Virgin Express, easyJet, Ryanair and other carriers. Price became a driving factor in the market. Intense competition and limited funds combined to break Debonair, which became insolvent and shut down on October 1, 1999. It was hoped that one of the airlines Debonair had operating agreements with (Swissair and Lufthansa) might be a "white knight" and come in and save the British carrier. That savior never arrived. Michael Cawley of Ryanair said "Essentially they're flying expensive, smaller aircraft, flying through expensive airports. There was never a question they'd go out of business." The running joke was Debonair was a high cost, low fare airline.

Mr. Mancassola seemed to acknowledge Mr. Cawley's statement. "It also helps if the plane doesn't keep breaking down. It was a choice dictated by our lack of finance. At the time they were available, they were parked in the desert. It turned out to be a poor decision on my part, and I take full responsibility. The aircraft proved to be unreliable, and in worse condition than anticipated. We had to spend tons of money to put it right. It was simply a mistake by yours very truly" (from the book "No Frills"). Even with reportedly high load factors (~ 70%), Debonair expanded faster than the cash was coming in, and as a result, its business could not be sustained. With 16 aircraft in operation (13 of which were BAe 146's), the airline shut down. But unlike other bankrupt airlines, which left employees 'high and dry', Debonair paid its staff in full. Mancassola believes 'our mistake was a poor choice of aircraft, they were very costly and not properly financed.' The irony in that statement is that for a poor choice of aircraft, why start with five and end with eleven?

The positive for BAe was that it had not backed Debonair with money or aircraft, so it became one operator that did not come back to haunt the manufacturer. But the aircraft's image had taken another battering.

Discovery's first BAe 146 at Midway Island on a fuel stop prior to arriving in Hawaii.

Thai Airways - 1989

Thai Airways placed an order for four BAe 146-300, which it would lease, at a cost of approximately $200,000 USD per aircraft per month. Initially the order, placed in January 1989, was for two aircraft but a second pair was added in May 1989. In the interim, BAe arranged to lease a single 146-100 (HS-TBO), and two 146-200s (HS-TBQ, G-CSJH) until the series -300s began arriving, going into service in April of 1989. The initial four -300s were provided on short term leases running from 1989 until 1991. In August of 1990 Thai Airways, satisfied with the leased 146s, placed a firm order for five -300s to permanently join the fleet. These -300 aircraft featured upgraded EFIS flight decks, and Thai Airways became one of the few airlines to operate all three passenger versions of the 146.

Thai Airways first BAe 146-300 to be delivered.

Thai Airways operated their -300s with 108 economy seats, while the -100 accommodated 83 passengers. The BAe 146 were used to replace small and slow Short 330s, partly because the 330s didn't have sufficient range but also because of local political pressure at some destinations where turboprops were regarded as 'old fashioned'. Boeing 737s were seen as too big and too fast for the routes the 146 flew. Thai Airways 146s were flown into short and difficult airstrips such as Mae Hong Son in the North West. Thai's Vice President of Domestic Flight Operations Captain Ataya Watanapongse said "So far we've had good service from them. Dispatch rate of 99.5% and the 146 has enabled us to open jet service into the north of Thailand where some runways are too short for the 737. I'd like more power though, but not necessarily new engines (an upgrade to the ALF-502R5). At present, power drops off above ISA +15 degrees Celsius, and we need more power for the hot, high, short strips." Thai believed the -100 was better suited to its needs as the bigger (and heavier) -300 could not lift a full load from some of the country's hot, high and short strips when the temperatures exceeded ISA + 15 degrees. As a result, at times Thai had to deplane passengers because of weight, limiting departures to no more than 50 travellers.

The chief complaint against the -100 was its limited capacity; it obviously could not transport as many customers as a -300. In extreme temperatures, the series -100 was the more appropriate aircraft despite the series -300 having the ideal capacity. Thai Airways had suggested the ATR as a suitable alternative, even though it was a turboprop and would be around 30 minutes slower over a typical flight sector. Thai Airways mentioned that Boeing saw the BAe 146 as a threat to the Boeing 737 in the region. While four engines were a good safety feature, Thai Airways having some shrewd negotiators played BAe off against ATR and Boeing to obtain the best price.

Thai operated BAe 146 aircraft between 1989 and 1993, after which the fleet was grounded until 1995. The aircraft were then reactivated again until 1998, and then left the fleet permanently.

Thai Airways leased a series 100 for immediate capacity until their -300s were delivered.

Thai Airways was operating their 146s to airports that hadn't been served by jets previously.

Thai
HS-TBM
HS-TBK
G-6-191

LAN-Chile - 1989

Founded in 1929 as Línea Aeropostal Santiago-Arica, LAN-Chile would grow over the decades to become the Chilean flag carrier. Based in Santiago, it was the dominant carrier and was state owned until 1989 when it was taken private. Shortly afterwards, LAN-Chile began looking for an aircraft with excellent short field performance to use at remote destinations including unpaved airfields. On September 29, 1989 LAN-Chile sent a letter of intent (LOI) to British Aerospace's sales division in Washington D.C., United States informing the manufacturer of its intent to acquire three used BAe 146 aircraft (one series -100, two series -200), and including with it a $100,000USD deposit. The aircraft were formerly operated by Presidential Air. LAN-Chile configured the aircraft with five-abreast seating and a maximum capacity of 85 passengers. Although LAN specified a single -100 aircraft, in the end only the series -200s were acquired. The aircraft were to be leased for nine and a half years, with first flights occurring in August of 1990. Jack Thomson, BAe's Regional Manager for South America, mentioned LAN-Chile had intentions of using the BAe 146-200 for flights to Antarctica as the type was ideally suited for the unpaved runways.

The first aircraft was handed over in a ceremony at Page Terminal at Dulles Airport (IAD) at 2:30pm on January 8, 1990. The next day, the flight crew that were part of LAN-Chile's team went through its acceptance procedures and the aircraft was formally taken on charge on January 10. Then began the long ferry flight from Maryland to Chile, starting with a flight to Miami on January 11. On the following day, a long-distance flight from Miami to Arica, Chile took place with fuel stops enroute. From there, the aircraft transited through Iquique and Antofagasta before arriving at Santiago. On Monday, January 15, a formal reception was held for the aircraft and staff of BAe. It wasn't long before LAN's first 146 undertook a proving flight to Antarctica and on January 26, 1990, it touched down at the Chilean Air Force base on King George Island. LAN-Chile was planning for the BAe 146 to become the first commercial airline service to Antarctica for tours through King George Island. It was one of the rare occasions when LAN-Chile's BAe 146s would service small fields with unpaved runways, such as King George Island.

LAN-Chile ordered a third 146-200 for delivery in September 1990, which was also an ex-Presidential aircraft. It had been engaged on a sales tour of Central and South America prior to delivery to LAN-Chile.

If the BAe 146 had performed as expected, LAN-Chile planned to acquire more, shifting its commitments to the BAe 146-200QC (Quick Change) variant so that they could operate as freighters at night.

By summer of 1992, reports were emerging suggesting that LAN-Chile was not happy with the aircraft and found it deficient in certain areas. Rumors of poor reliability, low load factors, and high operating costs began to circulate. On October 1992, LAN-Chile contacted British Aerospace requesting a modification to the aircraft leases, allowing them to be sub-leased to other airlines. This letter made it very clear that LAN-Chile, contrary to rumors, was quite happy

LanChile was the first airline to land a commercial jet aircraft on the Antarctic continent.

with the aircraft. It stated everything about the aircraft was in-line with the representations British Aerospace made before they were handed over. In fact, the customers and pilots loved the aircraft. But the letter went on to say the 146s where acquired as part of a business plan that was no longer economically viable, and the new plan didn't feature the 146 in future LAN-Chile operations. The 146 wasn't the only casualty of the reorganization as some of the Boeing 767s which the carrier acquired were also being subleased. With larger 737 aircraft in the fleet that could do stage lengths out of reach (and capacity) of the 146, LAN-Chile was grounding the small fleet of two aircraft because of marginal results (the third aircraft was lost in an accident the previous year). Both aircraft remained with LAN-Chile until 1997, when they were sold to Jersey European Airways.

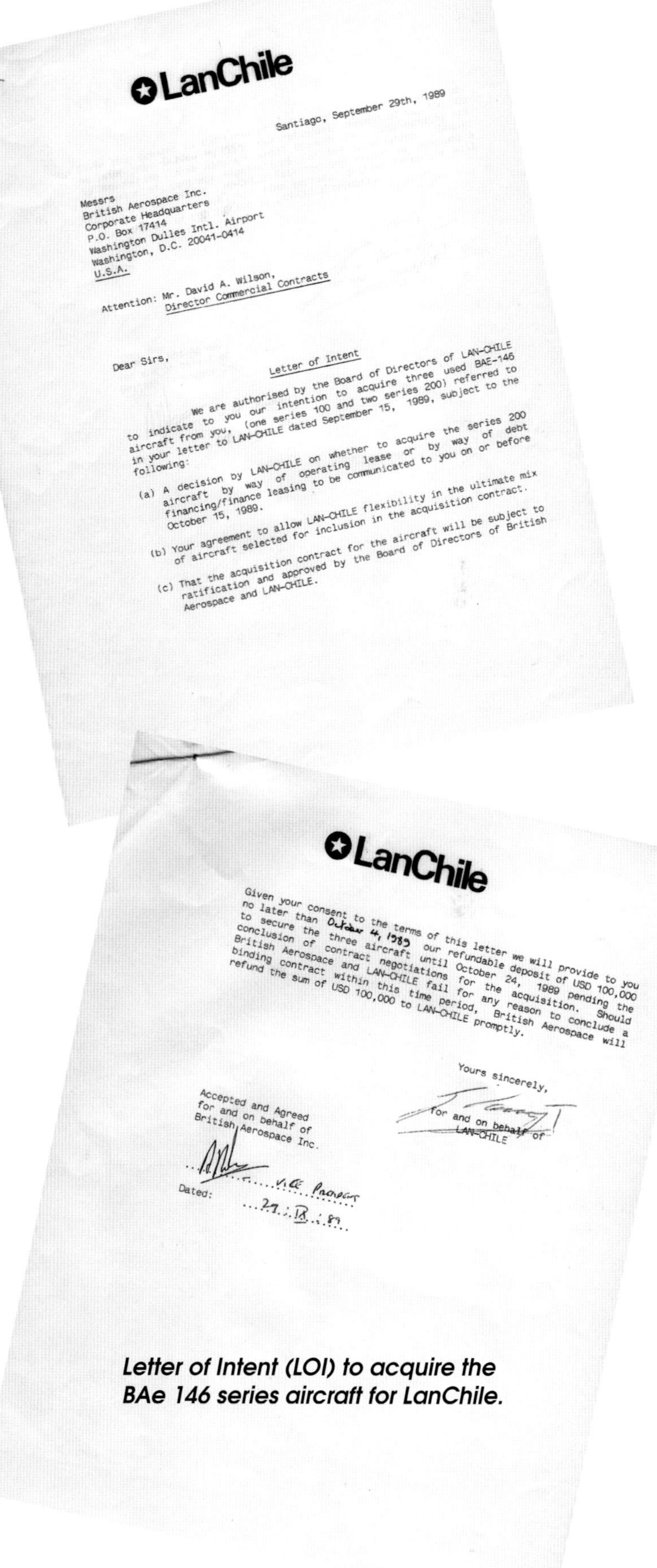

LanChile

Santiago, September 29th, 1989

Messrs
British Aerospace Inc.
Corporate Headquarters
P.O. Box 17414
Washington Dulles Intl. Airport
Washington, D.C. 20041-0414
U.S.A.

Attention: Mr. David A. Wilson,
Director Commercial Contracts

Dear Sirs,

Letter of Intent

We are authorised by the Board of Directors of LAN-CHILE to indicate to you our intention to acquire three used BAE-146 aircraft from you, (one series 100 and two series 200) referred to in your letter to LAN-CHILE dated September 15, 1989, subject to the following:

(a) A decision by LAN-CHILE on whether to acquire the series 200 aircraft by way of operating lease or by way of debt financing/finance leasing to be communicated to you on or before October 15, 1989.

(b) Your agreement to allow LAN-CHILE flexibility in the ultimate mix of aircraft selected for inclusion in the acquisition contract.

(c) That the acquisition contract for the aircraft will be subject to ratification and approved by the Board of Directors of British Aerospace and LAN-CHILE.

LanChile

Given your consent to the terms of this letter we will provide to you no later than October 4, 1989 our refundable deposit of USD 100,000 to secure the three aircraft until October 24, 1989 pending the conclusion of contract negotiations for the acquisition. Should British Aerospace and LAN-CHILE fail for any reason to conclude a binding contract within this time period, British Aerospace will refund the sum of USD 100,000 to LAN-CHILE promptly.

Yours sincerely,

for and on behalf of
LAN-CHILE

Accepted and Agreed
for and on behalf of
British Aerospace Inc.

Vice President

Dated: 29.IX.89

Letter of Intent (LOI) to acquire the BAe 146 series aircraft for LanChile.

CHAPTER 07

The BAe 146 Family Grows

The BAe 146 was developed primarily as a passenger aircraft for commercial operations, but since its inception British Aerospace had flirted with other variants that made use of the basic 146 airframe – everything from freighter and quick-change to military variants. Many aircraft programs go down a logical lifetime progression of airframe development over a substantial period of time, with the aircraft and the airframe being optimized to excel at one purpose initially, before further derivatives are considered. The evolution of the 146 took place in much the same way as Airbus, Boeing, and McDonnell Douglas have done with their airframes. Few manufacturers get it right from the start and have to adapt once they realize where the biggest market opportunity lies, much as Boeing did in the progression of the 737-100 series to the -200. BAe went down similar path with the 146, moving the emphasis from the -100 to the -200 as it became apparent there was a larger market for the latter. The 146 was struggling with slow sales, poor build quality, high operating costs and unreliability. Whilst trying to fix these issues, BAe also embarked on expanding the range to drum up further business.

Stretching Capability: The BAe 146-300

The first and obvious choice for a variant was to stretch the 146 again. By offering an even larger version, the 146 would evolve into a family of aircraft, in much the same way as Boeing and Airbus were offering aircraft families and giving operators flexibility. Airlines had been hinting that a larger variant would be preferred, with pressure coming from a key early operator: Air Wisconsin. BAe also lost an important sales campaign when Swissair ordered the Fokker 100, despite the British manufacturer offering the concept of a larger version of the 146. The Fokker 100 was an existential threat to the programme, so BAe had to act to avoid being further marginalized in future sales campaigns.

On September 3, 1984 at the Farnborough Airshow, BAe announced the launch of two new variants of the 146, starting with the stretched BAe 146-300. It was designed to seat up to 130 passengers in a high density 29" pitch 6 abreast configuration. The -300 also had improved wing aerodynamics (with winglets similar to those Airbus was proposing to introduce on the A320 series, in which BAe was a risk sharing partner) and a strengthened wing, improving overall operating economics. Additional changes and improvements were expected to come in the form of Avco Lycoming LF-502R7 engines (in early development stage), electronic flight deck glass cockpit, and a smaller stretch of 10 feet over the series 200. The aircraft's maximum weight would increase to 102,000-104,000 pounds. Reaction to the aircraft at the Farnborough Airshow from airlines was positive but there was no firm commitments or orders from a launch customer.

The prototype BAe 146-300 public presentation event May 1, 1987.

The new variant would cost more than £120 million to launch and develop. The major cost of the work on the initial -300 derivative came from a need to strengthen the fuselage and the wing to enable the gross weight to exceed 100,000 pounds. But the substantial cost, the lack of a launch customer, and a production run estimated of between 150 and 180 aircraft with a selling price ($18m) higher than the Fokker F100 and Boeing 737-200 Advanced doomed the proposals. BAe struggled to obtain funding and instead decided to pursue a "minimum change" derivative focused more on a simple fuselage stretch, limiting gross weight to 98,000 lbs. Upgrading the existing wing would be far less expensive for the minimum change aircraft than the original series -300 proposal, as powered ailerons and wing-tip fences would be dispensed with. As a result, development costs would drop to no more than £65 million. The revised forecast was for the derivative aircraft to have a maximum production run of 150 airframes, and for manufacturing to end in 1996.

Initial concept for the series 300 featuring winglets and an additional fuselage plug fore/aft of the wings with a higher row count.

The revised BAe 146-300 would carry 20% more payload, but with a negligible increase in fuel burn. Further changes planned were digital autopilot and avionics with an electronic flight instrument system (EFIS), using cathode ray tube displays, or conventional analog displays for fleet commonality. An updated flight deck was promised in the early launch days, but by 1987, sales material only discussed the standard analog instrument layout, which made crew cross-qualified nearly instant and reduced fleet integration costs for existing BAe 146 operators. The aircraft would have two fuselage plugs: a forward section of 8ft 1 inch and an aft plug of 7ft 8inch, bringing the total length to 100ft 9inches. This was sufficient for up to 15 additional seats.

Besides being able to carry additional passengers, the interior was be redesigned with rounded lines and wall wash lighting to offer a more spacious widebody feel. The cabin was widened due to modified frame profiles being introduced within the fuselage (shrinking by 2 inches), allowing more shoulder room and the raised seating to provide enhanced passenger comfort. The wing of the BAe 146-100/200 was so efficient it could lift the increased weight of the series 300 with no changes aside from a thicker fuselage skin at the mid-section where it joined the fuselage, and a stronger top skin and stringers attached to the existing wing.

E1001 was transformed into E3001, the demonstrator and prototype for the BAe 146-300.

Shortly before the launch of the Series 300, BAe sought launch aid for its participation in the Airbus A320 programme to the tune of £250m ($328m USD), which did not include another £200m that BAe had to provide itself. BAe was

spreading itself thin as it was committed to the Airbus consortium while funding its own regional jet growth.

After a protracted development schedule that ran later than expected, Air Wisconsin announced its order for five series 300 aircraft, which would enter service in United Express livery (Air Wisconsin was operating exclusively as a United Express affiliate). The announcement of a launch customer came on the day of the series -300 rollout. Air Wisconsin was no longer confined to its hub in Chicago O'Hare airport, but was also serving Washington Dulles. The order was revealed during the ceremony marking the opening the new BAe 146 assembly hall at Hatfield, which had cost £4 million. With the order book growing, the existing building which had been erected in

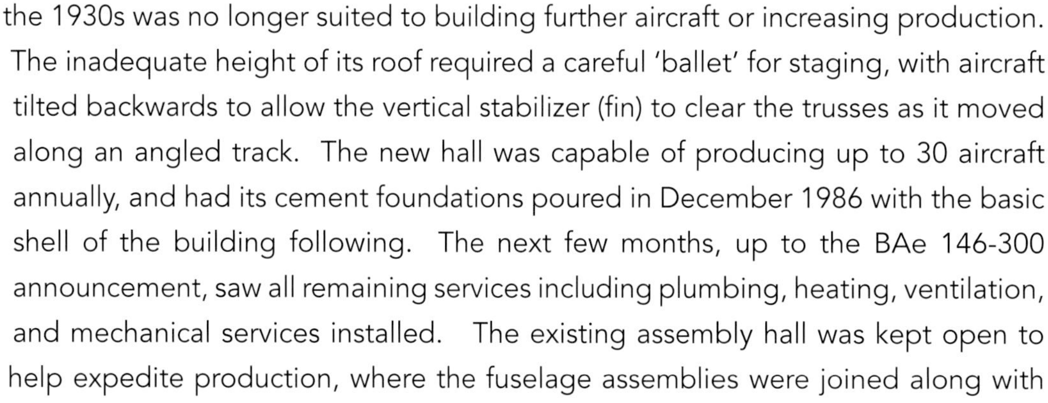

the 1930s was no longer suited to building further aircraft or increasing production. The inadequate height of its roof required a careful 'ballet' for staging, with aircraft tilted backwards to allow the vertical stabilizer (fin) to clear the trusses as it moved along an angled track. The new hall was capable of producing up to 30 aircraft annually, and had its cement foundations poured in December 1986 with the basic shell of the building following. The next few months, up to the BAe 146-300 announcement, saw all remaining services including plumbing, heating, ventilation, and mechanical services installed. The existing assembly hall was kept open to help expedite production, where the fuselage assemblies were joined along with

1025	Mr. Preston Wilbourne (President, Air Wisconsin is invited to signal the start of BAe 146-300 public unveiling ceremony
1115	British Aerospace Civil Aircraft Family flying display
1200	First flight of BAe 146-300
1230	Lunch
1400	Return of BAe 146-300
1445	Departure

Sound of Silence

wings, undercarriage, and fin. The aircraft was then moved to the new hall for final assembly where the engines and tail plane were installed, and ground testing took place. Once those tasks had been completed, each aircraft was towed to an adjacent hangar where final preparations for its first flight took place. The new arrangement saved BAe money and time, cutting the build sites from six to four and resulting in a 5% reduction in man hours, because build efficiency was increased and the flow of the line was sped up. Instead of 21 weeks from start to finish for a 146 build, the time was reduced to 15 weeks. In 1991, aircraft build times had been further reduced to nine weeks, enabled by crews with different skill sets moving between aircraft as needed instead of having the aircraft come to their stations. Ironically, despite Hatfield receiving its largest infrastructure investment for many years, the Board of BAe was already embarked on a programme of closing expensive sites around London for commercial re-development, including Weybridge and Kingston. Hatfield was another obvious target, as events were to prove.

In August 1986 BAe selected the prototype 146-100, E1001, which had completed nearly 1,240 hours of flying, and began to modify it to become the prototype series 300. The timeframe to modify the aircraft and get it flying again was short, as it needed to be ready to fly again within a year at the Paris Air Show in 1987. E1001 was positioned in the Technical Services hangar where the Comet 1 prototype had been built 37 years earlier. The -100 was cut into three sections, and fuselage plugs were inserted fore and aft of the wing. It was re-registered G-LUXE and given the construction number E3001, rolling out on March 8, 1987, and was painted in a new livery days later in preparation for its first flight.

The 146-300 was unveiled to press and workers under the banner "The Sound of Silence" on May 1, 1987, during which a special rendition of Simon and Garfunkel's classic song the "Sound of Silence" was performed by the Pontarddulais Male Voice Choir as a tribute to the world's quietest airliner. Preston Wilbourne was presented with a gold model of the BAe 146 by Civil Aircraft Division Marketing Director Brian Thomas as a token of appreciation. Lord Lieutenant of Hertfordshire Mr. Simon Bowes-Lyon presented the Queen's Award for Export to British Aerospace for the sales

achievement as a result of the overseas customers it had signed up for the BAe 146. Afterwards the aircraft rolled out of the hangar to the music of "Chariots of Fire" and took to the air at precisely 12 noon. After a test flight of just over two hours, E3001 returned to Hatfield where Chief Test Pilot Peter Sedgwick reported "it flies beautifully".

Boeing had been taking notice of the growing popularity of the BAe 146, specifically with North American operators. The larger 146 was encroaching into the 100-seat category, and the US manufacturer felt that it could no longer ignore the 146 – or the Fokker 100 for that matter. Boeing officially launched the Boeing 737-500 in May 1987 as a direct competitor to the BAe 146-300. It was the smallest variant of the 737 series, with seating for 110 (8 first class, 102 coach) in a typical mixed class configuration. It provided growth potential into larger 737 variants as the need arose, while maintaining fleet commonality in a family of aircraft. Boeing gained 51 orders plus 22 options for the 737-500 before it was formally launched by Southwest Airlines.

Appearing at the Paris Air Show as planned in 1987, the BAe 146-300 took to the air in the flying display. Following the conclusion of the airshow the aircraft entered a test flying programme, just like its predecessors. These tests would encompass flutter, minimum speed, stall,

Spacious interior for the BAe 146-300. BAe began presenting interior configurations showing a 2-3 5-abreast layout even though carriers could configure with 6-abreast.

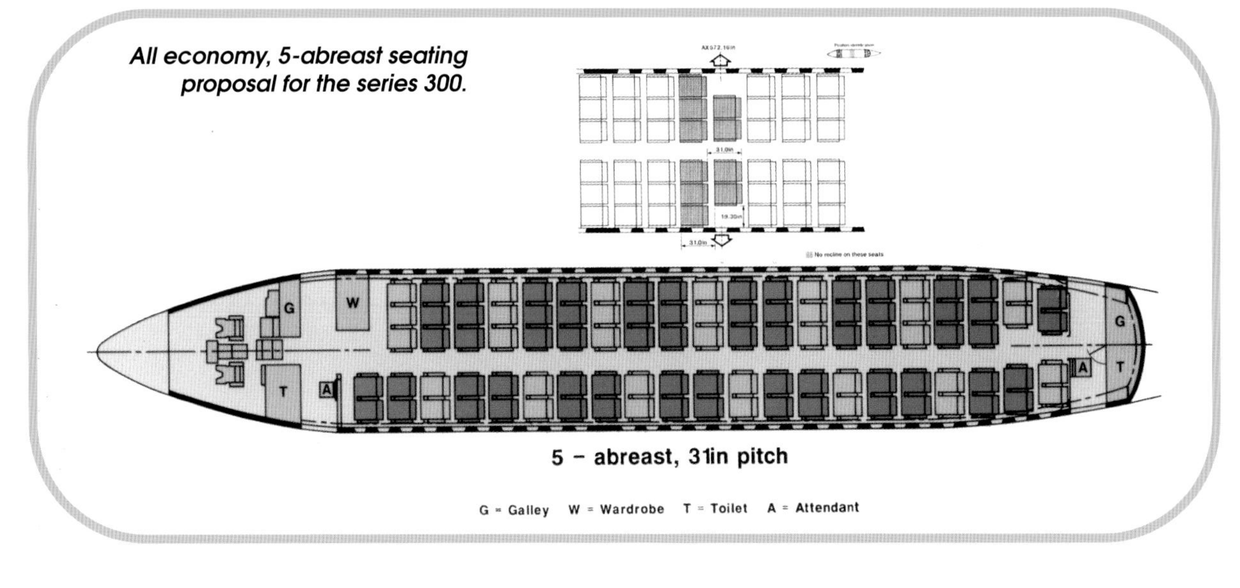

All economy, 5-abreast seating proposal for the series 300.

high speed and high-weight testing. After approximately 250 hours, the first production -300 (E3118 G-OAJF) joined the prototype for trials. The production series 300 incorporated further design changes and was used for testing at operational weights. On September 6, 1988 at the Farnborough Air Show it was announced that the series -300 had received certification from the UK CAA. G-LUXE took part in the flying display, while Air Wisconsin's first aircraft was on display in United Express livery (as G-BOWW) where it was joined by E1002 which had been reconfigured to become the Sideloading Tactical Airlifter (STA). Air Wisconsin put the first series -300 into service in December 1988 as more orders continued to pour in for the new and larger aircraft.

Further improvements to the BAe 146 series continued even though the series -300 did not launch with new bells and whistles such as an electronic flight deck. However, BAe soon realized that the marketplace was demanding EFIS and the first production 300 G-OAJF was quickly modified. Certification for the EFIS equipped aircraft came in late 1989, and by late 1990 20 -300s with EFIS had been delivered, marking a further evolution of the basic design. At the conclusion of manufacturing the BAe 146-300, a total of 71 had been built, more than twice the number of BAe 146-100s produced. This further demonstrated BAe's initial assessment that the series 100 would be the most popular version had been a fallacy.

Series 300 launch customer Air Wisconsin aircraft were delivered in United Express livery.

Roger Pascoe

QT
146-QT
QUIET TRADER

BAe 146QT – Quiet Trader

BAe announced another derivative of the 146, a Freighter dubbed the "Quiet Trader" (QT). Brian Thomas was appointed to run the newly formed Marketing Operations Center (MOC) covering all BAE's in-production regional aircraft. He saw variant aircraft, especially freight door equipped aircraft, as having the potential to widen the 146's appeal and underpin production and profit margins.

Thomas had taken over the relationship with Sir Peter Abeles who, as previously related, had seen the benefits of the 146 and against internal opposition had ordered the BAe 146 for Ansett Australia. Sir Peter Abeles had made his reputation and fortune in the freight business and in addition to his leadership of Ansett was still running TNT the international logistics freight forwarding company. Thomas sensed an opportunity for the 146QT and persuaded Abeles to come to Hatfield and see the concept for himself while he was in London. In due course Abeles and TNT 's Chief Executive arrived late one evening to see the outline of what would be the 146QT taped onto a hangar floor. Next to this 'tape mockup' were the containers that could be fitted into the aircraft. Although the economics did not immediately work, and with TNT managers against the idea, Abeles saw an opportunity to leap frog his competitors with the 146QT as it overcame the new noise curfews being implemented across the EU. Discussions continued, but whilst attracted by the concept Abeles wanted to see the reality before making a commitment. With this tenuous possibility Thomas persuaded the Board to fund a development aircraft.

Initially utilizing the series 200 fuselage, the freighter would provide for 11ft 2 in internal diameter allowing it to carry six standard 108"x88" (2.74m x 2.23m) pallets with space for an extra half pallet (53"x88"). The floor strength would be increased to allow a maximum payload of 22,000lb (10 tons), with each pallet having a maximum load of 6,000lb (2.7t). Optionally, industry standard LD3 containers could be used, with up to nine (9) on the -200 and ten (10) on the -300. LD1, LD2, and LD4 containers could also be accommodated in varying configurations providing that weight distribution rules were adhered to, with floor-loading limited to 70 pounds per square foot.

Freight could be loaded via a rear upward opening door to accommodate pallets on the port side of the aircraft. The door measured 131"x76" on the series -200 and -300, and the same size door was also fitted to the STA -100 aircraft. BAe did not market the series 100 as a freighter, instead promoting the -200 and -300 to airlines. Early concept drawings showed a series 100 with a freight door installed ahead of the engines, although this configuration was never pursued. BAe also offered the 146 as a combi that carried passengers in the front half of the aircraft and freight in the rear, or as a convertible which would fly a full load of passengers when needed, but with the ability to remove the interior quickly (seats were on pallets) for use as a pure freighter. Although it was discussed in depth with some potential customers, the combi variant was never put into production. Increasingly stringent fire regulations and the variant's development costs made the idea completely unprofitable for both the manufacturer and the operator.

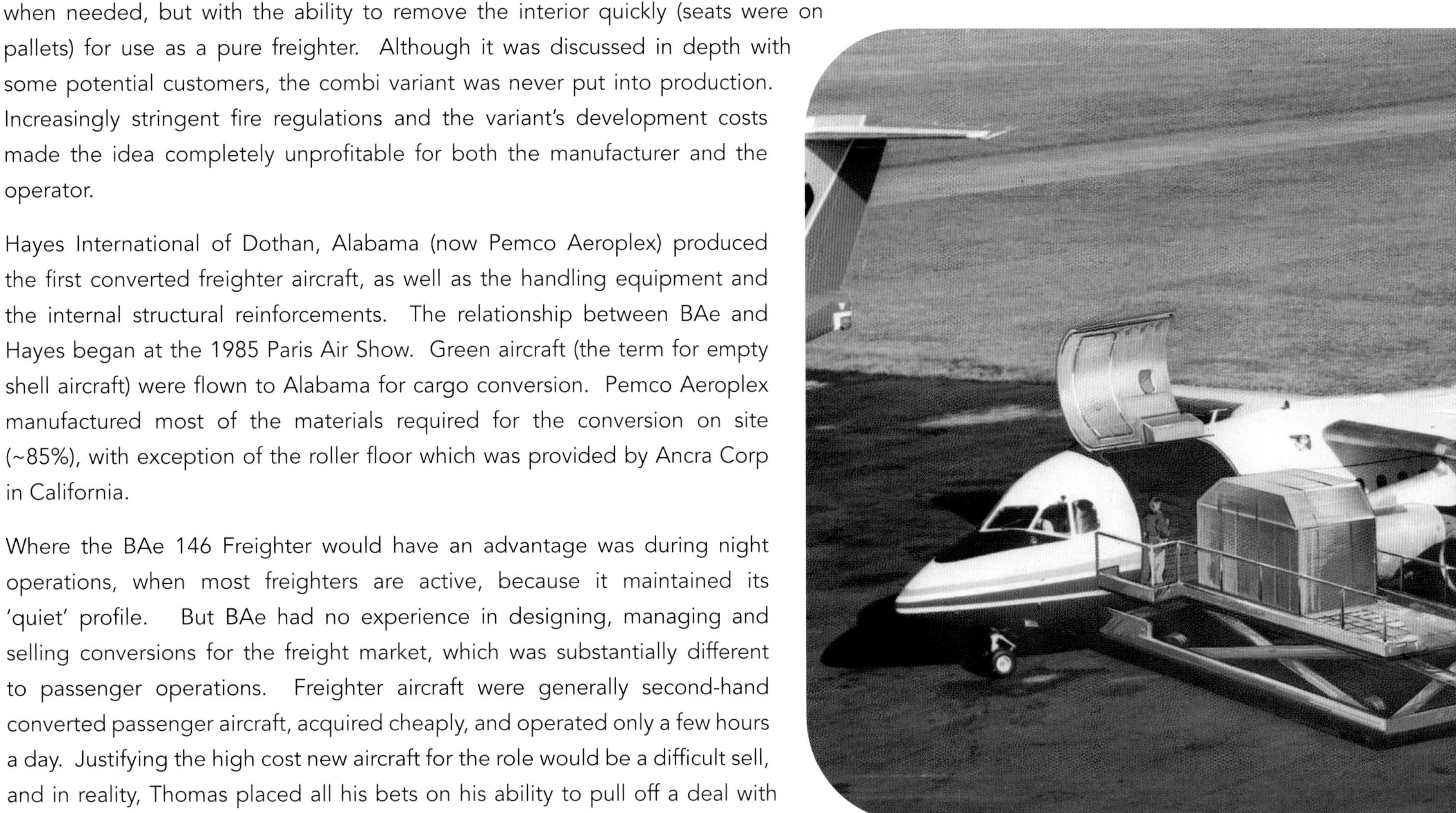

Initial concept with a forward loading cargo door.

Hayes International of Dothan, Alabama (now Pemco Aeroplex) produced the first converted freighter aircraft, as well as the handling equipment and the internal structural reinforcements. The relationship between BAe and Hayes began at the 1985 Paris Air Show. Green aircraft (the term for empty shell aircraft) were flown to Alabama for cargo conversion. Pemco Aeroplex manufactured most of the materials required for the conversion on site (~85%), with exception of the roller floor which was provided by Ancra Corp in California.

Where the BAe 146 Freighter would have an advantage was during night operations, when most freighters are active, because it maintained its 'quiet' profile. But BAe had no experience in designing, managing and selling conversions for the freight market, which was substantially different to passenger operations. Freighter aircraft were generally second-hand converted passenger aircraft, acquired cheaply, and operated only a few hours a day. Justifying the high cost new aircraft for the role would be a difficult sell, and in reality, Thomas placed all his bets on his ability to pull off a deal with TNT.

TNT put the QT to the test at BAe's expense in November 1986, before placing an order for the aircraft. The manufacturer offered TNT the prototype 146-200QT, along with two pilots, crew, and load supervisors for a series of flights that took place between the UK, Ireland, and Germany. The flights began each night at 8pm and operated between the following destinations: Dublin-Birmingham-Nuremberg-Hanover-Birmingham-Dublin (arriving at 5:30am). Each flight was approximately an hour in length, and turnaround times were kept under an hour. Adding an obstacle to the 'test drive' was Birmingham's runway length being temporarily reduced to approximately 4,300 feet due to improvement work taking place at night. The German cities had stringent night noise restrictions, so the flights tested the QT's short field performance and ability to operate within the noise parameters.

The QT trial proved a major success and then came the negotiations between BAe's Chief Executive, Sir

TNT proving flight tests November, 1986.

Raymond Lygo and TNT's Sir Peter Abeles that took place in a hotel room in Monte Carlo. Unsurprisingly Abeles thought the aircraft price was too high and there was a lengthy debate, with Abeles saying, 'he could get competitive aircraft from Boeing or McDonnell Douglas for less than BAe was asking for the 146.' It was rumored that the manufacturers were offering 737 and MD-80 aircraft at pricing below the 146. Lygo knew that one of the key traits for the 146, its low external noise, was a key advantage that neither Boeing nor McDonnell Douglas could match. Being given unlimited operations at night and with a noise rating below the limits imposed by European countries meant the 146 was really the only choice if TNT were to acquire new build aircraft. Lygo summed up his position on pricing by saying, "yes, the 146 isn't the cheapest aircraft, but I can't lose money on each one, and over time with economies of scale the cost can be reduced." Apparently, Lygo eventually agreed to sell the 146QT aircraft at cost, to which Abeles

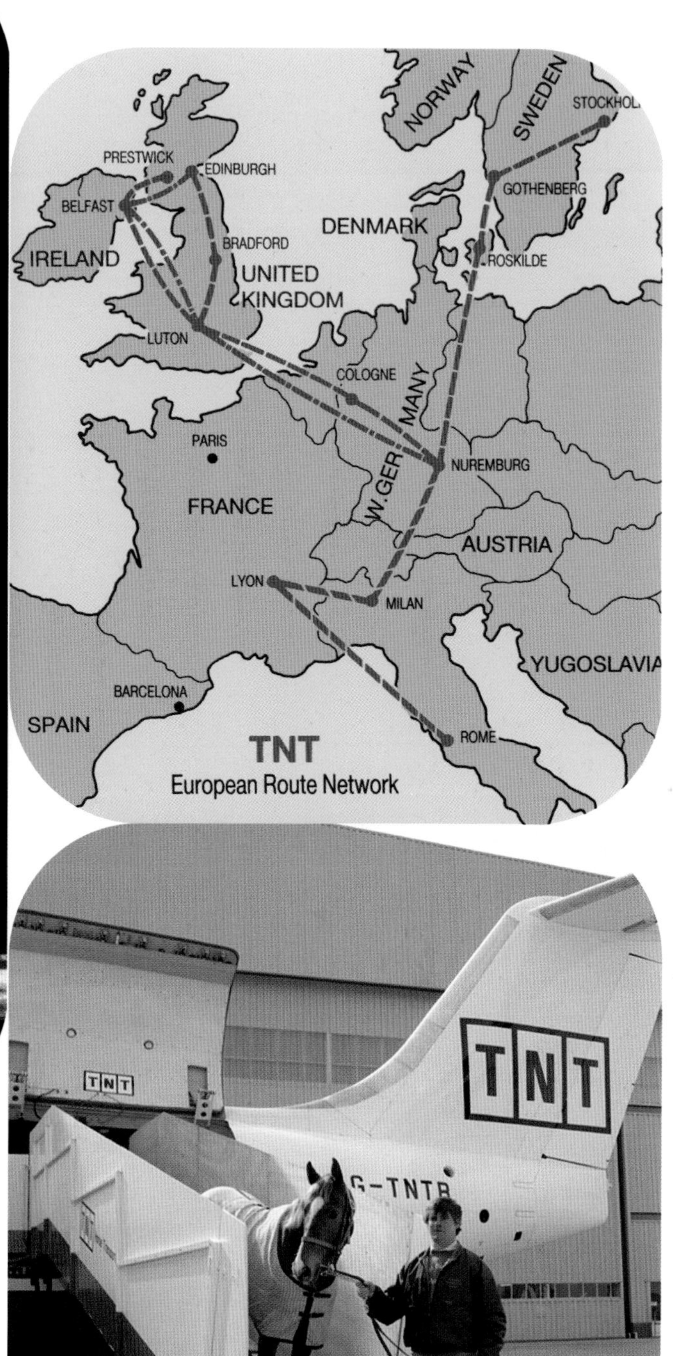

responded: "Well, I think we'll go for it" and that was it. It was not a formal agreement, and outside of undisclosed deposits for production positions on the aircraft being sought, it was just a gentleman's handshake.

Notwithstanding, when the deal was announced at the Paris Air Show, it was stated to be an order for 72 aircraft, notionally representing all of the QT production slots for the following five years. The reality was more prosaic. TNT signed a firm contract for just five 146-200QT aircraft. Subsequent orders for further -200QT's and the subsequently developed -300QT were placed but the total number actually purchased by TNT amounted to just 22 aircraft and the firm ended up being the sole customer for the type. The carrier's failure to take more aircraft probably had two main causes. The first was that Abeles had intended to buy US Carrier Airborne Express and saw many QT's being required for that market, but the deal never happened. Secondly the cost of each 146QT was booked against the account of the TNT country manager that it operated for. The consequence was that each had to bear new aircraft costs which made a serious impact on their financial results. Over time matters improved but in the short-term TNT managers squirmed every time an aircraft was allocated to them, creating a degree of resistance within the company to acquiring yet more aircraft.

BAe for its part was hoist with its own petard. Having elected to go big on the deal in public and adding 72 aircraft to its order book, it had raised both internal and external expectations. Efforts to get TNT to sign anything at all confirming the full extent of the nominal commitment failed and as time passed BAe was forced to admit the deal's true extent. By that time the damage had been done. This Faustian Pact effectively locked in QT production to one customer and left BAe unable to market the aircraft effectively to TNT's rivals, thus nullifying the benefits of the deal in the first place and condemning the QT to becoming another dead-end niche.

The aircraft served TNT very well for many years and as the acquisition costs were written down, its true virtues came to the fore. TNT Aviation Managing Director Neil Hansford fired a shot across other airlines bow's when he said "If they spent as much in looking at aircraft as they do on public relations, then they would choose the 146." He went on to say that the aircraft was giving TNT Aviation a time and cost advantage. Hansford continued by saying that it's difficult to be reliable with 30-year-old used aircraft. Even factoring in the engine problems the 146 had endured, Hansford said they were able to change an engine in two hours. With a £2,000 per block hour cost and the low noise footprint, a 10% reduction in landing fees at West German airports as well as unlimited night movements made the BAe 146 an economic powerhouse. And although a 20-minute turnaround was time allotted, the aircraft were often turned around in under 12 minutes.

TNT was not a European company, so it contracted with separate operators in the countries where the aircraft were registered. At its full extent this included carriers in the UK, France, Germany, Spain, Italy, Sweden and Hungary. For example, Air Foyle flew the UK based and registered 146s on its AOC on behalf of TNT, in whose livery all the aircraft were flown, while BAe performed maintenance. The 146QCs achieved a 99.13% dispatch rate according to Brian Ayling of TNT. Two aircraft were flown in Australia by Ansett Air Freight and two more operated for a period in the Philippines, but Europe was to be home for the majority of the fleet. When Sir Raymond Lygo retired from BAe, he became the Chairman of TNT until it was acquired, after the death of Sir Peter Abeles, by the Dutch Post Office.

The statist 1970's had been superseded in the UK by the more entrepreneurial business-friendly era of Margaret Thatcher. Sweating the assets (or asset stripping for the more cynical) became all the rage. BAe began to diversify beyond its aerospace roots acquiring a property development company, Arlington,

TNT
The Express

amongst other things. Why allow others to get the benefit of prime real estate re-development, as its facilities around London were closed, when it could keep things in house? When BAe subsequently sold Arlington back to its founders some years later after failing to reap the benefits anticipated, at a much-reduced price, it was easy to see who had really benefited and it was not BAe.

In line with this philosophy, Lygo had long wanted to shut down Hatfield and re-develop the site. Production would be moved to Woodford, a lower cost facility near Manchester. The orders TNT promised BAe allowed Lygo to do just that. Initially it was said that Woodford would just be a second production line with the QT as the lead aircraft. Perhaps that might have been so if the full TNT commitment had ever been realised, but it is hard to avoid the conclusion that Hatfield was living on borrowed time in any circumstances. In the end the Hatfield facility was shuttered and all production moved to Woodford, with the last aircraft to be built at Hatfield being the first production Avro RJ aircraft in 1993. It would be interesting to know just how cost effective the decision was in reality. The major infrastructure developments at Hatfield including the new Production Hangar and the centralised canteen (a state-of-the-art help-yourself facility that replaced the five seniority stratified restaurants that had previously existed) were written off. Thousands of employees were either laid off, transferred to Woodford, Airbus or elsewhere at huge expense and disruption. Hundreds of years of experience were lost, and the project memory bank substantially diminished.

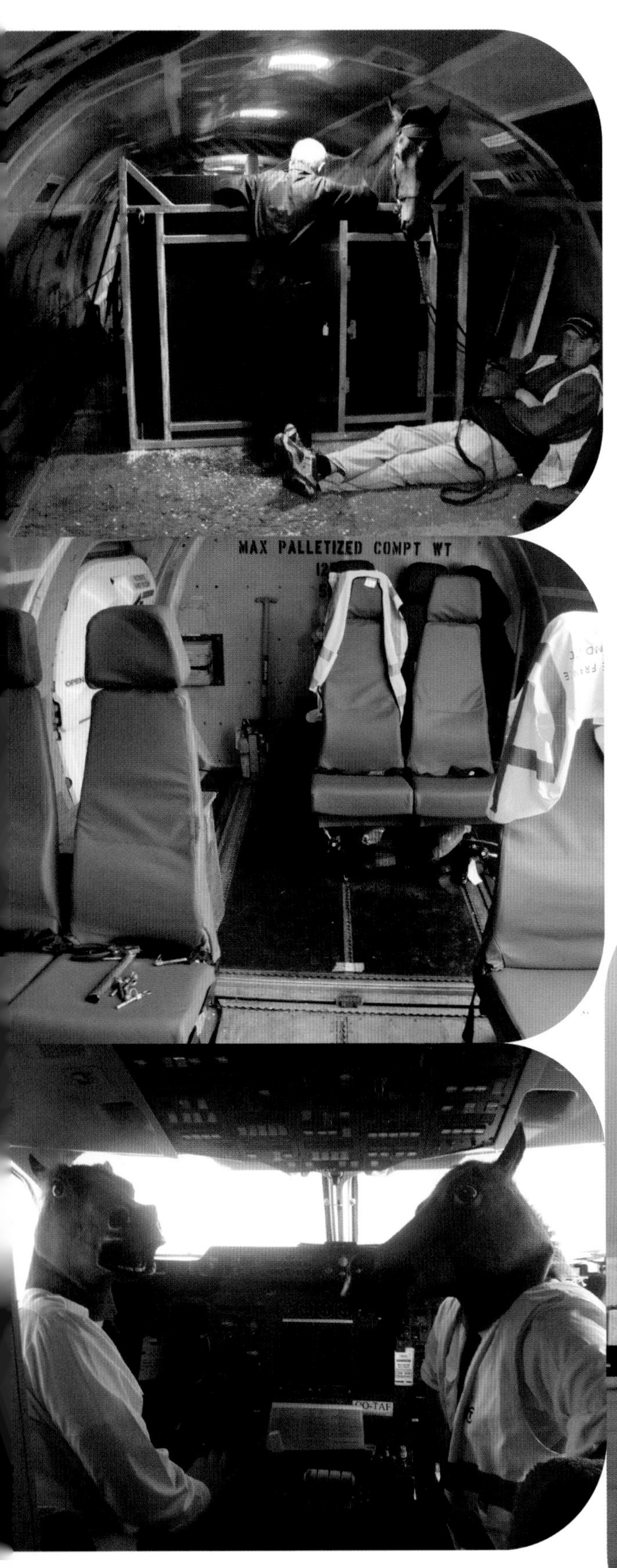

All photos this page provided by Kevin Davies.

All photos this page provided by Kevin Davies.

BAe continued to expand its presence in the United States to serve not only the emerging North American market, but also South America. The firm established a base just outside Washington Dulles Airport with offices for suppliers, spare parts and logistics warehousing, sales and technical support, and finally a simulator training bay for pilots. An affiliate company, Arkansas Aerospace, Inc., became BAe's principal centre for acceptance and delivery preparation for its complete line of aircraft in the Western Hemisphere. It also served as an alternative factory authorized maintenance center.

BAe 146QC – Quiet Convertible

The –QC (Quick Change or "Quiet Convertible") aircraft was essentially a –QT aircraft but with the passenger windows intact (vs. omitting them like traditional freighter aircraft). The overhead bins, galleys and toilets were not excluded from the -QC, and thus the aircraft could only hold six pallets versus the additional half-pallet series -200QT aircraft were capable of accommodating.

The launch customer for the QC version was Princess Air, which was announced at the 1989 Paris Air Show. It was a start-up carrier based at Southend in the UK and owned by Ivor Burstin, who was also the owner of

Burstin Travel. An interim -200 passenger aircraft (G-BRXT) was delivered on March 21, 1990 and was leased from BAe until the first of the two aircraft princess ordered were delivered. Burstin's plan was to operate the aircraft in passenger configuration during the day and as a freighter at night. He knew that he could utilise the 146s for his own travel business during the day, but the freight business was based on untested assumptions. The fist -200QC G-PRIN "Princess Allison" was delivered on June 8, 1990 and G-BRXT was returned to BAe on July 10, 1990.

During the configuration changes the cargo door would be opened, and the passenger seats which were installed on pallets would be removed. The aircraft was then ready to operate as a pure freighter during the night hours, when demand for cargo aircraft peaked. When being flown as a freighter, the 146 was operated by P.A. Cargo Ltd, a wholly owned subsidiary of Princess Air. Cargo flights took place between Southend and Cologne, Germany, and as well as Brussels, Belgium. Initially the trips took place between 1am until 4am each day, with a single flight running between Southend and Cologne. True to the noise cloak the BAe 146 operated, Princess Air operated under the call sign "Whisperjet".

Passenger operations ran to a number of tour destinations during the day with a high-density configuration of 94 seats in a 31" seat pitch. Not long after it was delivered, on June 16, 1990 Princess Air's first aircraft G-PRIN experienced a hard landing at Jersey Airport in the Channel Islands which resulted in fuselage damage and minor perforation of the pressure hull. The pilot in command was 36 years of age with over 8,000 hours, but only 63 hours on the 146. The aircraft was repaired and returned to service 21 days later. Unfortunately, from day one, the freight business did not meet expectations and the business model required freight to be an equal contributor to the bottom line. Additionally, the team at Princess Air did not have the experience to develop that aspect of the business. In the end, the airline failed to last a year and the aircraft was repossessed by BAe.

The concept of the QC was not just convertible from passenger to freight (and vice-versa), but also to carry freight plus passengers.

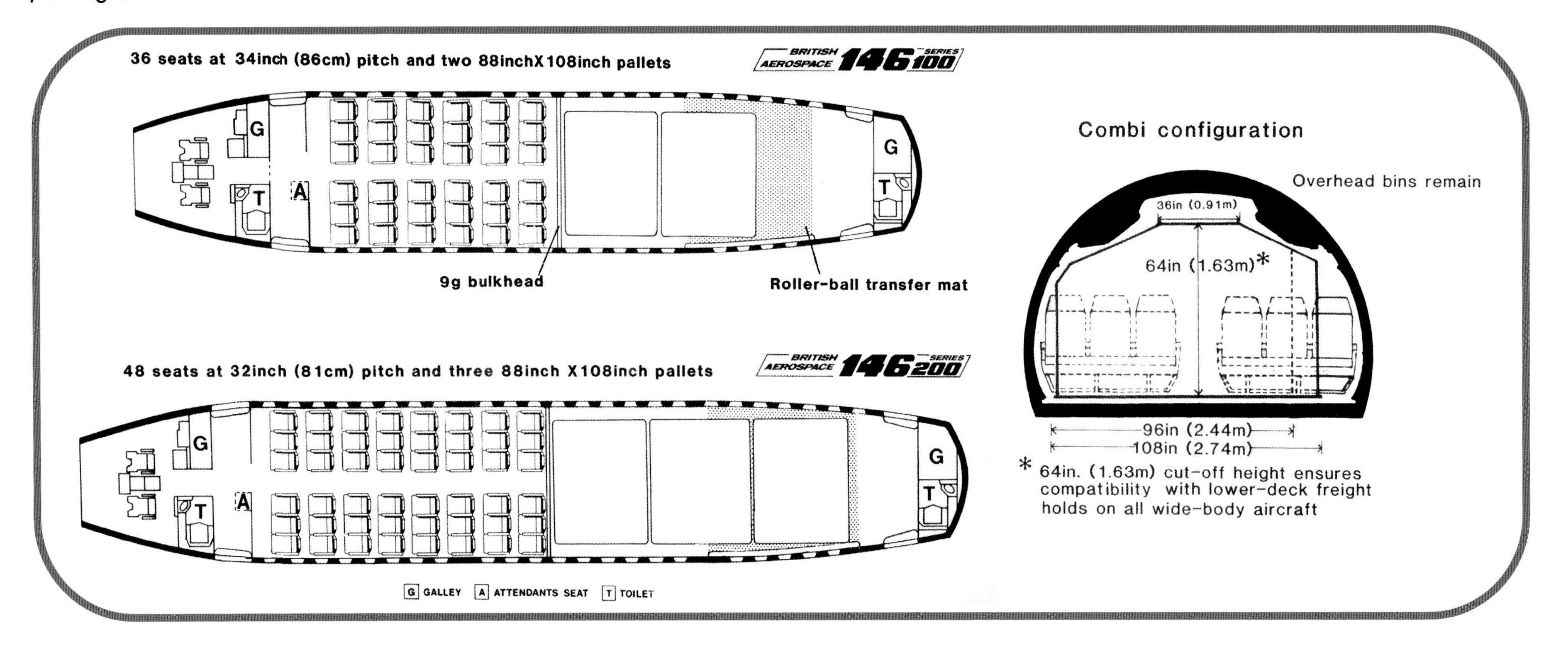

Princess Air
Princess Alison

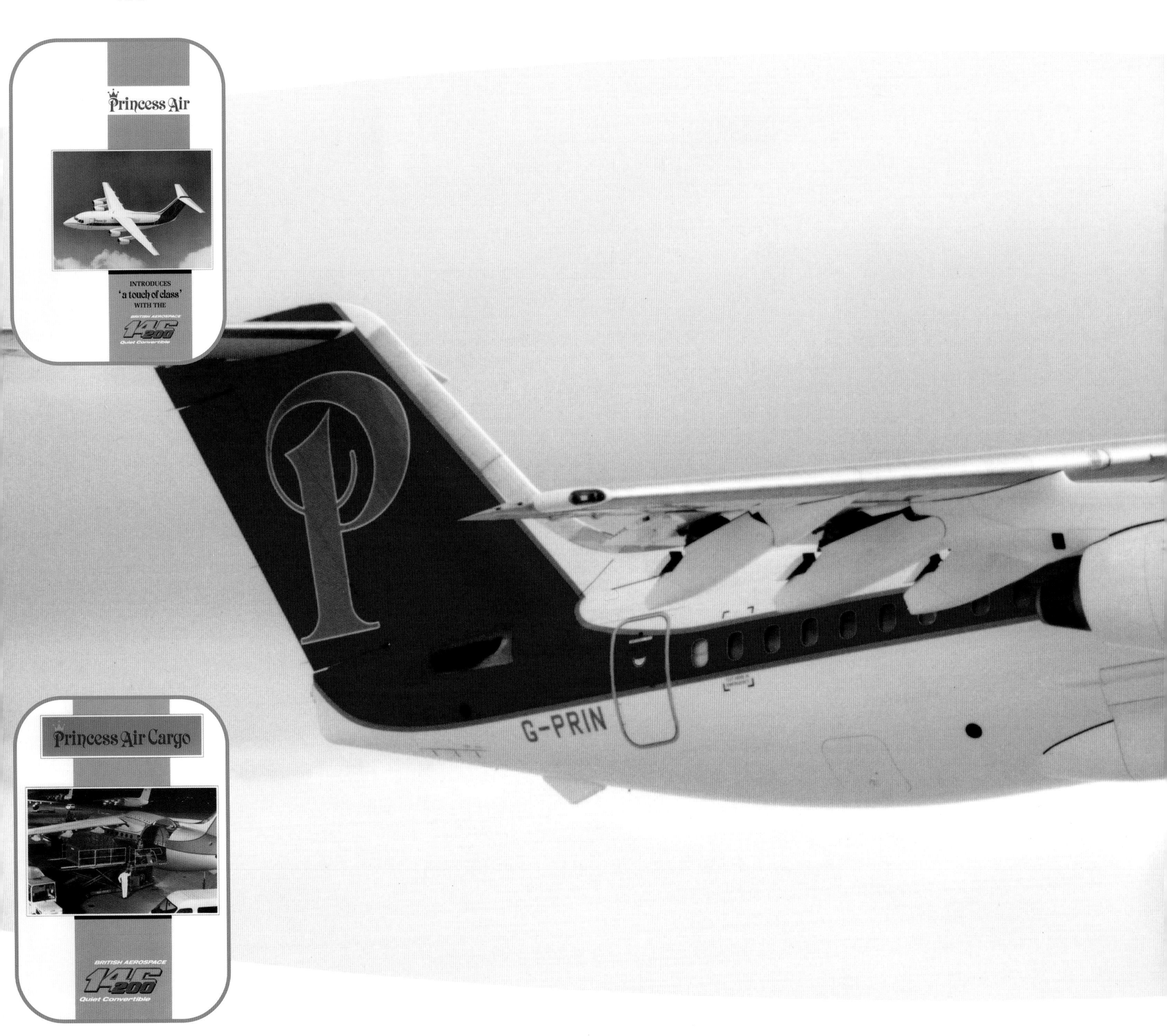

Princess Air
INTRODUCES
'a touch of class'
WITH THE
BRITISH AEROSPACE
146-200
Quiet Convertible
G-PRIN
Princess Air Cargo
BRITISH AEROSPACE
146-200
Quiet Convertible

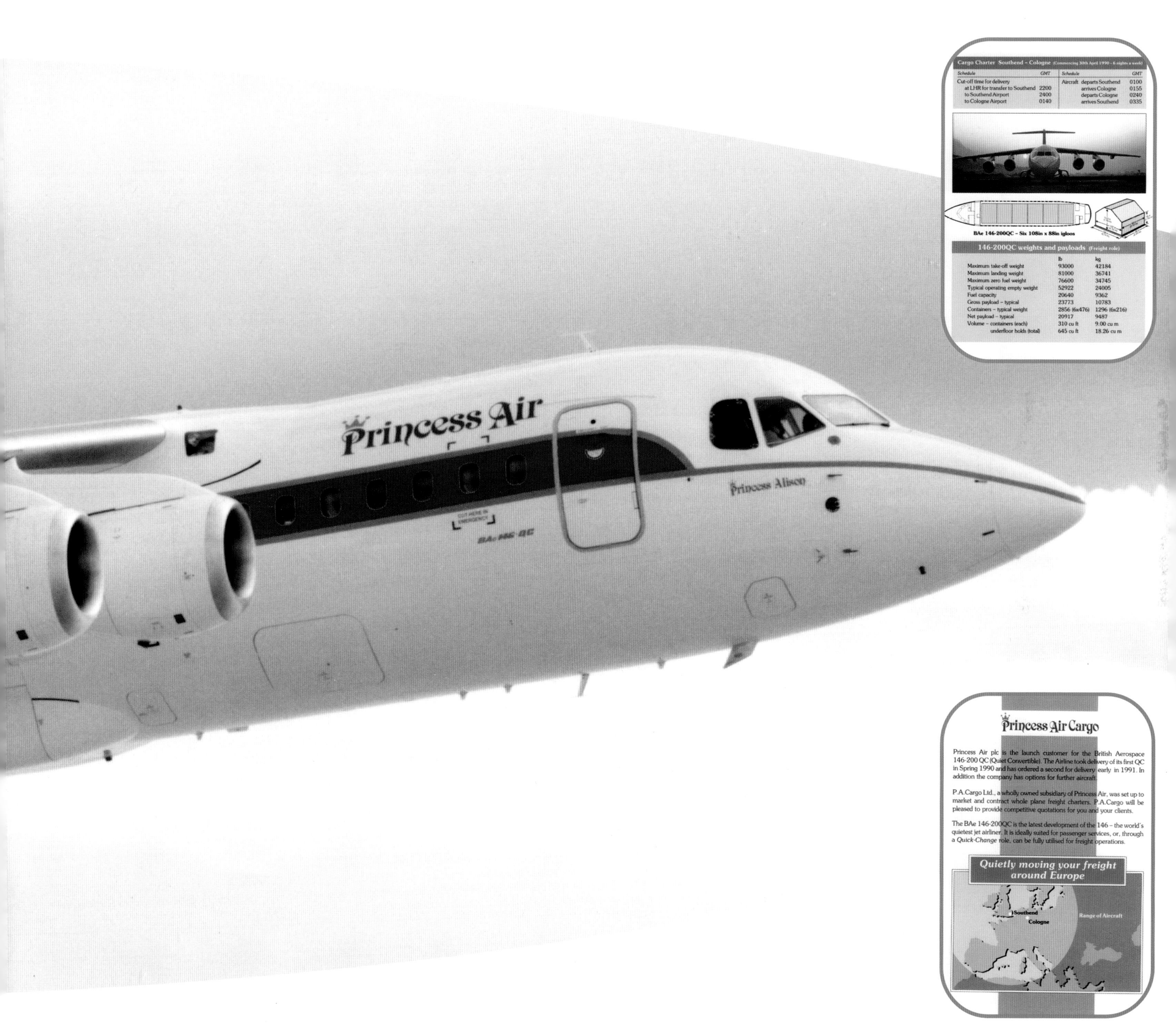

Cargo Charter Southend – Cologne (Commencing 30th April 1990 – 6 nights a week)

Schedule	GMT	*Schedule*		GMT
Cut-off time for delivery		Aircraft	departs Southend	0100
at LHR for transfer to Southend	2200		arrives Cologne	0155
to Southend Airport	2400		departs Cologne	0240
to Cologne Airport	0140		arrives Southend	0335

BAe 146-200QC – Six 108in x 88in igloos

146-200QC weights and payloads (Freight role)

	lb	kg
Maximum take-off weight	93000	42184
Maximum landing weight	81000	36741
Maximum zero fuel weight	76600	34745
Typical operating empty weight	52922	24005
Fuel capacity	20640	9362
Gross payload – typical	23773	10783
Containers – typical weight	2856 (6x476)	1296 (6x216)
Net payload – typical	20917	9487
Volume – containers (each)	310 cu ft	9.00 cu m
underfloor holds (total)	645 cu ft	18.26 cu m

Princess Air Cargo

Princess Air plc is the launch customer for the British Aerospace 146-200 QC (Quiet Convertible). The Airline took delivery of its first QC in Spring 1990 and has ordered a second for delivery early in 1991. In addition the company has options for further aircraft.

P.A.Cargo Ltd., a wholly owned subsidiary of Princess Air, was set up to market and contract whole plane freight charters. P.A.Cargo will be pleased to provide competitive quotations for you and your clients.

The BAe 146-200QC is the latest development of the 146 – the world's quietest jet airliner. It is ideally suited for passenger services, or, through a *Quick-Change* role, can be fully utilised for freight operations.

Quietly moving your freight around Europe

When the 146STA arrived after conversion and painting from Hayes International, the glossy clear coat made the camoflauge paint visible from far away.

In October of 1990, BAe delivered the 146QC demonstrator to Ansett New Zealand for a six-month trial. By mid-December, Ansett New Zealand announced it was going to buy the aircraft outright. The demonstrator saw 280 flights averaging 50 minutes each on eight passenger sectors in the first month. At the conclusion of passenger services each night, it took 20 minutes for the seats to be removed and the freight containers to be installed. In under one hour, the entire aircraft had gone from passenger to freighter complete with cargo loaded, refueling, and crew change. Turnaround for freight was under 30 minutes, demonstrating the expediency of the aircraft.

Unfortunately, very few regional airlines have an expertise in both passengers and freight. Consequently, there were only a few niche opportunities in the market and in the end just five 146-200QC were built

146STA – Military Side-loading Tactical Airlifter

The 146 had been as a candidate for a military derivative from the beginning, even when it was known as the Hawker Siddeley HS146. Carrying a CVR(T) Scorpion tank or similar vehicle with a rear loading ramp was the prospect, even though the RAF had no such aircraft requirement. In the early stages of planning after the re-launch in 1978, it was assumed that up to 100 examples of a rear loading version of the 146 for the military could be sold, although a prototype was never built. With a capability to airlift vehicles, paratroopers, a combination of troops and freight, BAe saw the options as limitless; even the lower cargo holds could be configured to accommodate additional fuel tanks for increased range. Early configurations of a military 146 included concept drawings for a heavily modified rear, eliminating the petal tail brake and instead featuring a ramp that lowered from the rear to facilitate loading/unloading. Planned for launch and management out of the Manchester office, these derivatives never came to fruition as the company was simply not prepared to fund the substantial development cost.

May 1986, BAe and Lockheed agreed to a joint venture to evaluate the market potential for a military version of the 146 aircraft. Lockheed produced a larger high wing aircraft turbo-prop military cargo, the C-130, of which over 1,700 had been produced by 1986.

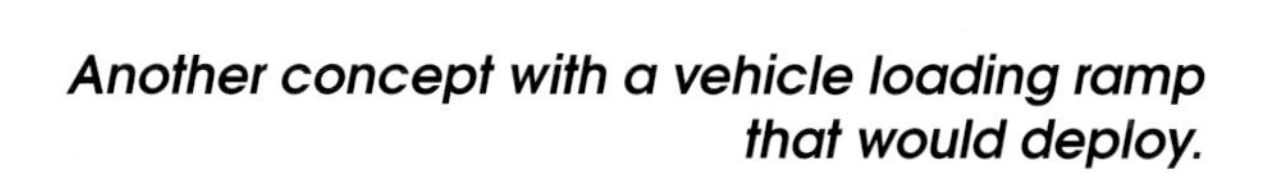

Another concept with a vehicle loading ramp that would deploy.

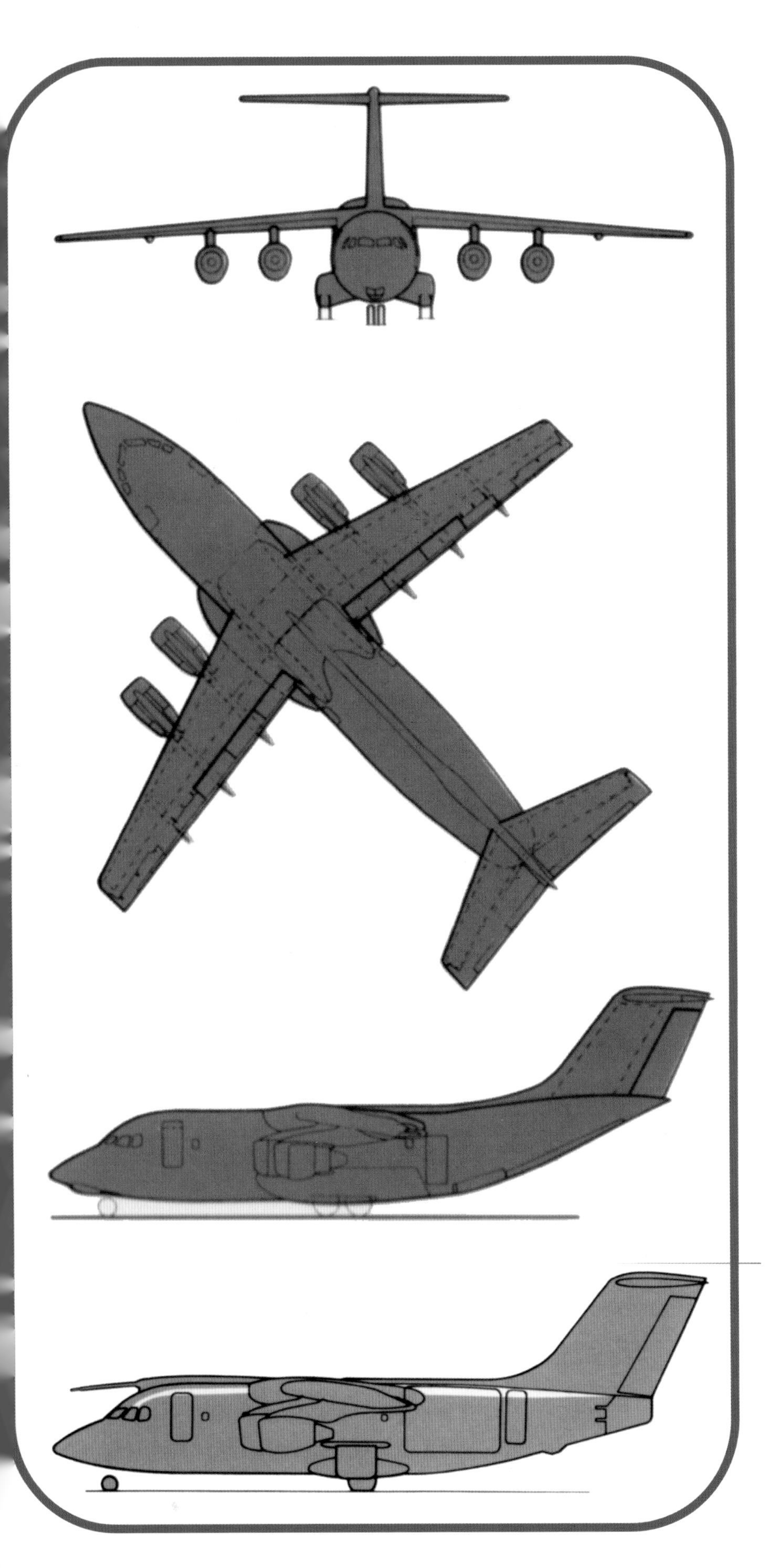

Concept artwork for a revised HS146 (and later BAe 146) military variant with a rear floor loading door.

It was also producing the C-5 Galaxy, one of the largest cargo jets in the world for the military, also a high wing aircraft. The appeal of the 146 for military applications was its low noise profile and ability to land on undeveloped airfields with a reasonable payload range.

When the initial draft was signed, BAe flew a series 100 aircraft (E1010 in Royal West livery) to the U.S. to demonstrate it to various branches of the country's military services. A tour was undertaken which included installations such as the US Navy's facilities at Patuxent River and the Air Force's Edwards Air Force Base flight test centre. At Pax River, carrier deck landings were simulated, after which the pilots regarded the aircraft favourably. Initially Lockheed saw the 146 as complimentary to the C-130, but BAe's cooperation with the U.S. firm did not last long. Based on the positive response to the 146 following the Navy's and Air Force's flight tests, Lockheed began to view the 146 as a competitor to future C-130 sales and the partnership was terminated quickly afterwards.

To offer a flexible aircraft, BAe needed to lower the floor to accommodate the rear door and allow vehicles to be transported. Additionally, the lowered floor meant the 146's landing gear would need to be redesigned. Other proposed modifications included a hydraulic lift system to speed loading and unloading. But the firm was not ready to invest capital into a 'maybe' aircraft, for which there was no clear business case. If the U.S. military or other military services issued an RFP (request for proposal), then BAe might have reversed

The Modular Personnel Unit (MPU) concept was presented as a solution to make the aircraft infinitely configurable for any mission.

course and made the investment to integrate and test these features. It would certainly have enabled the 146's role to be expanded into platforms beyond cargo, including possible opportunities for the type to be used for surveillance or tankering.

Despite the collapse of the very short-term joint venture with Lockheed, BAe still believed there was a need for a military 146 and that there was a market for it. Enter the STA: Side-loading Tactical Airlifter. The firm modified BAe 146-100 E1002 with a side cargo door, which carried the registration G-BSTA and was painted in a camouflage livery. It also had a non-functional refueling probe above the cockpit (which was removed for positioning flights). The aircraft was initially painted at Hayes International after conversion in March 1988, but with a glossy sheen rather than a typical military matt finish. When it arrived at Hatfield, everyone joked you could see it coming from miles away because of the shiny paint.

BAe removed the gloss seal to represent a true military aircraft.

BAe was trying to use the aircraft as a multi-role demonstration platform, capable and flexible of everything: troop transport, military freighter, and more. Marketing of the STA began in 1987 at the Paris Air Show and a year later it was shown for the first time at the local bi-annual Farnborough air show, with a vehicle ramp affixed and a vehicle half way up the ramp. The front cabin of the aircraft was fitted with sixteen seats in a VIP configuration, while the middle of the cabin was fitted out for paratroopers and the rear had a cargo floor with rollers to ease the moving of pallets. Paratroop tests were conducted in cooperation with the British Army's Red Devils at Cranfield on October 3, 1988. The 146STA was cleared to carry 60 fully equipped troops using the aft fuselage door for rapid departure, and by the Civil Aviation Authority for free fall jumping.

Proposals were produced for further variations such as medical evacuation, troop or passenger transport, as well as a tanker capable of carrying up to 25,000lb of fuel over a 200-mile radius of action. The refueling pods would be installed under the outboard wings making the 146STA into a two-point tactical tanker. Other uses such as a Special Forces support (because of low noise level on landing), or search and rescue carrying 32-man dinghies plus smoke floats, were also put forward.

The 146STA with the clear coat removed, and the BAe logo added to the tail.

BAe toured the aircraft between 1988 and 1990, attending the Farnborough and Asian airshows, not to mention private showings to military services all over the world. The 146STA went on a 36-day tour covering 30,000 miles in October 1988. It was later engaged in demonstrations to the Royal Australian Air Force (RAAF), flying in and out of the 1300m airstrip at Puk-Puk, the first time a jet the size of the 146 had done so and an ideal mission for the 146STA. It was then flown to Wellington and Auckland in New Zealand for demonstrations to MOD and Air Force personnel. The STA visited Queenstown following

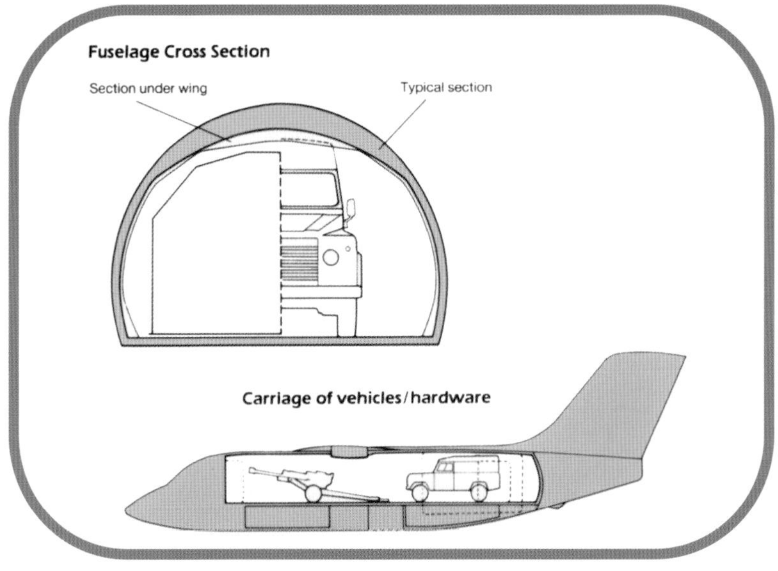

Concept artwork from Hawker Siddeley for the HS146 as a military transport.

Another concept with a vehicle loading ramp that would deploy.

Vehicle loaded into the 146STA.

The ramp was installed next to the aircraft (it did not deploy in this example), and the logistiics to get a vehicle in and out of the 146STA meant this design exercise would never manifest into an actual aircraft.

Mockup of an MPU being loaded onto the 146STA

an announcement by Ansett New Zealand that it had ordered two 146-200 aircraft. The 146 was the first commercial jet operated into Queenstown, and noise checks noted the 146STA was hard to detect above the wind noise. Subsequently the 146STA finished up a two-week tour in Southeast Asia which included Thailand, Indonesia, Malaysia, Brunei and Singapore. The aircraft also toured Middle Eastern countries in February of 1989, as well as the United States and Canada in May of 1989. The 146STA was further demonstrated in

the aerial display at the large bi-annual Farnborough Air Show in 1990, alongside a BAe ATP.

The 146STA was being pitched as a replacement for existing piston and turboprop military transports, the thought being that it would be a good replacement for ageing C-47s, F-27s, C-119s and C-123s. BAe even went as far as offering licensed production to a country that bought at least 30 aircraft. Hayes International of Dothan, Alabama would modify the 146STA by strengthening the floor and adding the rear loading cargo door, just like it did for the –QT aircraft. Further upgrades in the 146STA included fuel capacity increased to 3,384 gallons. The prototype was based on a -100 aircraft, but production airframes were likely to

The first test of the 146STA as a paratrooper aircraft.

have been based on the series -200 or -300 for greater capacity, and the 146STA would utilize the heavier reinforced wing of the passenger -300.

BAe took the 146STA to Air Shows complete with a mock-up non functioning fuel probe mounted forward of the cockpit.

BAe offered the 146STA in a typical conceptual fashion as "everything you want it to be", with the variations utilising customised containers, known as Modular Personnel Units (MPU), for each task. Quick connect devices would provide electrical, air, and other necessities making the aircraft incredibly flexible. For example, there was a VIP Modular Personnel Unit (MPU) that could be loaded through the side rear cargo door, transforming the aircraft (or part of it) into a VIP role. Other MPUs were proposed such as command centres, staff units, medical units, and more.

Although at one stage the STA was painted up in Austrian military markings, an anticipated order did not come through. Only the RAF Queen's Flight aircraft in VIP configuration, and subsequently a couple of modified –QC aircraft also flown by the RAF, were ever operated by military air arms. In the end, notwithstanding the imaginative efforts of the BAe sales team, the 146STA was not much more than a –QC version of the commercial BAe 146 series of aircraft with a refueling probe and some minor additions. Of course, it was pitched and billed as the Swiss Army knife of aircraft, but the reality was that each function would require significant investment to bring to reality. Anyone who watched the complex and protracted actions needed to load a vehicle into the STA could not but reflect on why you would want to go through such contortions when you could buy a far more practical ramp loading freighter aircraft instead.

No one bought the STA concepts, but BAe subsequently modified two BAe 146-200QC aircraft for the UK Ministry of Defense (MoD), to be used by the Royal Air Force in Afghanistan to supplement the Lockheed C-130s in 2012. Modifications included Defensive Aids Systems to protect the aircraft in hostile situations, resulting in them being designated as BAe 146 C Mk3 models. They were also equipped with a SIFF system (Successor Identification Friend or Foe), as well as high frequency and ultra-high frequency communications and SATCOM satellite communications systems that had been previously fitted to the two RAF 32 Squadron -100's. Following the conclusion of the Afghanistan campaign, the C Mk3s operated alongside the earlier airframes as part of a common fleet at RAF Northolt.

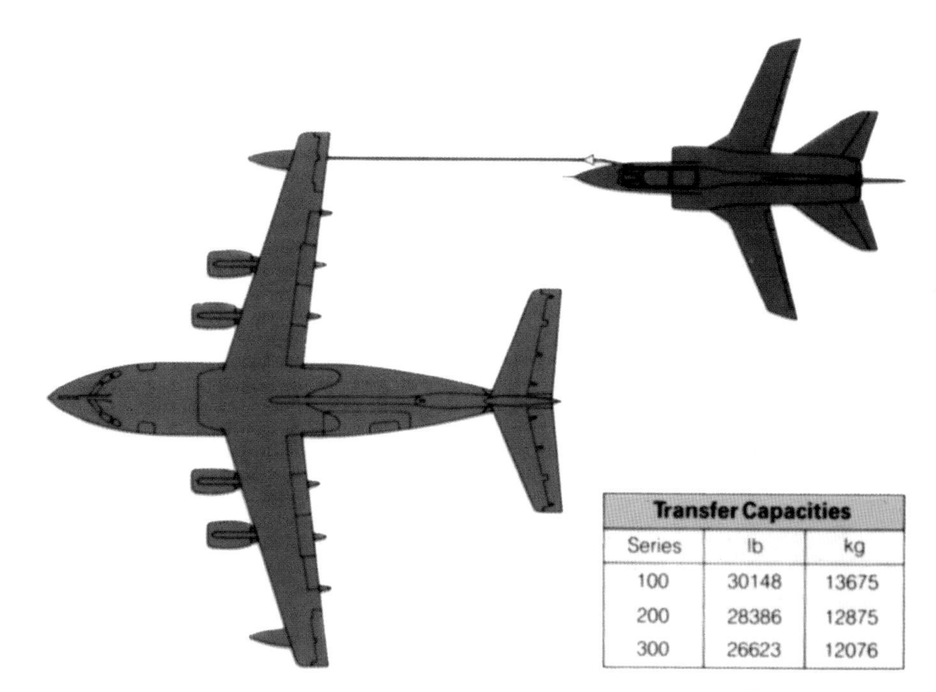

Transfer Capacities		
Series	lb	kg
100	30148	13675
200	28386	12875
300	26623	12076

BAe invested considerable effort in trying to develop a market for the 146 freight door variants but it is unlikely that the meagre addition of 27 conversions made financial sense. The sad truth was that BAe simply did not understand the markets it was aiming at and instead relied on a misplaced intuition that it could stimulate enough interest to make the project viable. The 146 is an incredibly versatile aircraft and has proven its potential subsequently in many roles. But BAe was never willing to make the serious investment required in capital and expertise to realise that potential whilst the aircraft was in production.

As the 1980s came to a close, there was a financial storm brewing for British Aerospace and the 146 program, and that storm would be so powerful it would bring the company to its knees.

Business Air was a carrier that started service out of LCY.

CHAPTER 08

Exploiting the Capabilities of the 146 - London City Airport

BAe was about to be given the perfect gift, one for which the 146 was designed, in its own backyard: London City Airport (LCY). Opened on November 5, 1987 and situated in the London Dockland redevelopment area, located very close to the developing Financial district and on the north bank of the River Thames, it was to become the ideal compact airport catering to business travellers. The irony was that its location, originally known as the "Royal Docks," was at one time part of one of the busiest seaports in the world. Destroyed in World War II, the area rebounded, and saw peak cargo shipments in the mid-1960s. Over the subsequent years, larger ships with containers and the rise in air cargo rendered the Royal Docks obsolete.

The London City Airport logo was updated to feature the BAe-146 as a design element.

After a long period of dereliction, the Thatcher Government established the London Docklands Development Corporation (LDDC) to regenerate the area. Special planning regulations were put in place, office buildings, housing and hotels started to appear, and the concept of a local airport began to be discussed. What emerged after design and debate was an facility built around STOL aircraft and business traffic. Only the De Havilland DHC-7 was certificated to land at the airport, owing to its short runway and the steep approach angle of 7.5 degrees which ensured noise was kept to a minimum for surrounding residents. The first air service was flown by Brymon Airways, operating to Paris, Amsterdam, Brussels, and other local European cities. But with a runway just shy of 4,000 feet which was too short to allow jet aircraft, a campaign to win around city planners and residents was started with the aim of permitting a minor runway extension to support jets.

BAe began preparing for the possibility of a demonstration into LCY, hoping to pave the way for future operations of the jet. The manufacturer set up two locations to validate its position that the 146 could fly into LCY, enabling testing of the 146 in visual flight rules (VFR) and using the instrument landing system (ILS) when visibility was poor. The runway at Hatfield was set up for VFR with a 5.5-degree approach. A steeper 7.5-degree approach using ILS was set up at BAe's Dunsfold facility to demonstrate that the autopilot could navigate the approach. Using a steeper approach meant using less engine thrust, while flaps and airbrakes were used to control the descent. The 146's high wing really optimized the STOL design for such situations, and the aircraft showed it could operate under either flight parameter.

BAe obtained permission from the LDDC for a demonstration day for local residents which took place on Sunday July 24, 1988 and involved not one but two BAe 146 aircraft. Flown from Hatfield by test pilots Peter Sedgwick and Dan Gurney, the first aircraft to land was a series -200 in Loganair colors (G-OLCB), prior to delivery to the airline who was a strong proponent of using the 146 at LCY. The second BAe 146 to arrive was also a series -200 aircraft, painted in BAe livery with a large 146 logo painted on the vertical stabilizer. After completing a series of takeoffs and landings, mixed with flyovers at 500 and 2,000 feet, the aircraft demonstrated the same approach pattern that the Dash 7 aircraft performed daily at LCY.

The point of the demonstrations was to gain the public's trust that introducing the 146 would have negligible noise or safety impact on the surrounding community. Every local household received information in the mail about the demonstrations and was invited to attend. The event was a success with large crowds, a carnival like-atmosphere including a children's play area, burgers and ice cream provided, and bands playing for the attendees. BAe brought in a partial cabin interior mockup, enabling residents to 'try out' the 146. Parked on the ramp were the two 146s along with three Dash 7s which helped alleviate fears that the 146 might be too big, when in fact its wingspan is smaller than the Dash 7. Yet it offered twice the speed, range, and passenger carrying ability as the Dash 7. Local interest in the aircraft was quite high and residents queued up to go inside the Loganair aircraft.

The aircraft in BAe livery was fueled up to simulate carrying a near full load of passengers, and a series of take-offs, fly-bys, and landings demonstrated to residents that the jet aircraft was in fact quieter than the turboprops then operating at the airport. In fact, the take-off used slightly more runway than the Dash-7 aircraft serving the airport. London City Airways, which had been considering the aircraft, flew one of its Dash-7s on the same pattern as the test 146 so there was an apples-to-apples comparison. Afterwards, the press reported the event had gone well and had residents not been notified of the event, they would not have known it was happening given how quiet the aircraft was. Even the local resistance leader (Mrs. Leisha Fullick), chairperson of Newham Education Committee, admitted she could not hear the 146 land. If there was a complaint it was that the Dash 7s were actually noisier than the 146, especially during taxi and take off maneuvers. The Times noted "A hundred local protesters made more noise than the planes."

London City Airport airport had been designed around the Dash 7 as the ideal STOL aircraft; everything from taxiways, runway length, and noise considerations were based on the aircraft's characteristics. Unfortunately for the Airport's pioneering owners the Dash 7 did not have the range, speed or drawing power necessary to attract enough high yield traffic to make the facility viable. Without the 146 the airport

146
LOGANAIR
Scotland's Airline
BRITISH AEROSPACE
146

would have faced closure. But the 146 proved it could do everything the Dash 7 could do and more while operating within the confines of the airport restrictions. The 146's cruise speed was nearly twice as fast as the Dash 7's (490mph vs. 259mph) and was heavily marketed to airlines as an advantage for customers. BAe used an example of a business meeting in Dusseldorf: with the time savings of a faster cruise speed a passenger could have a longer meeting and still return home the same evening in time for dinner. The 146 had already opened up noise sensitive markets such as the Swiss capital Berne, which had originally been turbo-prop only, but now permitted the 146.

Following the trials Steep Approach Certification was granted to the BAe 146 series 100 and 200 (even with one engine out). The 146 aircraft also received a minor update to the Ground Proximity Warning System (GPWS) which alerted the pilots that they were about to fly into terrain if they were not careful. All aircraft have a GPWS, but the steep approaches meant that the GPWS would not activate as designed.

The runway at LCY was eventually extended to 4,900 feet to allow more unrestricted 146 operations. It re-opened on March 5, 1992, a mere four years later, and with jet operation and a more extensive route network the airport's popularity with business travellers increased. The new runway length enabled the glide slope to be reduced to 5.5 degrees after sufficient data showed that noise parameters would not be impacted.

In 1990, Crossair purchased their first jets, taking three used BAe 146-200 aircraft (ex-US Air) after it was shown that they could operate out of Lugano where no other jet could operate. The 146 excelled on the steep approach and very short runway surrounded by mountainous terrain, and was capable of flying a full load without penalty while being incredibly quiet for the noise sensitive community. Beginning service to London City in June 1990, Crossair continued to grow and serve more destinations across Europe. Crossair dubbed the 146 the "Jumbolino" as a miniature jumbo jet, and even applied a flying elephant logo under the cockpit windshield on the port side. The carrier flew the first service with the BAe 146 from Lugano and Zurich to LCY, and went on to become one of the largest 146 operators. In terms of quality and prestige Crossair was to become a reference customer for BAe, as well as a real and effective cheerleader for the aircraft. It was also a launch customer for the Avro RJ85 development, ordering four aircraft initially and taking eight options.

The BAe 146-200 became a popular aircraft to operate into LCY over the next few years, and was eventually joined by the series -300. Many of the flights out of LCY covered short stage lengths, and therefore the majority of departures were not subject to payload limitations, lowering the risk of a tail strike on rotation. With the runway extended and Crossair's 146 services underway, British Aerospace began marketing the

BAe 146-200 demonstrator doing touch and go's, demonstrating how quiet the aircraft was to local residents.

146 specifically as the ideal aircraft for operations at LCY. Major carriers were initially reluctant to develop LCY services fearing the cannibalisation of their own Heathrow services, and were cautious about acquiring a niche jet that they felt could not be utilised elsewhere. This left the field comparatively clear for smaller airlines either operating independently or under franchise to the majors. Crossair's success forced a change of attitude and as time passed the major carriers began to see LCY's true potential, which had important consequences for the Avro RJ development of the 146. With its 10-minute check-in and good train links into London, the airport's throughput grew steadily and the number of travelers passing through it increased year over year. What started out at 230,000 passengers in 1990 grew to 500,000 passengers by 1995, and by the year 2000 more than 1,500,000 passengers were passing through the airport. The BAe 146 and Avro RJ brought frequency, greater passenger loads and above all profitability that the Dash-7s could not have achieved.

In another twist of fate, a similar airfield on the other side of the Atlantic was facing a similar issue. Toronto Island Airport, in Ontario Canada, was a struggling downtown facility that fought for years to allow jet operation. The data used from the noise standards agreement in 1983 (with the Dash-7 as the baseline) was no longer relevant by the mid-1990s. Attempts to bring the agreement for operations at the airport up to date using new data from current jet operations had been waved off. When LCY's managing director Bill Charnock attended a seminar held by Canadian Urban Institute, he felt a sense of déjà vu witnessing the arguments being made by Canadian officials. Regional city center airports are always experiencing pushback from the NIMBY (Not In My BackYard) syndrome with incorrect perceptions of noise, pollution, and traffic. The city and its residents, through taxes, had subsidized the airport that had seen traffic fall from 450,000 passengers in 1987 to 170,000 in 1993. It was as if Toronto was trying to kill off the airport for good.

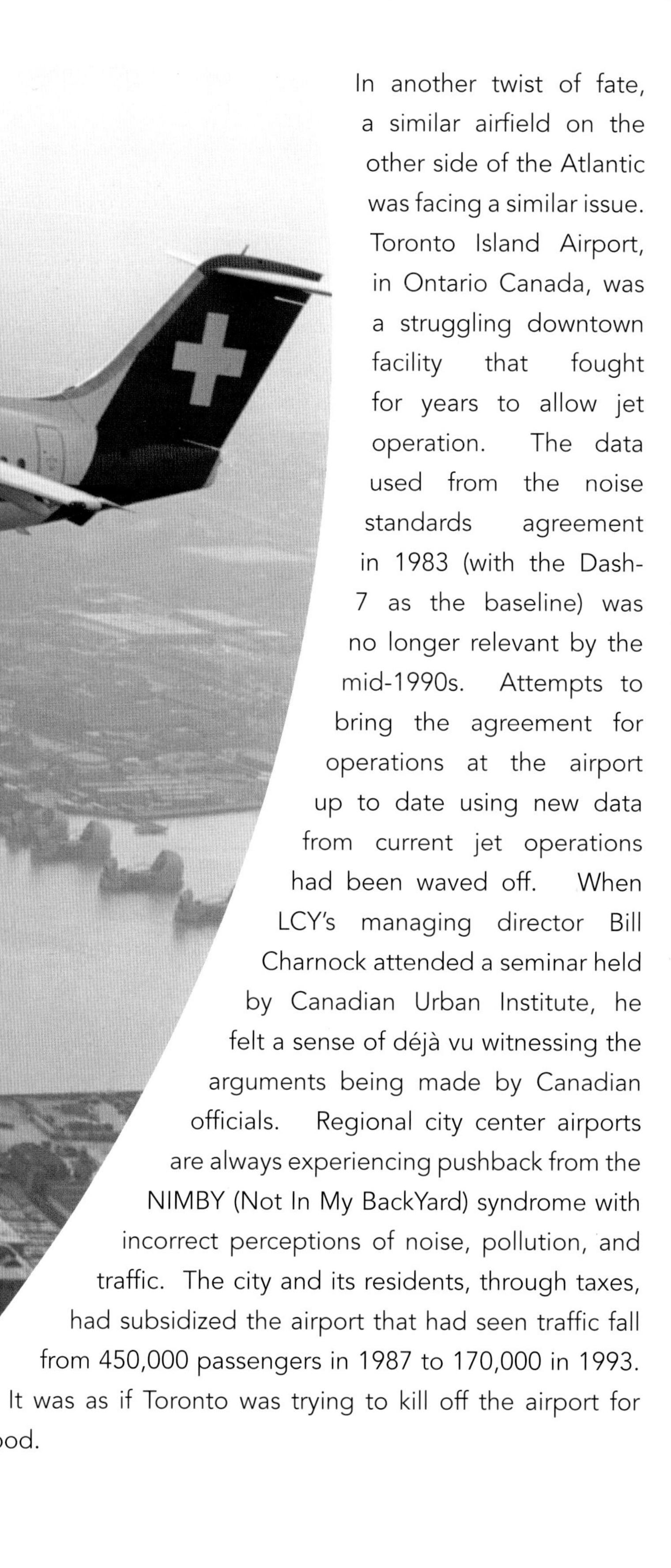

Crossair overflying London City Airport just before inaugural 146 service began.

In the late 1980s, local airline owner and proponent Victor Pappalardo arranged for the Vice President of Technical Sales for BAe, Ron Bustin, to provide data and 146 demonstration flights. Two of the three parties, Transport Canada and Toronto Harbor Commission agreed to revisit the issue, but the city council refused to even meet or discuss it. At the time jet aircraft were only permitted to operate into Toronto Island provided they were participating in the three-day Canadian National Exhibition (CNE). In conjunction with BAe, Air Atlantic flew one of their aircraft in for the show and performed a series of demonstration flights, each one noise monitored. A year later, BAe brought an Avro RJ70 to the CNE show for further demo flights and to influence VIPs from Toronto Island. Despite this and even supported by data from LCY operations, to this day, Toronto Island Airport still does not permit jets and it seems unlikely there will be a change before the ban on jets expires in 2033.

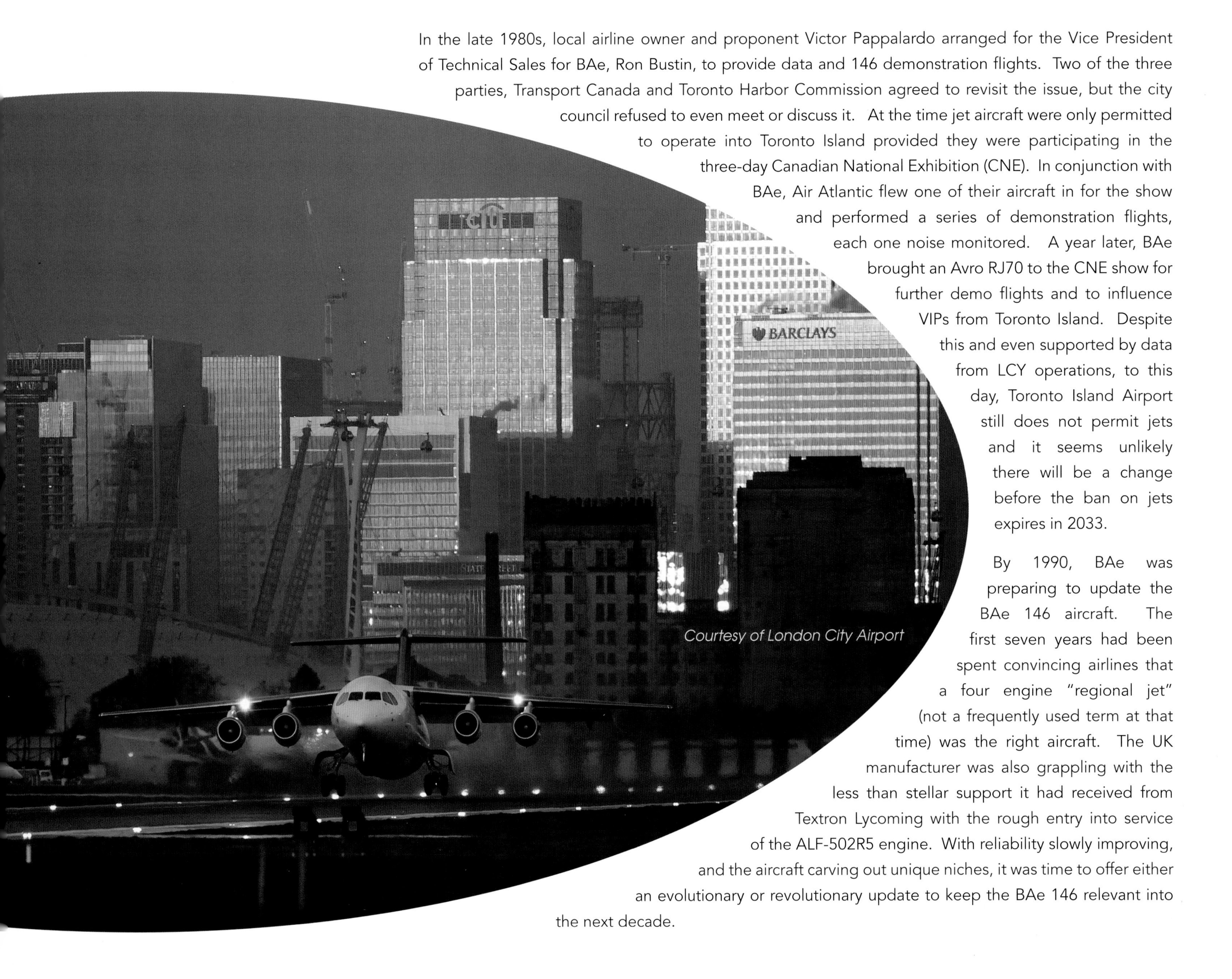

Courtesy of London City Airport

By 1990, BAe was preparing to update the BAe 146 aircraft. The first seven years had been spent convincing airlines that a four engine "regional jet" (not a frequently used term at that time) was the right aircraft. The UK manufacturer was also grappling with the less than stellar support it had received from Textron Lycoming with the rough entry into service of the ALF-502R5 engine. With reliability slowly improving, and the aircraft carving out unique niches, it was time to offer either an evolutionary or revolutionary update to keep the BAe 146 relevant into the next decade.

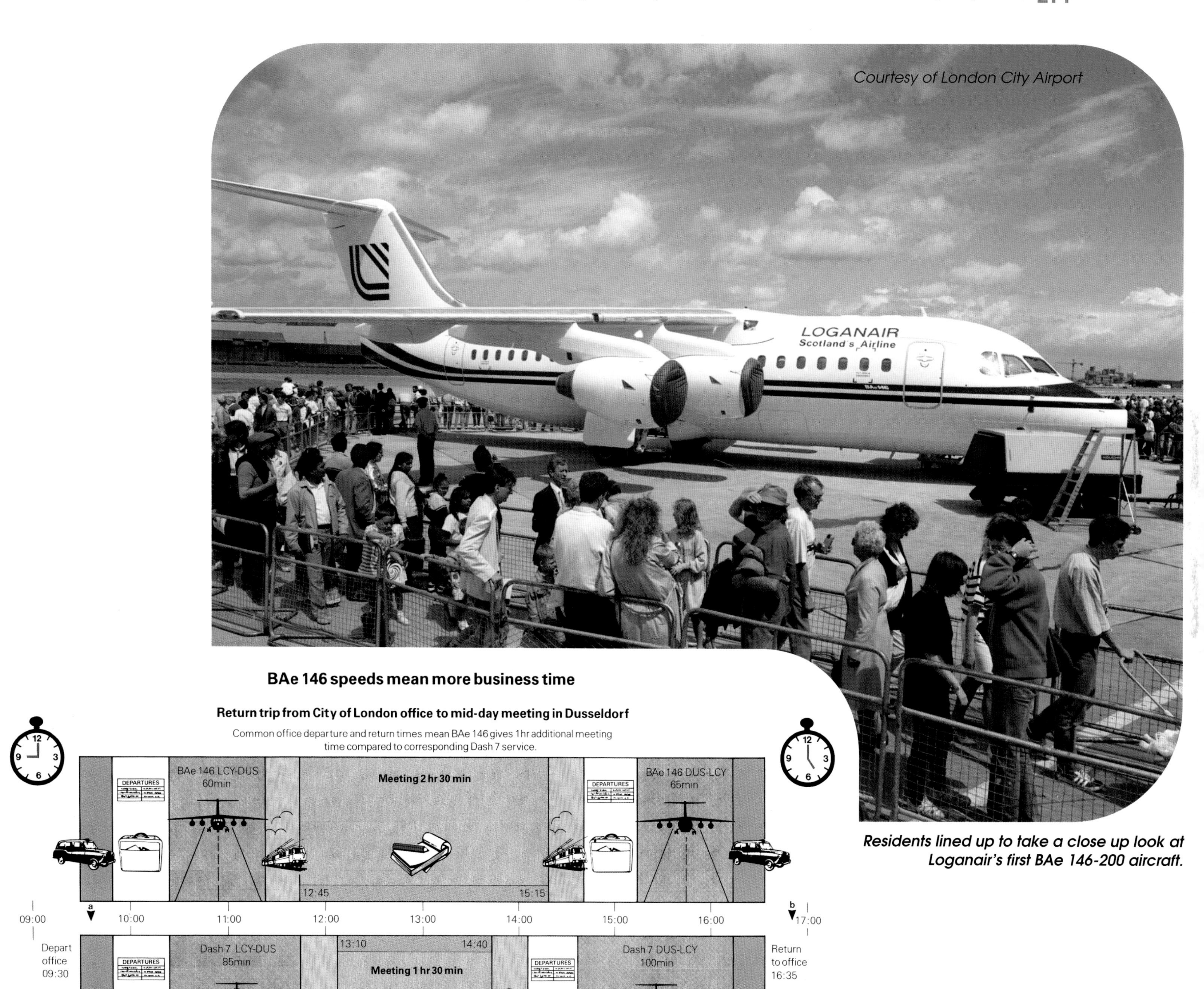

Residents lined up to take a close up look at Loganair's first BAe 146-200 aircraft.

Courtesy of London City Airport

CityJet was the last scheduled operator of the 146/Avro series into LCY, ending passenger flights in March 2020. Today, Embraer and Airbus aircraft have replaced the 146 for scheduled jet operations at LCY.

Courtesy of London City Airport

CHAPTER 09

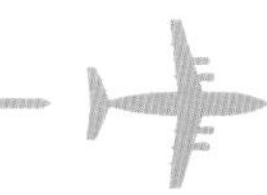

"Sales" of the 146 Come Home to Roost

Sales of the 146 were painfully slow until late 1983 when PSA placed a substantial order. Even after that order, sales were never brisk or high. By the latter part of the 1980s, sales had picked up with BAe selling quite a few 146s, which was a positive development for the slow selling jetliner. But was the 146 actually selling? BAe was so desperate that it was cutting deals left and right with customers that generally did not have the balance sheet or credit rating to support buying new aircraft. With the exception of carriers like PSA, Air Wisconsin, Ansett or AirCal, many of these carriers operated second hand aircraft that were not economically viable for their previous owners that were disposing of them. The chickens were getting ready to come home to roost, a phrase that signifies bad past deeds were about to come back to haunt the person or company responsible for them.

The leasing industry was gathering pace in the airline industry in the 1980's. Companies like GPA, GECAS and ILFC developed large portfolios of the most popular products from Boeing, Airbus and McDonnell-Douglas. This gave airlines at all levels more fleet planning flexibility and pushed new or nearly new aircraft into carriers that had hitherto not been able to afford to buy or finance them. Now all these airlines had to do was fund a security deposit and pay monthly rentals over an agreed term. This revolution largely passed the regional airline industry by as its economics, operators and the aircraft available were not attractive to the lessors. BAe tried hard to get a big lessor on its 146 order book, and it held discussions and made offers for a mega-order to GPA in particular, it did not succeed.

BAe's dilemma was that although it was not a high-volume producer it still needed to secure sufficient sales to keep production efficient and cost effective whilst avoiding the most dreaded phrase in the aircraft manufacturer's dictionary, "white tails". These were aircraft that had been produced but had to be parked painted white awaiting customers. White tails tied up substantial capital, required further investment when the aircraft were finally sold, and were a PR disaster. Whilst BAe generally managed to avoid building unsold aircraft most of the time, it was living hand to mouth, and forward orders were rarely enough to cover planned production. The harassed Sales Team then had to spend much of its time focused on finding immediate deals and not enough time on developing quality customers for the future.

Having failed to get an independent lessor to place an order, BAe almost imperceptibly drifted into the leasing business itself, first with the Jetstream 31 and then with the 146. The firm deluded itself that it was selling aircraft when in fact it was arranging the finance of its aircraft using its own credit rating to get the best terms, and then leasing aircraft to customers on varying terms. Sometimes these matched the terms of the finance, sometimes not. But the arrangements enabled the BAe sales teams to offer prospective customers attractive and flexible terms, similar to those available from the big lessors.

Unfortunately, many of the airlines that acquired the 146 on these terms were either not properly capitalized or had risky business plans, and in some cases both. If the airline stopped paying for the airplanes and eventually went broke, as most did, BAe was on the hook to cover those payments, interest included. If customers were blue-chip airlines it was not a problem, but BAe was not generally getting that kind of operator. The firm's short term need to place aircraft pushed it towards lower quality operators that were having trouble finding anyone to back them. The BAe sales team learned faster than the company's finance department and was gaming the system to achieve its sales targets.

As time passed BAe became more sophisticated in structuring the financing of such deals. Offshore companies were established in tax havens like the Channel Islands and complex leveraged lease finance structures were used that maximised BAe's return but all suffered from the same fatal flaw. They were only as good as the credit of the airline operating the aircraft, and if anything went wrong the complexity of the funding structure could in itself become a significant issue. It seems strange, with the benefit of hindsight, that no one at a senior level seems to have appreciated or even understood the mountainous liabilities that were being piled up.

After years of doing such deals, with customers going bust left and right, the financial noose began to tighten. The world economy was in a dive, the airline market was suffering, and BAe was having a hard time selling new aircraft or placing off-lease, even repossessed, aircraft with customers. The challenge it faced was exacerbated by a major airline, USAir, parking an entire fleet of nearly 20 BAe 146s in the desert. To make matters worse, values were depreciating faster than expected and placing off-lease aircraft at much lower market rates made the commitment BAe had to financial firms ever clearer.

A variety of superficially clever devices were developed to encourage and close sales

"Integration Funding" was notionally supposed to be a cash grant to help airlines deal with the inevitable negative cashflow as they went through the delivery, training and integration phase of placing a new type into service. Provision for these costs may have been made in the notional sale price, but if the deal did not run to term then such upfront payments were irrecoverable. Airlines like Westair and Meridiana Spain were beneficiaries, often to the extent of a seven-figure sum.

"Stepped Rentals" were used to aid cash poor airlines. BAe would offer escalating lease agreements where, for example, instead of a monthly payment of $150,000 USD, the first six months would be charged at $50,000 per month, and then the lease fees would rise in increments over agreed periods until the airline ended up paying $200,000 per month and thus the earlier shortfall would be recovered. But what if the airline ceased operations? What if the airline was on the wrong side of this curve, or was never able to afford the higher rentals? The result was BAe burned a lot of cash while being stuck with a massive financial liability: the aircraft.

Worst of all were the vanity projects. Senior BAe personnel allowed themselves to become too close to projects they were supposed to be evaluating. They became convinced that such prospects were strategic "no brainers", resulting in BAe becoming effectively investors in these projects. When things went wrong, which they nearly always did, too often additional monies followed in a desperate effort to keep the project alive until the promised day arrived, which it never did. Discovery and Air Brasil were classic examples that consumed not just money but resources, management time and effort.

BAe was seeking financing for the next iteration of the 146 via Taiwan Aerospace Corporation.

The fact was that BAe over extended itself substantially. This ranged from guaranteeing the engine performance from the poorly performing American engine maker Avco-Lycoming, to financing the entire purchase of aircraft by many of the carriers covered in the past few chapters; Carriers that ended up going out of business or being bought up, and the aircraft returned. BAe spent untold amounts of money chasing deals with carriers that clearly had no money or real prospects. As one former salesman put it "you could see the spin, you couldn't see the substance." As the decade was ending, BAe's liabilities and exposure were increasing at a rate that would bring the entire company to its knees. When the Gulf War began in the summer of 1990, fuel prices doubled overnight and airlines struggled and faltered as a result. This made operating a four-engine regional jet (with questionable reliability) an unwise proposition.

Efforts were made to try and exercise some control and to learn from the mistakes of the past. Every time discussions with a prospective customer reached the point where a formal proposal was going to be made, the matter had to be considered at a "Deal Committee". This was an ad hoc body consisting of senior managers from the commercial, finance, and sales departments before whom the sponsoring commercial and sales people would present their proposed deal. In advance of the meetings detailed proposals were circulated outlining the terms together with, if it was a lease deal, an analysis of the airline together with its business plan for the operation of the aircraft. The Deal Committee was not a rubber stamp process, and discussions could be extensive and prolonged with input from more senior management sometimes required.

There was a fatal flaw in the Deal Committee: it did not necessarily have the final word. At best, it could only try and rule out some of the wilder deals that clearly did not make

BAe was faced with financially unsound airlines collapsing, returning aircraft or not taking up delivery (like this former undelivered Discovery Airways BAe 146-300 that BAe had rebranded for demonstration purposes).

sense. If a deal that BAe Marketing Director Brian Thomas supported was rejected, he would go over the committee's head to more senior managers and often the deal would go through, perhaps with some cosmetic changes.

Company auditors who had to sign off the accounts with their bulging off balance sheet liabilities were becoming increasingly reluctant to do so. The Finance Director then brought in outside consultants to go through the deals in detail. Only when the analysis was complete and all the dots connected did the true financial horror emerge. BAe was imploding from within, and up to that point no one was doing anything to course-correct the path the company was on. BAe terminated Brian Thomas (who had been promoted as President and CEO of the company's US business, BAe Inc) and those that supported him in an effort to right the ship, but the damage was done. It was clear that senior management across the board had been asleep at the wheel, and focused on the wrong priorities. The manufacturer needed a more rigorous oversight process, and the divisional management should have highlighted to BAE's board that "this is where we are at, here are the options (and costs associated) we can go with." The aircraft was still underdeveloped, it was generating a bad undercurrent amongst operators, and consequently in the industry. No one within BAe thought they would get any major expenditure approval to rectify these issues, and thus the program just ploughed on with those involved hoping for the best. British Aerospace's sales campaigns and perceptions of the aircraft were not helped by photographs of 20 BAe 146-200 aircraft, each barely six years old, that were stored in the desert. USAir originally stated to the press and financial markets the aircraft were not economical and even though the carrier backtracked on its statement, the damage was done.

The Financial Noose Tightens – and so does aircraft development

With the financial stability of the company in jeopardy BAe was in a pickle, facing the decision to iterate on the BAe 146, make the jump to a twin-engine design tentatively called the "RJX", or even exit the business completely. It was clear that making evolutionary changes and improvements to the existing BAe 146 was the lower cost and thereby lower risk option. The alternative was to make the substantial investment (estimated at over $1.5 billion) in an aircraft that would be a hybrid between being 'all new' and a derivative. With liabilities stacking up, and orders slowing, it was clear that the RJX was the aircraft to pursue – but only if BAe could sign up risk sharing partners.

In January 1991 BAe began talks with the newly formed Taiwan Aerospace Corporation (TAC), aimed at forming a risk sharing joint venture to build the twin-engine RJX. The UK manufacturer also discussed launch aid for the aircraft with the UK Government. Marketing and sales

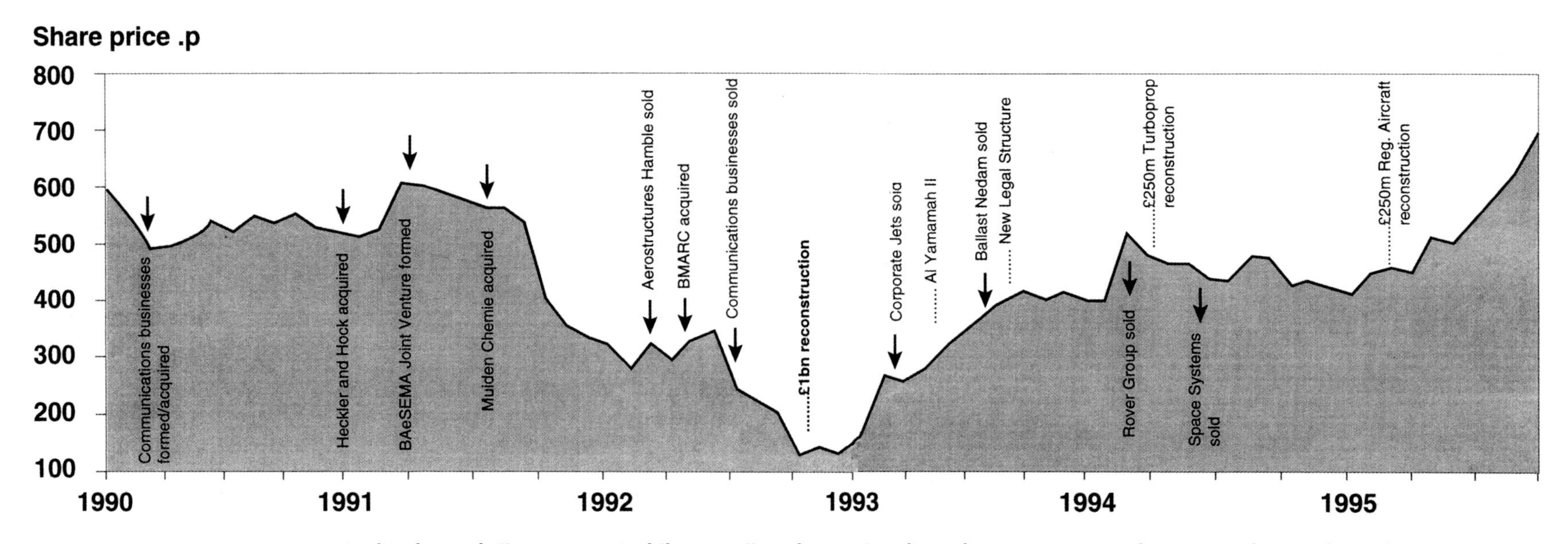

BAe was faced with financially unsound airlines collapsing, returning aircraft or not taking up delivery (like this former undelivered Discovery Airways BAe 146-300 that BAe had rebranded for demonstration purposes).

discussions with airlines were opened at high levels, while negotiations with TAC for investment and risk sharing continued. BAe was looking for TAC to take a 50% share in the RJX program. There was also debate with TAC about opening a second production line in Taiwan to build the aircraft (at a significantly reduced rate versus production in Britain), but this never went anywhere beyond high level discussions and a few diplomatic photo opportunities. BAe was desperate for a risk sharing partner to help maintain its relevancy as the regional jet provider while reducing its capital outlay for the new aircraft programmes that would keep it relevant and solvent. Internally, many BAe staff were scratching their heads at the Taiwan Aerospace discussions while some salesman were forecasting to TAC that there was local sales potential of up to 72 BAe 146 aircraft. Market research within BAe eventually presented the reality that none of the local airlines' route structures would support even half that many airframes. Although TAC wanted to get heavily into aerospace, there were other countries were a risk-sharing would have made more sense including China, Korea, India or Japan.

BAe had been running a parallel development track aimed at updating the BAe 146 aircraft, while investigating the two-engine RJX aircraft as an alternative means to expand the family of regional jets on offer, and the TAC partnership would have permitted both an updated BAe 146 and the new RJX to be developed. In the end, it was decided that the cost of developing of a twin jet RJX derivative that internal studies showed would offer customers only a minimal reduction in direct operating costs did not justify moving forward with the project. No airline had shown major interest in a more expensive aircraft offering negligible improvements over the existing line of regional jets, and a deal with TAC never came to fruition after two years of discussions. By late 1992, Taiwan Aerospace disappeared quietly into the night.

Furthermore, British Aerospace did not have any interest in developing a new airliner that would cannibalize sales of the existing BAe 146 line. Instead, the firm put its effort into an updated 146 series that included CATIIIa approach capability that airlines had been asking for, as well as refreshed engines with improved reliability and better fuel burn. Faced with an investment of over $150 million for the aircraft improvements and TAC completely out of the picture, it was clear that going ahead with the revised aircraft was an all-or-nothing moment for BAe.

The revised aircraft (G-ISEE) flew in April 1992 for the first time, while internally BAe was still grappling with its development costs, lower build times of just 12 weeks (later dropping further to nine weeks), and the impending liabilities of BAe 146s coming off lease. The situation was growing dire, and BAe needed to find a way to not only ensure its relevancy in the market, but its very own survival. BAe would soon find it was no longer the only regional jet available.

The Chickens had come home to roost

At the end of 1992, the financial house of cards had come crashing down. BAe had to face its mounting liabilities and declining position in the newly competitive regional jet marketplace. The firm was struggling with a substantial number of aircraft on which it had signed guarantees with the banks that were being leased to less than financially viable airlines. Just six months earlier, with a heavily depressed stock price, BAe had considered shuttering its regional jet division. The business was in serious financial trouble and the company was teetering on the verge of insolvency. The chickens had come home to roost. On one hand, BAe's exposure on aircraft leases was in excess of £3 billion and was still growing. But BAe's net worth as a company had been reduced to £1.7 billion and was a mere £100 million away from insolvency which would allow the banks to take over. BAe's assets were less than its liabilities. The share price of nearly £6 in 1990 crashed to £1.20 in early 1993, and BAe had to act quickly if it was to avoid being bank-owned and made irrelevant.

BAe took a £1 Billion charge which initially alleviated the liability. This was not just for the aircraft on record, but to cover other areas of the company's performance. The first step was to get the forty-plus grounded aircraft off its books as they were a financial noose that was tightening by the day. A new business, Asset Management Organization (AMO) that was also based out of Hatfield, was set up and all the leases and unsold aircraft BAe was responsible for were transferred to this new organization. AMO would be responsible for getting these aircraft back into operators' fleets, and it would live-or-die by its success or failure to do so. Richard Thomasson was responsible for setting up the Used Aircraft Group, which eventually was absorbed within AMO, and he and his team were responsible for aircraft storage, conversions, flight tests, modifications, and so on before eventually being redelivered to new customers. AMO also became a barrier between sales of new "Avro RJ" aircraft, and the BAe 146s of the past would be forgotten in the new organization. AMO took on a total of 118 BAe 146s that BAe was on the financial hook for, along with a fleet of Jetstream 31 and ATP aircraft, and tried to find new homes for them as quickly as possible with financially viable carriers.

The second part of ensuring BAe's survival was mass consolidation, with nearly 10,000 jobs eliminated in the organization. These were not limited to commercial aircraft and cuts were made across the board in all divisions. Hatfield accounted for approximately 15% of the total and as a result, the factory where the BAe 146 was built was wound down. All future production and final assembly would be initiated out of Woodford, where the rear fuselage was already being produced, and nearby Chadderton where other portions of the 146 were also manufactured. Hatfield was shut down after general assembly of its final jet aircraft (Avro RJ85 E2208 G-ISEE), which occurred on March 23, 1992, even though £4 million had been spent to overhaul the assembly line just five years earlier in 1987. Production of Avro RJ/146 parts including the nose continued but finally ceased on April 4, 1994. This closed a chapter in British aircraft manufacturing, which had been taking place at Hatfield since 1934 and where more than 8,000 aircraft had been built.

The last and final piece of the restructuring was the resurrection of the Avro name, a former manufacturer that built famous warplanes such as the Lancaster and the Vulcan, as well as the 748 turboprop. Avro had been absorbed into Hawker Siddeley, which itself was became part of the nationalized British Aerospace. Rebranding the redeveloped aircraft as an Avro RJ was an attempt by BAe to get the industry to "forget" about the BAe 146 and treat the revised and refreshed model as a brand-new aircraft. The sales team reported that nomenclature generally worked and many believed the Avro RJ was in fact a new aeroplane. Aircraft would no longer be 'sold' the way they were previously, with hybrid lease agreements not being pitched to customers that were not blue-chip airlines or could not finance the aircraft themselves. Production would be cut nearly in half, regardless of demand. BAe could not afford to find it in such a precarious position again moving forward, if it was to survive and remain relevant. It was taking a low risk gamble with its next aircraft and engaging in an evolutionary update. The enhanced BAe 146, the Avro RJ, was to be manufactured at Woodford, just as Sir Raymond Lygo wanted to do years earlier. Ironically, Sir Raymond had retired from BAe in 1988, before the proverbial roof fell in.

This near-death experience was a seismic event that changed BAe's whole approach to its civil airline business. The number one priority became to manage its existing liabilities, initially to ensure that no further financial provisions were necessary and then to try and maximise the possible return from the assets and ultimately to dispose of them altogether. Production of new aircraft was allowed to continue because closing it down completely would adversely impact the value of the used aircraft portfolio. Few thought it would continue for very long and that when the used aircraft liabilities had been stabilised an opportune reason would be found to close it down. As it turned out that decision was eight years away and no one could have foreseen that it would be 9/11 that delivered the coup de grace.

During the 1980's the 146 had been an important part of BAe's business that was supported and promoted at the highest level. As the company grew and diversified that interest and enthusiasm diminished, exacerbated by the seemingly never-ending problems that the programme suffered from. The financial crash in 1987 moved the dial on from indifference to active hostility. The whole programme became seen as a disaster area and any future investment would be made solely to milk the asset to its greatest extent, and not with any thought of it remaining in the business any longer than was absolutely necessary. It may have been unstated but that was the paradigm shift that had occurred.

A major cultural change also emerged within the whole regional aircraft business. The emphasis changed from a management that only wanted to hear what it wanted, to one that wanted to know the truth, however unpalatable that might be. Unsurprisingly this resulted in a more positive atmosphere where problems were faced and dealt with, and initiative and achievement were rewarded.

Digging the company (and the 146) out of a deep hole

Enter Asset Management Organization (AMO), formed in January 1993 and based out of Hatfield. Its sole responsibility was to manage the fleet of aircraft (not just 146s but any civil aircraft in which BAe had any liability) that were on-lease and coming off of lease, in order to keep aircraft flying and revenue coming in. In short, the goals were simple: Get the aircraft flying again, increase revenue, and reduce BAe's exposure. It would be staffed initially with 55 persons from the commercial group, including marketing who were responsible for re-marketing aircraft. The new management of AMO cherry-picked the best people, which was not difficult because many did not want to move to Woodford, 168 miles north of Hatfield.

AMO was backed up by a secondary company also formed in January 1993, JSX Capital Corporation, that was based in Washington, D.C. in the United States and had a focus on managing the Jetstream line of aircraft which was primarily with North American operators. By the time AMO was formed, it already had 21 idle BAe 146 on its books, another 40 expected by the end of the year, and in all liabilities on a total of 118 out of the 221 BAe 146 aircraft that had been built at the time. Nearly 20% of the total production run was going to be idle, without customers. Additionally, the Avro team was not permitted to market new build Avro RJ aircraft to existing AMO customers to help get existing aircraft off the books. Ironically AMO did, on occasion, try to lure Avro RJ prospects to used 146s as a cheaper alternative.

PUBLIC SALE

BAe 146-100,
N 801 RW, S/N 1002,
Man. 1982

and

BAe 146-100,
N 803 RW, S/N 1011,
Man. 1983

each including (4) Avco Lycoming ALF 502R engines

To be held on May 28, 1987 at 10am, P.D.T. at the Offices of Jones, Jones, Close & Brown, 700 Valley Bank Plaza, 300 South Fourth Street, Las Vegas, Nevada, 89101, USA.

To be sold to the highest bidder, with the exception that the secured creditor reserves the right to credit bid.

Aircraft located at or near Las Vegas, Nevada, USA, and are available for inspection upon prior notice along with historical records.

Please contact:
Mr SAUD SIDDIQUE
Tel: (212) 574-3552
Telex: 232 432/420 520

Aircraft began to be reposessed by banks and sold off.

The formation of AMO was a trailblazing move which other manufacturers would subsequently copy. The separation of liabilities on the existing fleet from the production and sale of new aircraft not only made matters transparent but also ensured better and focused management with clear and accountable goals. It is easy to forget that BAe was not alone in getting into this kind of mess; every manufacturer in the regional business had joined in this lemming like activity and some including Fokker ended up being bankrupted by it. BAe had survived and now the experience was going to turn AMO into one of the most focused and effective units in the business.

AMO was responsible for taking off-lease aircraft back, remarketing them to other airlines and refurbishing them. It also began changing the way it leased out aircraft. Instead of offering them to any airline, or finding a way (creative or otherwise) to underwrite aircraft for an airline that had a poor credit rating, AMO started to become more selective about the carriers it was prepared to work with. It meant that it would turn down deals where risks outweighed benefits. It also managed the portfolio so that when a point was reached where leasing no longer made sense, it would sell the aircraft opportunistically.

Another plan initiated through AMO involved beginning to entice airlines with weak financial status to return aircraft earlier than contracted. This had the potential to avoid the hassle of impounding and repossessing aircraft that more than likely would be in a state of disrepair, which would incur additional expenses for AMO. Receiving an aircraft back often involved refurbishment for the next customer, with costs running at approximately $1million USD per unit billed back to the previous operator. This included everything from paint and interior to maintenance. But new customers had the option to take aircraft as-is (after any maintenance was performed), such as keeping the same interior as the previous carrier, to keep costs low.

AMO was proposing lease terms only to solid, financially viable airlines for at least five years or longer. Not only did the prospective customer have to comply with financial requirements just to lease the aircraft, it also had to provide BAe with financial and operational disclosures periodically during the lease. This put BAe on the offensive, ready well before an aircraft came off lease, or in the worst case, before a customer encountered financial difficulties which risked blindsiding AMO.

Being proactive, AMO defined five financial categories for its prospective customers

Those in category "one" were flag carriers or blue chip airlines that were profitable and had healthy balance sheets. Category "two" incorporated major regional and national airlines with a reasonably strong balance sheet. A category "three" airline was independent or small and had a checkered past of profits and making payments on time. Operators with a consistently weak balance sheet and poor profit record were placed in Category "four". Finally, category "five" included airlines in default. AMO was primarily interested in pursuing, understandably, Category One airlines while those in any other category assigned a risk value.

By 1997, AMO had more than recouped the liabilities that were at hand with BAe. It had sold 22

Roger Pascoe

First tests of the EFIS equipped BAe 146.

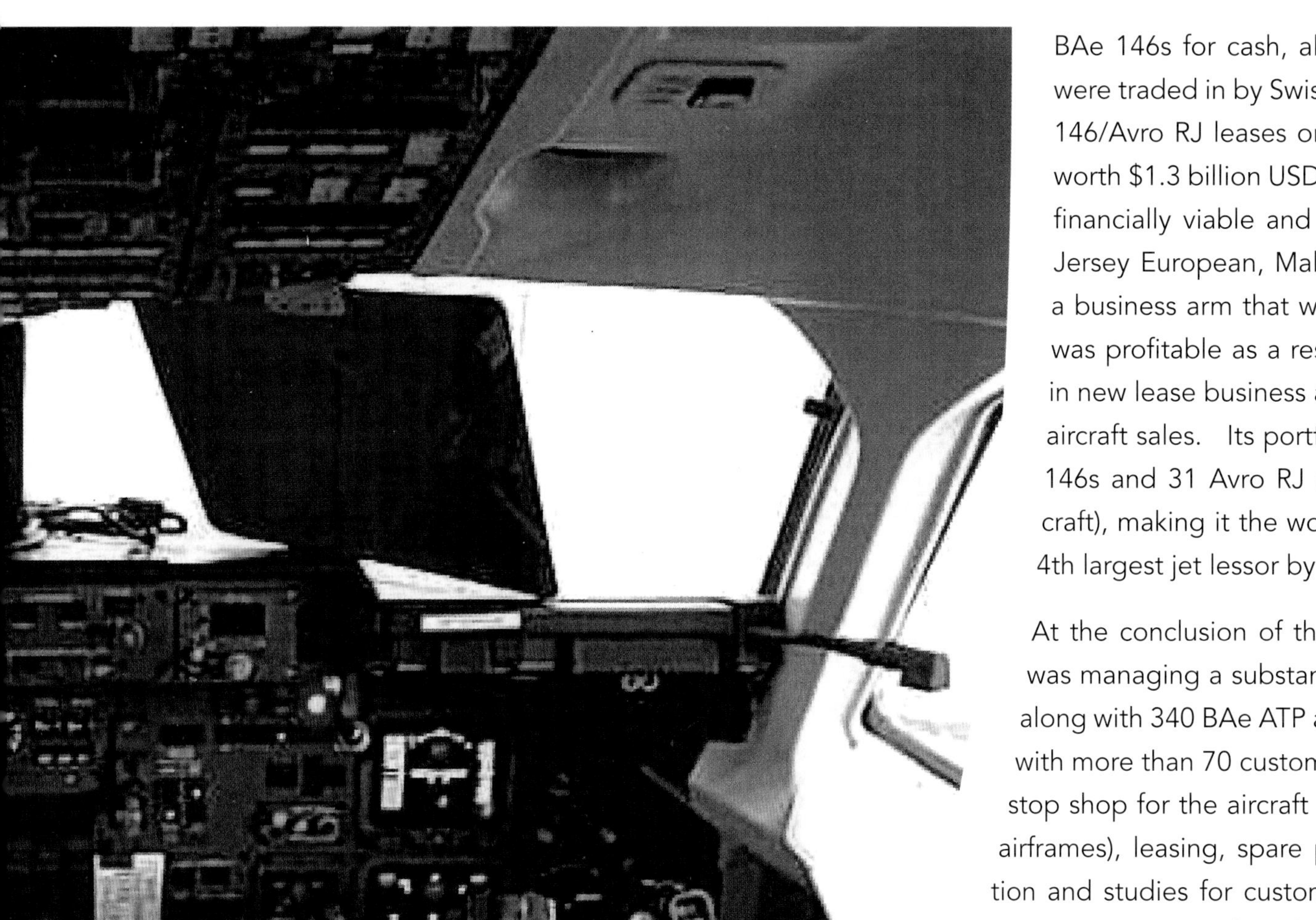

BAe 146s for cash, along with ten Fokker 100 aircraft that were traded in by Swissair. It had an additional 235 new BAe 146/Avro RJ leases or extensions signed, all of which were worth $1.3 billion USD (£999 million). AMO was signing new financially viable and stable customers such as Eurowings, Jersey European, Malmo, and Qantas. AMO had become a business arm that was solving BAe's earlier problems and was profitable as a result. In 1998, AMO won $150million in new lease business and made more than $100million from aircraft sales. Its portfolio had expanded to include 75 BAe 146s and 31 Avro RJ aircraft (as well as 260 turboprop aircraft), making it the world's largest Turboprop lessor and the 4th largest jet lessor by fleet numbers.

At the conclusion of the first year of the new century, AMO was managing a substantial portfolio: 101 BAe 146/Avro RJs, along with 340 BAe ATP and Jetstream 31/32/41 aircraft placed with more than 70 customers worldwide. It had become a one-stop shop for the aircraft it specialised in: sales (used/off-lease airframes), leasing, spare parts, training, refurbishment, evaluation and studies for customers, and even supported further improvements to aircraft such as ALF-502R XRP modifications which reduced the removal rate of engines. For the first time in BAe's regional jet history the sales program was running properly.

At the end of 2010, BAe had renamed AMO as BAE Systems Asset Management. By then it was responsible for 151 aircraft placed with 16 customers in 11 countries. It had also generated more than 1800 leases and/or sales of aircraft worth in excess of $3.1billion USD. But May 2011 marked BAe's exit from the regional aircraft leasing and sales business. The manufacturer sold the remaining assets and organization to Fortress Investment Group LLC for $187million, and was also able to redeem $514m Series G Equipment Notes and Series B Equipment Notes issued under the Systems 2001 Asset Trust financing, a private placement of debt securities completed in 2001. The assets were now managed by Falko, which was formed in July 2011 to handle the former portfolio of BAE Systems Asset Management, and is wholly owned by Fortress Investment Group. On October 29th, 2020, Falko announced that the they had sold the last BAE asset from their portfolio, an Avro RJ bringing an end to their 146/RJ management.

CHAPTER 10

Finding Its Way - The Avro RJ

RJ70 & 80 – False Start

Prior to realizing it was in a dangerous financial position, and before full-blown evolution of the aircraft updates were complete, BAe tried to put lipstick on a pig by offering a "new" aircraft to regional airlines. Unveiled at the Regional Airlines Association in Washington, D.C. in June of 1990, they were dubbed RJ70 and RJ80, and presented as variants, not replacements, for the existing BAe 146 based on the series -100 fuselage with de-rated LF-503 engines. They were presented as very short haul aircraft compared to the BAe 146 family, with up to 850nm of range. Whilst BAe did not foresee the success of the Bombardier CRJ, its actions were designed to try and forestall the momentum of that programme. It is a reflection of the echo chamber style of BAe's senior marketing management that they indulged in the fantasy of believing that such products could seriously compete with a far more optimised 50 seat design.

Airlines weren't impressed, nor did any of them buy the aircraft. The common theme was that the RJ70 was just a BAe 146-100 with fewer seats, and de-rated LF-503 engines that might (or might not) have lower maintenance costs because of the lower engine core operating temperatures. BAe was remarketing essentially the same aircraft with negligible changes as if it were a new design while sales of 146-100s had been declining as airlines chose the larger -200 and -300 variants. Airlines also questioned why BAe was pitching an aircraft with nearly the same capability as its poorly selling 64-seat ATP (Advanced Turbo Prop) aircraft. Even competitors like Canadair which was building a 48-seat RJ countered BAe's figures, stating that the RJ70s breakeven compared the Canadair RJ was not three or four additional seats filled, but more like 8-10 extra seats sold at "typical North American" yields. BAe claimed strong interest in the revised aircraft from Canada and South America, although no orders were ever placed.

Almost as quickly as it offered the RJ70 and RJ80, BAe reversed course, this time 'soft launching' what would become the Avro RJ and began a marketing tour to airlines. The new RJ70 and RJ80 were offered with a choice of analog or digital flight decks, full CATIIIa approach capability as a future upgrade, but with de-rated LF-507 engines to improve fuel economy and reliability. The new designations had significance: RJ70 would offered five-abreast seating for 70 passengers and was marketed to North American customers, while the RJ80 accommodated 80 passengers six-abreast and was promoted mainly to European carriers. BAe further relied on the wide cabin, low noise footprint, higher passenger capacity, and fleet commonality with existing BAe 146 aircraft to compete against the new segment of upcoming regional jets.

British Aerospace began taking the RJ70 with derated LF-503 engines on tour in the United States. No airline was interested in this variant, resulting in an updated LF-507 equipped version with a soft-launch of what would become the Avro RJ series.

At the time BAe 146s were offered with a phase I hybrid EFIS flight deck, using digital instruments displaying data from analog-derived flight data. From the third quarter of 1992, new build aircraft were offered with a phase II EFIS including an ARINC 700-standard digital databus. The RJ70 and RJ80 were marketed as baseline aircraft with simple galleys and larger overhead lockers for short haul flights, and would have lower price points of $18 million vs $21 million USD for the BAe 146-100.

Unfortunately with the world economy languishing and an ongoing war in Kuwait and Iraq, interest in the revised RJs was limited. Airlines perceived the offerings to be minor upgrades to the earlier aircraft. BAe began to realize that the rebranded regional jets, the RJ70 and RJ80, were not enough to attract interest and CATIIIa was still not quite ready on the BAe 146. Airlines did not buy and British Aerospace stopped marketing these incremental versions of the BAe 146, and instead went back to the drawing board to determine what to do with the arguably fourth iteration of the BAe 146.

Introducing the Avro RJ

By 1991 BAe was moving forward with an evolutionary update to the BAe 146 which included substantial improvements. With the loss of the potential partnership with Taiwan Aerospace and the company's precarious financial situation growing direr by the day, the money was not available to produce a radical new aircraft. Instead the 'new' Avro RJ Regional Jetliner was an aircraft that had all the advantages of the original BAe 146 and many of the improvement's airlines had been asking for; namely reliability, CATIII capability, and reduced fuel burn. In 1993 Charles Masefield was leading the revamped Avro sales and marketing team, based at Woodford and entirely separated from AMO and the used aircraft business. It was Masefield who had made the branding change to Avro International Aerospace in an effort to distance the company and the new aircraft from the BAe 146.

Avro was offering the following versions, all with designations that reflected the seating capacity in five-abreast configuration: RJ70, RJ85, RJ100. It also added another variant, the RJ115 which was an RJ100 with six-abreast seating that featured mid-cabin emergency exits. These mid-cabin exits were required because the increased passenger capacity meant that the evacuation time limits would be exceeded using just the front and rear exits. There were also QC and QT variants offered, and a retouched image showed the RJ85 as a QC and RJ100 as a QT. The de-rated LF-507 engines were carried over from the initial RJ70 introduction, with an option to select fully rated FADEC (Full Authority Digital Electronic Controlled) LF-507 engines for the new RJ70 variant alone – the other variants were only powered by the fully rated FADEC LF-507 engines. The family was marketed as having true commonality across each variant, including the older BAe 146.

Flight Deck

The new features touted for the Avro RJ family were the ARINC 700 digital avionics system with four Honeywell CRTs for display, and the CATIIIa digital flight deck. CATIIIa (Category III instrument landing system) allows for an automated landing in poor visibility down to 700 feet when manual landing is not permitted. The Honeywell CATIIIa autopilot and auto-throttle, which were further developments of those fitted to the McDonnell Douglas MD-11 and Boeing 737 Next Generation, were chosen based on in-service experience and reliability. Dual GNS-X Flight Management systems were also added along with LED displays for engine and fuel system information, reducing maintenance intensive electro-mechanical instruments.

Bittersweet ending: The last aircraft to be built at Hatfield was the very first Avro RJ series.

Paul Roberts

Engine

In designing a successor to the ALF-502, engineers studied the in-service history of 800 ALF-502 engines which had collectively accumulated more than six million hours of flight time. Textron-Lycoming also addressed issues raised by operators of the ALF-502, ensuring the new engine would have a smoother entry into service. The FADEC (Full Authority Digital Electronic Control) system controlled the Allied Signal LF-507 engine putting out 7,000 pounds of thrust, while running up to 100 degrees cooler, which was the major change in the new aircraft. BAe touted this feature on the exterior livery of the demonstration aircraft, with the phrase "new cool" LF507 painted on the engine cowlings.

Test flying the new BAe 146 (Avro RJ) with full digital flight deck, LF507 engines, and CATIII landing capability. All aircraft would now be built out of Woodford.

The FADEC computerized engine control ensured that the engine ran at maximum efficiency and automatically adjusted parameters such as fuel flow and temperatures in any flight condition ensuring the engines ran at maximum efficiency. It connected into the Engine Life Computer, which analyzed engine data. In the event of a failure, there was a full hydromechanical back up system, so a flight could depart with an inoperative FADEC computer. Unlike the ALF-502, the LF-507 prevented the pilots from exceeding safety limitations on the engine such as over speeding and over temping – the FADEC ensured the engine could not be run at undesirable parameters. The end result was that unlike during the early days of the BAe 146, greater reliability was demonstrated during entry into service of the Avro RJ family.

The LF-507 had other improvements over the ALF-502 such as a steel compressor casing (versus magnesium in the earlier design), which was more resistant to corrosion and increased reliability. The new case heated up and expanded at a rate similar to the compressor, keeping tolerances close and optimising efficiency and fuel burn. An additional supercharger was added to boost the flow of air through the engine, reducing turbine entry temperatures and widening temperature margins. This feature required a nominal increase in fuel burn which was cancelled out by all the other improvements. The end result was 11% more thrust in climb and cruise while extending the life of the turbine. Fuel capacity remained unchanged at 3,099 gallons, but with provision for up to 3,409 gallons when extended-range wing root pannier tanks were installed.

Digital flighteck of the new Avro RJ.

Under the skin, the payload and range were increased with the additional power of the LF507. Seat-mile costs in 1994 were shown to be as low as $0 .0975 USD per seat for the RJ115 (at six-abreast) and as high as nearly $0. 13 USD per seat mile for the RJ70 (at five-abreast). In comparison to other aircraft, an RJ flight would breakeven with approximately 30 passengers, while a Boeing 737-500 would breakeven carrying 37 passengers. Using data from 1991, BAe marketed the 146 as achieving 99% average dispatch reliability, while the Avro RJ was averaging better than 99%. Unscheduled removals for the ALF-502R5 ultimately dropped from 0.45 per 1000 hours in 1987 to 0.20 per 1000 hours in 1991, while the LF-507 was achieving better than Allied Signal's target of 0.15 per 1000 hours. The newer engines were just slightly longer (by about 1½ inches) than their predecessors and thus could not be swapped with older BAe 146 ALF-502 engines (although many of the improved parts could be added to the ALF-502 to increase reliability and service life).

Avro also began offering new customer support warranty packages (called JetStart and JetKey). JetStart was aimed at airlines taking on their first jets and covered the complete aircraft with advisory information, training, flight operations and technical support, plus additional spares and maintenance support. JetKey was for customers operating a range of larger aircraft but wanted support for a smaller fleet (while the airline itself managed the larger fleets). JetKey offered a guaranteed fixed cost package that covered all RJ Avroliner operations and maintenance. This included spares, heavy and line maintenance, basic management and administration, and Textron-Lycoming engine 'power-by-the-hour', which meant customers only paid for the aircraft when it was operating (and not when it was being serviced). In theory, the airline payed for use that included all engine maintenance, no matter how severe or intensive.

These services, especially JetKey, were innovative products that have become widespread in the industry today. However BAe's motivations were not, initially at least, to develop a new business line that would be profitable program. It was driven by the need to counter the widespread belief that the maintenance costs of the aircraft were very high. In offering these services BAe was able to offer airlines a cap on future maintenance costs. Even for the customers that did not adopt the new programmes, the fact that BAe was prepared to put its money where its mouth was gave the manufacturer's assurances extra credibility. Needless to say, there was always some internal dispute between the sales team wanting the lowest possible numbers and the support team wanting enough contingency to keep their program viable and profitable.

RJ70

Interior

The interior featured a new widebody look dubbed "Spaceliner", with newly designed larger, but compact looking, overhead bins with an integrated handrail system for passengers. A redesigned lighting system provided for ambient illumination, and largely white-trimmed and contoured side walls with dark colored seats gave a more spacious look. Of course, airlines were free to fit and change the interior to match their requirements.

APU

The Sundstrand APS1000 was a new APU that became standard on the Avro RJ, with Air Malta being the first customer selecting it for their RJ70 aircraft. It is a small constant speed single shaft gas turbine with an overhung back-to-back radial compressor. Like the LF-507 power plants, the APS1000 is a modular APU for ease of assembly and disassembly, lowering maintenance time and costs. The construction is simply three main modules: gearbox, turbine and combustor. The APS1000 was available as a retrofit for existing BAe 146 aircraft, and Crossair upgraded all of its aircraft to this new APU. Improvements included extended preventative maintenance actions, increased electrical power capability, lower operating temperatures, and no hot day bleed air limitations.

Assembly

Even though Hatfield had gone through a multi-million-dollar expansion in 1987, Avro RJ production was moved to Woodford, with Hatfield kept online to build the last BAe 146 aircraft on order and the prototype Avro RJ85. The relocation to Woodford presented another chance to reduce production costs and distance the new Avro aircraft from the BAe 146. Hatfield was shuttered in April 1993 with all sales, marketing, and manufacturing moved to Woodford. Only Asset Management Organization (AMO) remained at Hatfield as a distinct and separate business entity.

First flight of the Avro RJ85.

There was great uncertainty about the future of the programme with the move to Woodford and the financial mess the 146 left the company. The parent company British Aerospace had come to regard commercial aircraft as toxic, and the 146 as the biggest problem. While the Avro RJ and its success (or failure) would dictate the future of regional aircraft sales for BAe, AMO was the company's priority as it was a massive and deadly weight around the neck of the business at the time. With a battered team relocated to Woodford, and the remaining cherry-picked group for AMO at Hatfield, there was clearly no love lost.

First Flights & Certification

The first Avro RJ and the last aircraft to be built at Hatfield, RJ85 registered G-ISEE, made its initial trip aloft on March 23, 1992 flown by Dan Gurney. The test flight carried out two Category III landings (automated), including one in a 14-knot crosswind. Dan commented "the four- and half-hour flight was fault free, remarkable for a new development aircraft." Not far behind, RJ100 G-OIII took to the air from Woodford on May 13, 1992 and the last aircraft, RJ70 G-BUFI flew on July 23, 1992. All three carried a brand-new livery including "The Regional Jetliner" markings, with a similar primary blue colour while the secondary colour varied according to the aircraft's series (red on the RJ70 and RJ100, green on the RJ85, and yellow-orange on the RJ115 which was just an artist rendering). A photo session including all three aircraft inflight over the Isle of Wight was undertaken, and the results formed the primary image printed in sales literature and circulated to the press. Early photographs showed the RJ115 as well, but they were retouched images and no RJ115 aircraft were ever built.

First set of wings produced at Prestwick.

While BAe was marketing the Avro RJ as a new aircraft, it was essentially a derivative of the BAe 146, with the primary changes being the engines, glass cockpit, improved avionics, and the Category III landing certification. The general construction and design of the aircraft

All three test aircraft inflight.

INFO	RJ70	RJ85	RJ100	RJ115
Maximum Range	1,600nm	1,500nm	1,400nm	1,330nm
Minimum Runway	3,490 ft	3,660ft	3,880ft	3,950ft
Maximum Take-off weight	95,000lb	97,000lb	101,500lb	101,500lb
Operating empty weight	51,700lb	53,100lb	55,100lb	55,800lb
Total hold volume	470 cu. Ft	645 cu. Ft	812 cu. Ft	812 cu. Ft
5 Abreast (seats/pitch)	70/31	85/31	100/31	
6 Abreast (seats/pitch)	82/31	100/31	112/31	116/31

RJ QUALITY
QUALITY RJ
Fire
Lufthansa
sabena
crossair

Crossair's first Avro RJ85 on a test flight, with deployed airbrake.

had not changed. As a result, testing of the Avro RJ series went fairly quickly with the flight test team directing their focus to the updated systems. Certification of the RJ85 by the CAA and FAA was achieved on April 23, 1993, and Crossair became the first customer to fly the new Avro RJ. The RJ100 was certified July 2, 1993 for launch customer Turkish Airlines. The RJ70 was the last of the new series to be certified by the CAA on August 23, 1993 with FAA certification following on September 3 in conjunction with first delivery to Business Express (operating as Delta Connection). The BAe 146 and Avro RJ shared the same type rating, allowing pilots to jump back and forth between the two variants. With the type certified, Avro allotted times to take the aircraft on sales tours.

Revised overhead bins and bright colors give the Avro RJ a more spacious looking interior.

Air Malta had shown interest in the aircraft, and in an attempt to persuade the airline to move forward with a purchase, Avro RJ70 (E1228) was painted in Air Malta livery and registered G-OLXX (LXX is '70' in Roman numerals) in January 1994. It was also given the name "Jean L'Eveque de la Cassiere" who was the 51st Grand Master of the Order of Malta. The aircraft was flown to Malta and took a party of airline officials to Rhodes and back. After a hard-fought campaign against the Fokker 70, Air

Avro RJ85 Launch Customer Crossair preparing for its first test flight.

Malta ordered four Avro RJ70s with pannier tanks for longer range missions to London. The demonstrator continued on a sales tour of the Middle East and the Far East for the next month and a half.

In May of 1994, Allied Signal purchased Textron-Lycoming's powerplant business, and the engine became known as Allied Signal LF-507. Avro took the Avro RJ85, painted in Lufthansa Cityline livery on another tour in September 1995 to the United States, starting in Washington DC. It made visits to Air Wisconsin and ASA (Atlantic Southeast Airlines) as well as Air BC, Air Canada, Air Toronto, and Air Nova. It was shown to more than 19 airlines across the country, with the final stop in Halifax, Nova Scotia on September 27.

Stats

The most popular of the Avro series was the RJ85, which has a payload capacity of 17,000 lbs (85 passengers six-abreast) with a 860nm range (or 555nm at six-abreast and 100 passengers, a 20,000 lb payload). The RJ70 can lift 14,000 lbs (70 passengers, five-abreast) with a 1,330nm range (six-abreast is cut to 1080nm and the load is 16,400 lb). The RJ100 carries 20,000 lbs 1,355nm or 22,400 lbs for 1315nm with six-abreast seating, or 23,200 lbs with 116 pax over 1250nm range (RJ115).

Avro developed a 33-degree flap take off setting to improve the Avro RJ100's take-off performance and increase payload from

Crossair E2233 RJ85 with Paris Air Show markings.

operationally restricted airports. As a result, the RJ100 was certified for LCY operations by the UK CAA in 1995.

By 1995, further improvements had been made to the Avro RJ including solid-state cockpit voice and flight data recorders (vs. analog tape), and a shunt HF antenna fitted in the leading edge of the fin. Moving to solid-state was one aspect that helped reduce weight with an overall target of 1,500 lbs. being the overall goal, and a 1,000 lbs. reduction being identified and integrated with aircraft on the line. Modular toilets saved 25 lbs., a waste water system revision removed 10 lbs., new wire looms took off an additional 10 lbs., and more. Each pound counts in an aircraft, as every ounce of weight requires fuel burn to provide lift to carry the additional weight. The later aircraft would end up being lighter because of a more efficient application of Thiokol adhesive to the fuselage on the assembly line versus the earlier BAe 146 days.

More improvements came in 1996 and later airframes included exterior reductions in drag, notably derived from inclusion of a fairing at the joint points of the fin and tail plane, as well as improved sealing and the deletion of the tail bumper. Collectively these reduced fuel burn by 2%. An increase in cabin differential pressure enabled higher cruise altitudes, increasing the service ceiling from 31,000 feet to 33,000 feet, which offered another slight decrease in fuel burn. All of these improvements provided for up to 600nm additional range on the RJ70 or 200nm on the RJ85/RJ100. Unlike most aircraft, the BAe 146/Avro RJ family does not need a "D" check (heavy

maintanence) which reduces overall ownership maintenance costs considerably. Instead an "A" check is performed after 400 flights and a "C" check is required every 4,000 flights. Even the older ALF-502 power plants were seeing improvements derived from the newer LF-507, which lowered maintenance costs by over 20%

Special 75th Anniversary livery applied to the first Avro RJ100 delivered to Sabena.

Additional enhancements or options included a full width transverse galley, developed originally for Air Malta to provide commonality with its Airbus A320 fleet. The rear toilet was moved just forward of the rear passenger door to accommodate an optional rear galley. By late 1995, Textron Aerostructures (formerly Avco Aerostructures) started to transfer production of the wings from Nashville, Tennessee to the BAe factory at Prestwick in Scotland. During 1995, Textron completed 12 wing sets with Prestwick producing six. In 1996, Prestwick took full responsibility for wing production, increasing the UK content of the Avro RJ to more than 70%.

Configurations

The RJ70 with 70 seats @ 5 abreast and 31″ seat pitch (or 82 seats @ 6-abreast)

The RJ85 with 85 seats @ 5 abreast and 31″ seat pitch (or 100 seats @ 6-abreast)

The RJ100 with 100 seats @ 5 abreast and 31″ seat pitch (or 112 seat @ 6 abreast and 32″ seat pitch)

The RJ115 with 116 seats @ 6 abreast and 31″ seat pitch (and mid-cabin exits)

The RJ100 was capped at a maximum of 112 seats, as a higher seat count would require additional emergency exits. One airline did take delivery of two British Aerospace 146-300s with Type III mid-cabin pop out emergency exits installed (MSN E3161 & E3174). Makung Airlines wanted to outfit its aircraft with the maximum number of passenger seats, and this required the fitment of additional exits, and the modifications were completed by Fields Aviation at East Midlands Airport in the UK. British Aerospace used employees to demonstrate that the aircraft could be evacuated in the time required for certification with E3161 used for the full trials. A third aircraft (E3193) was built with Type III exits but Makung did not take delivery of it. Subsequently the exits were inhibited, and the aircraft was delivered to Dan Air.

Orders / Notable Customers

Crossair was the launch customer for the RJ85 aircraft, initially ordering four aircraft. At the time, the only true Avro RJ commitment was from Crossair. A further five aircraft were notionally on order for Malmo Aviation but this was solely a mutual PR exercise funded by raiding the security deposit BAe held against 146 leases with the airline. Ironically, under new ownership, Malmo would later become a long-lived and substantial Avro aircraft operator. The first Crossair aircraft entered flight testing in November 1992 with delivery following in March 1993. It carried a brand-new livery and the "Jumbolino" (with a flying elephant logo under the cockpit) would be joined later by 16 RJ100 aircraft.

The Jumbolino was so popular with passengers that some would ask for it by name when reserving a flight. Crossair redesigned the galleys in its RJ85s in conjunction with BAe in order to enable its renowned hot meal and champagne service, even on flights of only one hour. On March 29, 1995, Swissair Group confirmed an order for 12 RJ100s with 12 more options. Crossair eventually adopted the name of Swiss Airlines after the collapse of Swissair, and all Swissair operations under 100 seats were transferred to the rebranded airline.

Swissair acknowledged it could not, with its cost structure, make money with its 85-seat Fokker 100s, and the Crossair order for 12 RJ100s would replace the Fokkers one-for-one. Aircraft sales campaigns are usually protracted affairs and often culminate in a final negotiation led by senior management from both sides. The teams then have to endure a tense and sometimes frenetic bout of negotiation as the last key issues are hammered out. Few CEO's can resist the temptation of making the final grand gesture to close the deal, however irritating and unnecessary that might seem to the "sherpas" who have brought matters to this stage. The CEO of Swiss, Otto Loepfe, managed to peel off Avro's CEO from his minders for a one-to-one meeting. The net result was Swissair being given a gratuitous additional $2million in trade-in value on each of its Fokker 100s. Crossair returned two BAe 146-300s as the RJ100s joined the fleet, and the new aircraft were configured with 97-seats, along with long-range fuel tanks enabling them to perform sectors up to 1,300nm. The RJ100s complemented the four 82-seat RJ85s in Crossair's fleet. The carrier later determined that ten of the 12 new RJs would operate on Swissair routes and would have a two-class layout with the first seven rows being a four abreast business class configuration, for a total of 89-90 seats.

In a complete turnaround from the BAe 146, and to avoid the financial liability the company had taken on in the past decade, Avro focused on airlines that were considered 'blue-chip' and could obtain their own financing for the aircraft. As a result, the types of customers that began to buy the Avro were more financially stable and had better operational and engineering resources. The most notable often flew as flag carriers or as affiliates of major flag carriers.

Delta Connection affiliate carrier Business Express (BEX) placed an order for up to 20 RJ70 aircraft in December 1991, becoming the launch carrier. The carrier operated five leased BAe 146-200 aircraft before its own RJ70s began delivery in 1993. BEX ended up operating the five former Discovery Airways aircraft and took delivery of three RJ70 aircraft for a total fleet size of eight BAe aircraft. Not long after BEX began taking delivery of the Avro RJ70, the entire management team was replaced. One of the new executives was Gary Ellmer, who arrived after having just left WestAir, a former BAe 146 operator.

With the scope clauses in force at Delta, and declining revenue, from day one Gary knew the Avro RJ would not work in the BEX fleet. Sure enough, BEX cancelled the remainder of its RJ70 order and began withdrawing all BAe aircraft at the end of April 1993. With these reductions BEX could not sustain jet operations and would revert to operating solely as a turboprop carrier. BEX ran flights into hub-and-spoke airports for Delta on the East Coast and was based out of Portsmouth New Hampshire. It was unable to get landing slots at Washington National (for flights to Boston) due to a restriction on the number of slots for four-engine aircraft. At the end of the '90s BEX was acquired by American Airlines and branded American Connection.

Handover ceremony for DAT's (Sabena) first Avro RJ85.

BAe's financial implosion led, as we have seen, to the formation of AMO, JSX and Avro. A further consequence of the break-up of the centralised Marketing Operations Centre (MOC) was that responsibility for the turbo-props manufactured by BAe (Jetstream 41 and ATP) was passed to the Prestwick site in Scotland. ATP manufacturing was also moved from Woodford to Prestwick. There had always been tensions between site General Managers, who had hitherto controlled sales and marketing for their products, and MOC. With MOC being blamed for the financial problems BAE had suffered, the site General Managers wrested back control. As so often in BAE's civil aircraft history there was much wishful thinking involved. Whilst there were advantages in co-locating sales activities beside the production unit it was delusional to believe, as Prestwick did, that the failure to sell more J.41's and ATP's was due to MOC's incompetence and 146 focus. The problem lay more with BAe's own strategic mistakes. The J.41 and ATP were stand-alone products that simply did not have an adequate market to support them. The J.41 could not be produced at a low enough cost to be afforded by the airlines flying in that sector of the market. The ATP was being skewered by Bombardier and

ATR which had genuine 50/70-seat families that offered operators commonality and flexibility whereas the ATP was marooned in a 64-seat segment that was too small to support it. Unfortunately, it was to take some years and another big loss before Prestwick understood this.

The reason for highlighting this is that two crucial early Avro deals – Air Malta and Turkish Airlines – involved replacing existing ATP operations. The personnel responsible for those sales, done under MOC auspices, had subsequently moved to Avro and saw the opportunity for bigger Avro deals. BAe backed their judgement, given the importance of underpinning the whole quad jet universe to the company's balance sheet. With considerable dexterity given the complexities involved ATPs were maneuvered out and replaced by Avro RJ aircraft. Unsurprisingly Prestwick was not happy to see two of its very limited operator base jettisoned in this way.

Turkish Airlines placed an order for the Avro RJ100 series at the Paris Air Show in June 1993. The deal was structured for a ten-year lease, with an option to break after seven years. BAe was involved in supporting the finance arrangements but a crucial difference – compared to earlier years it was the quality of the airline's credit, which was supported by a Turkish Government guarantee. Initially ordering five RJ100s, after only six months in service the carrier ordered an additional three configured with a 99 seat five-abreast layout. A further four RJ70 aircraft followed along with the purchase of a simulator that was installed at the airline's Istanbul training base. Turkish operated the RJ100 domestically out of Ankara to destinations east, and from Istanbul to southern cities. Out of both Ankara and Istanbul there were select routes where both the RJ100 and other aircraft were flown. Turkish eventually also operated the RJ100 internationally to Rome, Bucharest, Athens, and Cairo.

Air Malta took delivery of its first Avro RJ70 on September 21, 1994. The carrier required 260 modifications to the aircraft prior to placing the order, which included the transverse galley and extended range fuel tanks. The first RJ70 was received during a lavish ceremony at Air Malta's main hangar at Luqa, following its delivery flight from Woodford. The very next day, the first revenue flight took off from Malta bound for Tunis. Less than a month later the second RJ70 arrived and went into service the same day. The orders from Turkish and Air Malta were crucial in stabilising the Avro operation, creating both a backlog of work and PR momentum. They gave Avro a platform on which to build its aircraft story.

BAe sales became aware that there was an opportunity developing at Lufthansa Cityline which operated an extensive fleet of Bombardier CRJ-200 and Fokker 50 aircraft. The German carrier was aware of the inroads being made by Crossair/Swissair, especially at London City, and could see the competitive advantage being provided by the Avro RJ. In theory this should have been Fokker's deal to win, given its existing extensive relationship on the F.50 and the fact that the company was now owned by Daimler-Benz, one of Germany's largest and most prestigious companies. Given the cozy inter-linking relationships that prevailed amongst the German business elite, it seemed unlikely that there would be any opportunity for Avro. In reality there are occasions, and this was one, when such a position induces a sense of smug complacency in the incumbent, and there was certainly a feeling at Cityline that Fokker was taking it for granted. By contrast Avro lavished attention on Cityline and Lufthansa, secretly building up its relationship to the point where, by the time Fokker realised what was happening, Avro's momentum was unstoppable. The campaign culminated in a demonstration to Lufthansa's board at Frankfurt where a Crossair aircraft was brought in for inspection and a flight. With that successfully accomplished, approval to proceed soon followed.

The terms of the first deal were brutal: just three aircraft taken on a lease structured so that Cityline could return them after just three years without penalty. In addition, BAe took three Fokker 50 aircraft in trade at prices well above market value, dictated primarily by the value of the assets on Cityline's books. Superficially this looked like a return to the bad old days of high liabilities and rotten terms.

Lufthansa's fifth Avro RJ85.

Interior for Lufthansa Cityline Avro RJ85.

It was a tough deal to get through the Deal Committee but the difference was the quality of the operator and the bigger prize on offer if the operation succeeded. Lufthansa did not exercise its return option, but added to the fleet in a series of further deals on better terms for BAe, including eventually purchasing five aircraft for cash. Ultimately Lufthansa Cityline operated a fleet of 18 Avro RJ85 aircraft. The Lufthansa order had an effect on the worldwide perception of the product. The German airline was a byword for quality and technical prowess and an order from it demonstrated clearly that the repetitional damage BAe suffered from US Air/ American was a thing of the past that was well and truly buried.

Lufthansa took delivery of its first RJ85 on October 17, 1994, and the fleet was based out of Munich. The carrier was operating a single "City Class" cabin with 5-abreast full leather seating and high-quality meals served on full Arcopal tableware with linen tablecloths and napkins. The RJ85 flew domestic and European points, including two of its longest flights to Helsinki and Minsk. Upon introduction, dispatch reliability was lower than expected, with a 98.48% rate but quickly increased to 99% with a target of above 99% the goal. In 1995, Stuttgart airport had to briefly close its main runway for resurfacing, with the only alternative being a 5,000-foot secondary runway.

As a result, only the Avro RJ (and BAe 146) could fly full loads out of the airport. For a brief period while the primary runway was closed seven airlines operated Avro RJ/146s to the German city. Crossair would handle Lufthansa Cityline's LF-507 maintenance out of its facility in Basel, Switzerland while Cityline handled other maintenance at Cologne, Germany. Every time Lufthansa placed a further order for the Avro RJ, it had to have its own internal airline scope clause adjusted and approved by the pilots' union.

Delta Air Transport (DAT) was a regional carrier that was owned by and flew services for Sabena. It operated a small fleet of 146-200 aircraft which it successfully flew for a number of years. Following the takeover of Sabena by Swissair, plans were soon put in place for a major expansion of the regional operation. The Swiss involvement gave Avro the advantage and although competition from Fokker was fierce, there was little doubt as to the outcome.

Meanwhile, in the Netherlands, Avro was conducting another secretive campaign with KLM affiliate, KLM Cityhopper. Once again, Fokker was asleep at the wheel and seemingly oblivious to what was happening on the other side of Amsterdam Airport.

Updated interior for Lufthansa Regional Avro RJ85.

Using three available former BEX RJ70's as the bait, Avro was in the process of negotiating an MOU with KLM Cityhopper. The CEO of Avro then took matters into his own hands. Meeting his opposite at Fokker, he struck an understanding whereby Fokker would pullback from the Sabena deal, and in return Avro would do the same with KLM. This unwarranted and unnecessary intervention worked out as planned, but is a fine example of what happens when the high paid help takes no notice of the advice of its own team.

Sabena committed to purchase 23 RJ85s in August 1995 in what would be a short-lived largest single order for the four-engine jet. The airline subsequently converted some of its RJ85 orders to the RJ100, and flew the RJ85s in an 82-seat five abreast layout. Sabena owned 49% of DAT (KLM owning the other significant percentage),. In 1996 the national carrier purchased all of the outstanding shares in DAT, which became known as Sabena, and two of the RJ100s were flown in a special anniversary livery for Sabena. Sabena collapsed in 2001, but was merged into the newly formed SN Brussels Airlines. SN Brussels became the second largest operator of the aircraft with 32 in operation, of which 26 were Avro RJ85 and Avro RJ100 and the balance the six original 146-200s.

Mainline U.S. carrier Northwest Airlines placed an order for 12 Avro RJ85 aircraft on November 6, 1996 with options for 24. This was the first (and only) order for the Avro RJ from a major North American carrier. The RJ85s were to be subleased to Mesaba, a Northwest JetLink carrier operating from Minneapolis and Detroit, and would replace DC-9-10 aircraft. However, due to Union scope clause regulation for aircraft with seating capacity of 70 or more passengers, the new fleet was configured for 69 passengers in a two-class layout: 16 first class in a 2-2 configuration, and 53 economy seats in a 2-3 configuration. Northwest marketed the aircraft to flyers as the only regional aircraft in the United States with a first-class cabin. They proved to be essential and Northwest/Mesaba exercised the options for 24 aircraft, making it not only the largest operator of the Avro RJ but also the largest operator of the BAe 146 (and derivative aircraft) with a total of 36 airframes.

SAM Columbia provided the first and only orders for the Avro RJ from South America, with nine RJ100s selected for its routes. The most critical, Bogota to San Andreas, imposed payload restrictions for the RJs because of the hot and high conditions. Modifications to the brakes by Dunlop (increasing brake energy by 10% with approval by aviation authorities) gave SAM the ability to carry up to seven more passengers. The 100 seat RJ100s filled a seating gap between the airline's Fokker 50 and its 150-seat MD-80s. The Avro RJ100 was evaluated against the Fokker 100 and Boeing 737, and SAM Columbia chose the Avro RJ to replace its ageing Boeing 727s. The ability to take off from short runways and high elevations was an important factor. The other was that Avro demonstrated that the space available per passenger to accommodate baggage as well as cargo was far greater than the competition. Four engines providing safety margins plus aircraft ferrying capability also worked in the Avro RJ's favor. After discussing Avro RJ operations with established customer Crossair, SAM Columbia was convinced the RJ100 was the right aircraft. The carrier's blue and green livery included a stylized bird logo on the vertical stabilizer, "Hace Amigos Volando" (making friends flying) of Sociedad Aeronautica de Medellin. The new logo paid homage to pre-Columbian artists who left in their designs the symbolic fauna.

Unfortunately, the SAM operation did not fare well, and the RJ100 struggled in the high temperature, high altitude environment. The engine was a particular area of difficulty, not helped by the Columbian maintenance staff being used to more robust equipment. Aircraft were soon being grounded and crisis talks followed but despite all efforts to turn the operation around, problems persisted. In the end BAe took the aircraft back and the least successful Avro RJ operation came to an end. With hindsight the RJ100 was clearly the worst variant to use given that it was the most performance constrained, but sometimes sales campaigns push sellers to outcomes that rely more on blind hope than reality.

1995 marked the best year the Avro RJ program had, with 50 aircraft placed demonstrating the new and effective sales team, and production at Woodford humming along continuing to improve the speed each aircraft could be built.

Loss Of A Competitor

Fokker had been a formidable competitor to the Avro RJ series with its Fokker 70 and 100. Back in the mid-1980's, U.S. carriers (and former BAe 146 operators) American Airlines and US Air both selected the Fokker 100. BAe had pitched the 146-300 against the Fokker 100, but American's negative view of the 146 (and planned retirement) pushed American to the Fokker 100, and USAir hadn't purchased PSA at the time to form an negative opinion on the 146. American Airlines went on to operate 75, and USAir 40, Fokker 100s. In Europe, three of the five larger carriers (Sabena, Swiss, Lufthansa) had chosen to operate the Avro RJ series over the Fokker 100. Fokker found itself in 1996 with a substantial number of white tail (unsold) aircraft, and less than a year later the firm went into receivership. After completing 283 Fokker 100s and 48 of the smaller Fokker 70, Fokker was gone.

Fokker closing up shop was a victory for Avro, but a short lived one. Avro had to worry about competing aircraft offered by Bombardier and Embraer, which were making significant inroads with inexpensive regional jets. Avro needed a follow-on aircraft to continue its relevancy into the next century if it was to continue producing regional jets. Four engines, no matter the explanation of operational benefits or economics, was being frowned on from new and existing operators. Even larger commercial jetliners with more than two engines, including the McDonnell Douglas MD-11 and Boeing 747, were falling out of favor and being replaced with twins such as the Boeing 777 and Airbus A330 series.

AI(R)

In yet another disruption, and in an effort to again consolidate the regional aircraft division to counter increasing competition, in January 1995 a consortium named Aero International (Regional) – or AI(R) for short – was created that brought together Avro, Aerospatiale and Alenia. After the agreements were signed by the companies in Rome, the new group was established on January 1, 1996 under French law as an S.A. company (public company). AI(R) made its public debut at the Paris Air Show at Le Bourget, introduced by Aerospatiale Chairman Louis Gallois. The business consolidated sales, customer support, and marketing effort at Toulouse, France, with 900 employees (460 coming from Avro & Jetstream) and in Virginia, United States for marketing operations. The management structure would be identical to that used by Aerospatiale and British Aerospace to manage Concorde, with the presidency traded every second year between Britain, Italy, and France. The board would be made up of two representatives from each of the three companies.

The combined efforts would leverage each other's products by offering airlines a 'family' of aircraft including both turboprops and jets, as well as providing a one-stop shop to address customers' needs. This meant that each division (within the consortium) would obtain efficiencies, with lower costs resulting from having a combined sales & marketing force (rather than each manufacturer having a dedicated team). This would theoretically enable lower prices to be offered to customers as an indirect benefit. Additionally, market research previously produced would be shared amongst the consortium members, and new research pursued jointly. For example, Aerospatiale and Alenia were involved with the "Regioliner", a 120-seat new generation regional get that was pursued with Deutsche Aerospace prior to DASA acquiring a 49% stake in Fokker. In the interim, Avro planned to pursue improvements to its existing family as sole ventures outside of the joint venture. Moving forward, the Avro name would be dropped from AI(R) marketing material, which would refer to the aircraft specifically as the "RJ".

There was no guarantee that the consortium would pursue an all new jet given the costs involved. BAE agreed to phase out production of the 70-seat Jetstream 61, successor of the BAE ATP. The regional aircraft market was becoming more crowded, with a variety of new players that previously were not a threat to the Avro RJ. AI(R) was hoping the new consortium would attract the Asian investment that it had been courting in past ventures and it was determined to reduce costs across the board for all three manufacturers to allow the organization to compete with other manufacturers. The group planned support existing aircraft lines, but would jointly work on new turbo prop and regional jet initiatives.

Northwest Airlines on behalf of regional airline Mesaba ordered 36 of the Avro RJ85.

Northwest Airlines wasn't the first airline to have a true first class on the BAe 146/ Avro RJ, but advertised that even their regional aircraft had first class. First class was the result of scope-clause limitations limiting the aircraft to just 69 seats.

Multinational joint ventures are difficult entities to create and manage. The ATR partners were state owned or effectively state controlled. Their employees felt they had a job for life, because in practice it was extremely difficult to make redundancies. There was a very autocratic management style with the Director General taking all major decisions and often being involved in closing sales deals. The incumbent, Henri-Paul Puel, had run ATR from its start and with a praetorian guard of other Frenchmen in key positions tried to carry on running AI(R) in the same style.

For many of the BAE personnel (200 people plus another 400 dependents), especially those from Avro, being moved to France was like going back to the worst excesses of the old BAE regime. ATR was pumping out more aircraft than it could sell, was accepting significant liabilities in supporting finance or leasing directly, and it was grotesquely mishandling its used aircraft activities, with huge refurbishment costs each time an aircraft was re-cycled back onto the market. The back-office support functions were in shambles with byzantine systems buried in paperwork and months behind events. There were two key differences to BAE. Firstly, ATR could bury its bad news in the labyrinth of bureaucratic structures that existed with its parent companies. The French state in particular took the long view and saw aerospace as a strategic industry that had to be supported through thick and thin and was prepared to allow mistakes to be buried in the greater glory of overall progress. Secondly, the ATR product was a first-class money-making machine equipped with reliable Pratt & Whitney PW.120 series engines so there was a good market underlying the situation, almost despite the mismanagement of ATR.

As may be imagined, the influx of BAE personnel were used to operating in a more transparent and open way: risk averse and used to a paper trail for each decision, and a lean but efficient back-office function. However this operating style caused problems from the start. Doubtless in time these difficulties could have been overcome and a new culture melded, but the fatal flaw was that the partners of AI(R) simply had diametrically opposed visions of the future.

Aerospatiale and Alenia saw the future as being the continuation of the ATR, dropping all the BAE legacy products and the launch of a new regional jet family. In those circumstances it was understandable that, with very limited exceptions, ex-ATR sales staff showed no interest in pursuing Avro RJ sales campaigns.

It is really quite difficult to discern what BAE was expecting to achieve. It seems to have seen the joint venture as a neat solution for its floundering Jetstream turboprop business, allowing it a graceful exit from those programmes. It may have felt that the Avro business could survive for a few more years on the backlog that had been created, supplemented with a few additional orders, and then similarly be allowed to die. In the meantime, overall costs would be cut, with a spiffy new regional entity they could either sell off their share or see the whole thing folded into Airbus.

BAE allowed an AI(R) regional jet design to be developed and for preliminary marketing to be done, but balked at any serious investment. Their partners became increasingly exasperated at British prevarication and after two years the joint venture was dead and buried along with the new regional jet programme. By the late 1990s, following the collapse of AI(R), Avro International Aerospace was be rebranded again, this time as British Aerospace Regional Aircraft (BAE).

The Avro RJ programme emerged from this debacle intact but severely battered. After two years of benign neglect the remnants of BAE's sales team had to try to pick up the pieces again. Having spent millions moving people to France it was decided to leave them there and run the new sales, marketing and support team from Toulouse.

The decade was ending, and the market was filled with competitors to the Avro RJ. Fokker was gone, but there were twin-engine regional jet threats from Embraer and Bombardier. British Aerospace Regional was at a crossroads: offer an evolutionary update to its RJ airframe and engines, retaining the four-engine layout? Or proceed with a revolutionary update which would be a twin-engine configuration, starting a new family of aircraft for the new century? BAE, as history has shown, would go down the path of least resistance.

SCM Colombia
RJ100
G-6-242

SCM Colombia
RJ
SCM Colomb
RJ
The Regional Jetliner
AVRO

Prototype Avro RJX85 and Avro RJX100.

CHAPTER 11

The Avro RJX – A Bittersweet Ending

By 1999, the wind in the sales and marketing department of BAE had begun to fade. New entrants into the "regional jet" market were active promoting quite a few new two-engine jet options, putting immense pressure on the four-engine Avro RJ. In what would be its swan song, BAE avoided the much-needed expense of designing a brand-new airliner with twin engines and went with yet another derivative of the Avro RJ. Although changing its name to Avro International Aerospace helped the regional jet manufacturer, it did not completely erase the past transgressions of BAE nor the pain of the BAe 146. BAE just did not have the stomach or interest to embark on a substantial redevelopment of its jetliner, but sales were slowing for the Avro RJ in its current form. Many customers were not interested in another four-engine regional jet as there were other two engine options available.

Twin engine aircraft had been around since the dawn of aviation. So were four engine aircraft which had found favour for overwater operations and high-demand markets. By the late 1980s medium-range twin-engine jets had begun to take over where four-engine Boeing 707 and DC-8 used to dominate the skies. With the introduction of the Boeing 777-200, a widebody twin-engine jetliner, the writing was on the wall even for the venerable Boeing 747 series. At the end of the century, the airline market had shifted, and aircraft with more than two engines had begun to fall out of favor. By the time the Avro RJX was getting off the ground, McDonnell Douglas (under Boeing ownership) was winding down production of its three-engine aircraft, the MD-11. Boeing and Airbus were focused on four-engine aircraft only for ultra-long haul widebody service, and newer regional jet aircraft were all of the two-engine variety.

Meanwhile, developments at BAE had marginalised the regional aircraft division even further. In 1999 BAE took over Marconi, a substantial part of the long-standing British conglomerate GEC. The new company which became BAE Systems quickly saw the effect not just of former Marconi managers but also of Marconi's culture which had legendarily been seen as parsimonious and spreadsheet oriented. For people coming from that background the once mighty civil airliner business was seen as a tedious hangover from bygone days. Continuing a growing trend in BAE since the financial crash, more and more senior managers had finance rather than a sales or operational background. The age of the accountant was never going to rest easily with the high-risk airliner business.

The Avro RJX represented the next improved Avro RJ series aircraft, utilizing the three airframe sizes of the older RJ. The RJX designation was often used internally at BAE to denote experimental versions of the BAe 146/Avro RJ series. In fact in the late 1980s and early 1990s it was used almost simultaneously for very different versions of a two-engine derivative of the 146. Development of the new four engine RJX began in February of 1999, by when offers were already being made to select airlines for launch customer status for the new aircraft. Honeywell, which purchased Allied Signal, had signed on to bankroll most of the reengineering cost of the aircraft, as it would in theory produce sales for the new AS977 engine and select flight deck avionics enhancements. BAE only proceeded with the RJX because it was a low-cost opportunity to extract the last juice from the quad jet market and underpin its still considerable used aircraft and support business. The investment to update and certificate the new aircraft would be a mere £65 million ($100 million USD), although BAE Systems was also trying to secure risk sharing partners for the RJX venture. Sales campaigns began in 1999, aimed at convincing current Avro RJ and BAe 146 operators as well as new operators that the RJX was the optimum aircraft. First flights and deliveries were scheduled for 2001.

Although work was already under way on the RJX, it was officially launched on March 21, 2000 with approval from the BAE Systems board and an order for two aircraft. Unlike previous internal designations of 'RJX', the updated aircraft would use the RJX designation officially to differentiate it from previous iterations of the regional jetliner. The market for regional jetliners was becoming quite crowded, with offerings from Bombardier (the CRJ700/900) which would enter service before the RJX; the Embraer E-Jet (170/190) aircraft that would join the existing smaller ERJ family; Boeing's 717 (formerly McDonnell Douglas MD-95), and even the Airbus A318 to some extent.

Avro was clearly no longer the only regional jet manufacturer to offer a family of aircraft. BAE Systems was going to try and get further mileage out of the BAe 146/Avro RJ line, in an age of airline operations where a two-engine variant would have made more sense. However, BAE Systems did not have enough faith or interest to invest the nearly £375 million ($500 million USD) required to transition to a two-engine airframe. Such changes would have necessitated an all new wing and engine pylons, as well as additional testing. In looking back at other RJX concepts, BAE might as well have proceeded with modifying the aircraft further to extend its reach into the 120-130 seat range as well.

Instead, BAE Systems updated the Avro RJ with a new engine, the Honeywell AS977, as the primary improvement over the Allied Signal LF-507 series. The engines featured a new design, which demanded a new engine nacelle as well. Promising up to nearly 19% lower maintenance costs and up to 15.3% improved fuel burn over the LF-507, serviceability would also improve over the previous engines. The power plants were also air started versus the BAe 146/Avro RJ engines electric start, which had higher power to weight ratio and did not suffer from overheat issues during startup attempts. Brief consideration was given to moving to the Pratt & Whitney PW308, a small engine with similar thrust, but with Honeywell funding the transition, BAE Systems stuck with Honeywell. Honeywell was active in soliciting feedback from customers using the LF-507, and formed a technical advisory committee with BAE as well as service centers such as Garrett Aviation and select customers to establish what they wanted in engine evolution. The driving response was cost; everyone wanted a reduction in the cost per-pound of thrust. Honeywell expanded the FADEC capabilities to include self-diagnostic and predictive software to monitor performance and planned maintenance requirements. The dual-channel system would be fully computer controlled, far more than the Avro RJ which had mechanical backup controls.

Avro were still offering the same fuselage lengths and designations, with the RJX70, RJX85, and RJX100 all with matching the previous aircraft's seating capacity at 5-abreast at a 31-inch pitch (and higher density seating with 6-abreast). Even though a higher-density passenger variant (the RJ115) was never produced, Avro went back to offering a very high-density seating option of the RJX (with mid-cabin exits), accommodating up to 128 seats at 6 abreast with a knee-pinching 29-inch pitch. Increased thrust allowed VMO/MMO to be reached more quickly, along with a 15% reduction in noise footprint. The RJX was projected to be up to 30% cheaper to operate than a BAe 146 and offered 20% better economic than the Avro RJ. The range was increased by as much as 20% over the Avro RJ series.

With the improvements in manufacturing lowering build times drastically, the aircraft was ready to fly much sooner than previous derivatives. Because the largest change to the RJX series was the powerplant, BAE was able to begin construction of aircraft quickly, with E2376 becoming the first RJX85 in March 2000. The initial flight did not happen as scheduled, due in part to Honeywell and GKN Aerospace experiencing performance issues with the power plant, which caused a delay of nearly a year. The first engines did not arrive until January 2001, and the remainder in February 2001. Installation and the first test runs began on March 18, 2001 followed by taxi trials

in preparation for a first flight in April 2001. Additionally, the APU was tested before the engines were installed, and unlike the Avro RJ even the APU was FADEC controlled. The day before the scheduled first flight, RJX85 engaged in high speed taxi tests in order to ensure that there was no disturbed airflow into the new nacelles; encountering this on the ground as opposed to during rotation would naturally be a safer option.

RJX85 G-ORJX (E2376) first took to the air on April 28, 2001 at 12:16pm, completing a two-hour 54-minute maiden flight from Woodford at an altitude of 20,000ft. Alan Foster, RJX Project Pilot, noted that it flew just like the previous Avro RJ with two exceptions: improved fuel burn and increased thrust at altitude. Aircraft Services Group's Managing Director Mike O'Callaghan noted that in just over two years the aircraft had gone from a paper-based project with Honeywell to actually flying. The plan was for the first aircraft to be joined by the RJX70 and RJX100 later in the year, with the RJX100 prototype scheduled to make its first flight in July (G-IRJX msn E3378), while the first production RJX100 (E3391) was to be used for cabin noise, electromagnetic capability and handling evaluations from October.

The RJX85 was launched with an order for two from Druk Air, an existing BAe 146-100 operator, followed by an order for 12 of the larger RJX100 variant along with eight options from British European in July of 2001. British Airways subsidiary CityFlyer Express also placed options for six RJX100s. U.S. carrier Aloha Air was interested in as many as 20 aircraft, noting the Boeing 737NG economics were not the best for inter-island travel but was also looking at the Boeing 717 as an alternative. The carrier put down a $250,000 deposit for a handful of delivery positions for Avro RJX85 QC variants, as it had a robust cargo business and could operate passenger flights during the day and cargo flights at night due to the low noise profile (versus the noisy Boeing 737-200s it was operating at the time). Woodford was not keen on having to embark on a QC derivative so early in the RJX programme, and Aloha subsequently withdrew due to the financial risk involved in going for a new aircraft.

In May 2001, the RJX85 painted in British European livery was shown off to specially invited guests at Duxford. Other airline executives like Antonis Simigdalas, COO of Aegean Airlines which was an Avro RJ operator, were also present. The sales team extolled the virtues of the RJX, promoting its lower fuel burn, increased range, and reduced operating costs. What was not planned for was for a fuel leak in the number 2 engine the next morning after the brief flight. It was suspected to have occurred during flight but was confirmed while taxiing into Woodford. Lucky for everyone, the engine did not catch fire, but it demonstrated why aircraft are tested to ensure any design or manufacturing issues are caught before entering service with airlines.

By August 2001, the RJX85 completed testing that enabled full and final speed and altitude limitations to be communicated to customers. Climb was even better than the last engine update, with a reported 6-11% improvement, and testing of the aircraft had exceeded the 120-hour mark. Performance of the aircraft was improved in comparison to the previous two iterations, owing largely to the powerplant performance. The AS977 was driven by a two-stage high pressure turbine, with 38 single-crystal SC180 blades, and the second stage driven by 34 nickel-alloy blades. The three stage LP turbine contains 56 blades in the first stage, 62 in the second stage, and finally 74 in the third stage, all rated for a 15,000-cycle lifetime.

AVRO
RJX
G-ORJX

Derek Ferguson

Marketing produced photo showing full hot meal service inside the Avro RJX.

Interior of the Avro RJX100, fitted with 6-abreast seating.

Testing with the AS977 was showing great progress, but certification was moved from September 2001 to the first quarter of 2002. The RJX program was continuing to slip because Honeywell was having issues with the integrated powerplant system supplied by GKN Aerostructures, and the lost time was not being made up. AS977 General Manager Mike Redenbaugh said "The engine is meeting its specific fuel consumption and temperature margin specifications, but we have identified a handful of areas where we need to incorporate durability and other improvements before it goes into service." Areas of improvement identified after 4,500 hours of testing included individually replaceable fan stator vanes; combustor cooling changes to increase durability; and bleed duct optimization to reduce losses. By that time, Honeywell had completed 6,000 hours of the planned 10,000 hours of testing, including flying the engine on its Boeing 720B with the AS977 mounted just behind the cockpit. BAE Systems own flight test engineer reported that the later 'block 2' AS977 engine had improved over the block 1 engine, but they were still not meeting the noise or fuel targets and were

Derek Ferguson

Water ballast tanks are equipped inside test aircraft to simulate passenger loads.

running about 2% higher in fuel burn. The aircraft was actually proving to be noisier than with the LF-507 engines, giving a more jet-like sound instead of the whine from the earlier geared turbofans.

All aircraft tests are designed in conjunction with past aircraft performance as well as CAA requirements. The sequence of the trials was determined based on need or around dependencies that had not quite been resolved, while sometimes changes were weather related. Many weeks (and sometimes months) went into planning the flight test schedule, and twice weekly a planning board determined when specific test flights would take place. On August 25, 2001 RJX85 prototype was flown to the United States for performance trials while RJX100 (E3378) G-IRJX was being prepared for its first flight that was planned for late September. The RJX85 was based out of Mesa's William Gateway Airport, just outside of Phoenix, Arizona. The location was chosen because of the predictable weather and the facilities available, as the primary task was hot weather performance and systems tests. The aircraft departed Woodford and overnighted at Keflavik, Iceland. It flew on to Goose Bay, Newfoundland where it was refueled and continued on to Bangor, Maine (United States) for

Paul Roberts

The very first production Avro RJX100 assigned to British European (G-6-39), E3391.

overnight crew rest. On Monday August 27, 2001 the aircraft continued from Bangor to Topeka Kansas for another refueling stop and then flew onward to Arizona. The weather at its destination was exactly what had been expected – hot – and the daytime temperatures peaked at between 100 degrees and 108 degrees Fahrenheit (38-42C).

Flight trials began on Wednesday, August 29 with a series of take-off and climbs to altitude. They continued with approximately eight missions every day and 79 runs were completed. These evaluations measured performance including take-off speeds, center of gravity, one engine out, and more. For example, on one trip, the aircraft's brakes were set on and the engines run up for three seconds. Upon brake release at a pre-defined engine speed, one engine would be throttled back to idle with applicable rudder applied. Once the aircraft rotated, all instrument readings including climb rate were logged. The test parameters varied with some take-offs at heavy weight, and the hot temperatures resulted in shallow climb outs. At 400ft, the aircraft was turned downwind to fly a full circuit followed by a landing on the same runway as the departure after which the flight team would discuss whether the run was acceptable or not. The criteria used assessed the achieved versus the target speeds as well as the rotation rates, using the onboard instruments in the analysis.

The take-off performance tests were completed just prior to the aircraft being grounded due to the events on September 11, 2001. The programme eventually resumed at Toluca Airport, Mexico where high-altitude trials took place, but it was abruptly cancelled before completion and the aircraft was summoned back to Woodford on November 22, 2001.

In the end, the RJX never went into production, nor was the testing gauntlet completed that would have enabled the RJX to receive certification. On September 11, 2001 the world changed with the terrorist attack in the United States. Overnight, flights were grounded, people were afraid to travel, economies began to unravel and in the days that followed unrest provoked by the terrorist group Al Qaeda spread across the globe. The world entered a brutal recession, followed by war and airlines were hit hardest by it. Many carriers furloughed or laid off employees, slashed schedules, canceled aircraft orders, and parked aircraft in the desert. This adverse impact on the industry would continue for several years.

British Aerospace had considered withdrawing from the aircraft manufacturing business more than once since the mid-1980s. It had increasingly viewed the regional aircraft business as an albatross, a giant weight around its neck that would never deliver the success originally envisioned. Sales had been slow until PSA stepped in, but even after the PSA order, sales were never brisk until a brief period in 1995. Convincing airlines to operate a four-engine regional jet was not an easy sell, and by 2000 such aircraft had lost their luster. But British Aerospace's management realized that pulling out of the regional business was not a simple matter. There was the ongoing financial liabilities from the company's recourse to its leasing obligations and the obligation to support airframes for the 10-20 years they were expected to fly. Without a new aircraft production schedule it was determined that residual aircraft values would fall, potentially opening the company up to needing to make further financial provisions for the difference. In other words, it was cheaper for the firm to keep "kicking the can down the road", hoping never having to swallow that bitter pill or at least having mitigated it significantly when it did. This strategy was successful and by 2001 Asset Management was well in control of its used aircraft liabilities and even a worst-case scenario was no longer an existential peril for BAE.

Derek Ferguson

Within days of the 9/11 attack and the subsequent grounding of all flights in the US it was clear that the airline industry and its suppliers were facing a major crisis.

With the first RJX100 having made its maiden flight on September 23, 2001, its fate had already been sealed. Sales for the RJX were already off to an incredibly slow start, with only fourteen orders and a handful of options. BAE put together a team (without any representation from Regional Aircraft) to conduct a thorough review of the RJX project. In reality this was no more than a fig leaf designed to support the decision the company wanted and intended to make, namely to terminate the programme forthwith.

In the context of 9/11 and the subsequent economic fall-out, the decision to pursue such a hatchet job was not difficult. The project had a weak backlog and further orders were unlikely to materialise until a recovery was underway. Competition was getting tougher and even with the makeover that was underway, the RJX was going to struggle. Other options such as riding out the storm and continuing production whilst searching out a buyer for the programme were given scant consideration. The 146 simply had no residual goodwill with BAE's senior management who just wanted as clean and final a break as could be managed. By contrast, as Bombardier has recently demonstrated, it is possible to engineer an exit without destroying the business and the livelihoods of the employees, but BAE had long since ceased to have any real long-term strategy for the programme and the crisis only proved the paucity of its insight into the business.

On November 27, 2001, at 8:30am just as the RJX85 was being prepared for a flight to test the avionics systems, the staff at Woodford were summoned to a darkened Hangar 3A. The announcement that BAE Systems was exiting the regional jet market was made, and redundancies at Woodford occurred almost immediately with 1,700 jobs eliminated. John Weston, CEO of BAE Systems said "the outlook for regional aircraft has deteriorated sharply. Regrettably it has been concluded that our regional jet business is no longer viable in this environment."

BAE Systems leasing arm (Asset Management Organization - AMO) was still managing a portfolio of over 100 BAe 146/Avro RJ aircraft and would continue supporting those in service. But any chance of the RJX being delivered was dead.

BAE Systems had struggled for decades to make the 146 a success, both operationally and financially. Even with the cost cutting and the closure of Hatfield, the aircraft maker could never move the programme into the black. The events of 9/11 had a major impact on the airline business, with vital competitions for revised regional fleets at British Airways and Qantas postponed. It was rumored that the BAE Systems Avro RJX100 was the front-runner for a big expansion of British Airways' network out of Gatwick. The market had changed, and prices that could be demanded for aircraft had gone down with increased competition. Vice President of Marketing for BAE Systems Nick Godwin made a statement that the increased competition and economies of scale from new competitors such as Canadair and Embraer would require BAE Systems to sell each Avro RJX at a loss just to compete. This was not a sustainable option.

While it might have been expected that flight testing would cease, it actually continued just in case the launch customers (Druk Air and British European, formerly Jersey European) wanted to force BAE Systems to fulfill its contractual commitments. Both airlines were operators of the BAe 146 series aircraft. Flight trials continued, but no high-risk tests required to certify the aircraft were carried out. Risking aircraft and air crew for a cancelled program did not make sense, but until negotiations with the two launch carriers were concluded the program was not completely shut down. With the writing on the wall though, the flight test team decided to do an air-to-air photograph session with all three RJX aircraft (two -100s, one -85) on January 10, 2002, while engaging in additional performance testing.

WARNING
HOT AIR
EXHAUST
WARNING
HOT AIR
EXHAUST

AVRO
RJX
G-ORJX
AVRO
RJX

Druk Air agreed to cancellation fairly quickly, and ended up buying Airbus A319s to replace both BAe 146-100 aircraft in its fleet. British European initially declined to cancel its order, instead wanting the new RJX aircraft. It was planning to take its order over the course of five years, but BAE Systems pushed back: the manufacturer wanted British European to take all of its aircraft at an accelerated rate. In the end, faced with the prospect of operating an orphaned fleet for which spares and support may have become an issue, British European decided it was safer to move to a different type rather than stay with the current RJX. British European reached an agreement with BAE Systems in January 2002, receiving 'commercial consideration' (i.e. penalty payment) in exchange for cancelling its order for the RJX. British European (now Flybe) went on to operate a mixed fleet of ATR-72s, Bombardier Q400s, and Embraer 175/195 jet aircraft. Although customer contracts were cancelled, Honeywell was not as forgiving and filed a $20 million lawsuit against BAE for cancelling the programme. Despite this, in the fourth quarter of 2001 the US firm wrote off nearly $100 million in development costs for the AS977 powerplant, navigation, and avionics systems for the Avro RJX

On January 23, 2002, BAE Systems assembled what was left of the team and grounded the RJX fleet, suspending further flight testing. The test specimens had accumulated 528 hours over 242 flights, and were put into store. The RJX series was only a few tests away from completing certification. Three RJX airframes were completed, and after careful consideration by BAE Systems including negotiating with Honeywell and the CAA, one was donated to the Runway Visitors Center at Manchester Airport. On February 6, 2003, BAe Chief Test Pilot Allistair McDicken was the pilot at the controls who took G-IRJX into the air for the last time. Prior to heading to Manchester Airport, G-IRJX made a series of low fly-bys for the remaining staff at Woodford. The aircraft then lined up for short finals at Manchester Airport with project test pilot Alan Foster becoming the pilot to land her for the very last time. The aircraft made a low pass before touching down. Upon arrival, G-IRJX made a trip around Terminal 1 and received a fire service water cannon salute before coming to a halt on Stand 1. No abnormalities were noted when shutting the RJX down for the last time, even though it had sat for nearly a year before its final trip aloft. The flight crew signed the flight deck bulkhead before departing. The weather was gray and wet, perhaps symbolic for what was the end of the line for a fully British-built aircraft.

The Manchester Airport Viewing Park is where G-IRJX RJX100 is on display today. A dedicated group of enthusiasts work tirelessly to keep the aircraft well maintained, including former RJX (flight test) engineer Derek Ferguson and other members of the Woodford production team. The other RJX100 (E3391) which was the first production aircraft for British European, was scrapped after only four flights, and the engines returned to Honeywell. The last Avro RJX, RJX85 (G-ORJX) was kept at Woodford and used as a training aid at BAE Systems Customer Training Center (which later became Oxford Aviation Academy). In 2011, the wings were removed and it was transported to BAE Systems Regional Aircraft Division at Prestwick to be used as an engineering design aid. There were seven additional Avro RJX airframes in various stages of completion at Woodford: all the components for E2396 RJX85 for Druk Air were available but it was not fully assembled; E3397 RJX100 for British European also included all the components for a complete aircraft but was not fully assembled. Various parts for E2398 RJX85, E3399 RJX100, E2400 RJX85, and E3401 had been delivered to Woodford, but not enough to have assembled complete aircraft. E3402-E2404 had provisional assignments only and did not include sufficient components for their construction to begin.

G-IRJX arriving at Manchester Airport on its last flight ever.

The last two Avro RJ85 aircraft on the manufacturing line did not have customers at that time, and were stored as white tails briefly. Support for the aircraft would continue, while an evaluation of the support groups (Aircraft Services Group) would be reviewed around the world. A £250 million-pound charge was taken in 2001 related to the closure of the production line. A £150 million-pound restructuring fee was instituted along with support costs of approximately £120 million pounds. Ironically, the RJ program contributed a £10 million operating profit in 2000, but that was not enough to save the program. Sufficient employees were kept on to support the existing aircraft in operation around the world and provide customer service. But no less than 1,000 posts were made redundant due to the cancellation of the RJX with a £370 million-pound charge against 2001 accounts. Nevertheless, £29 million was invested in manufacturing spare parts for existing aircraft and customers.

Asset Management Organization (AMO) continued to be managed out of Hatfield, with Prestwick becoming an engineering and support hub, and Weybridge supplying the spares. Woodford still handled customer training and engineering, and support and asset management centers continued operating in Washington, D.C. (United States) and in Sydney, Australia. BAE Systems had to provide spares, training and service for more than 300 BAe 146/Avro RJ aircraft it had built, not to mention all the other aircraft it had produced such as the Jetstream and ATP series.

BAE Systems as an end-to-end provider of aircraft was officially over. It went on to derive revenue from the vast amount of regional aircraft on its books, leasing them for as long as it could. It also continued to sell parts, as well as service and training for new airlines acquiring used BAe 146 and Avro RJ series aircraft. It even engaged in working on life extension services for the in-service fleet, extending the airframe life to up to 80,000 cycles (a cycle is one take-off and landing). Doing so helped prop up the residual values of the aircraft on its books, which was key to getting out of the financial mess of the 1980s that resulted from bad deals with airlines that had poor balance sheets.

The remaining white tail Avro RJ aircraft without owners were eventually sold and delivered to Blue 1 on January 25, 2004. The last aircraft to leave the British Aerospace factory was E1001, by then stretched and assigned construction number E3001. It was reconfigured as an atmospheric research vehicle and delivered on May 10, 2004. The program ended where it started, with aircraft E1001, the very first BAe 146 to be built.

CHAPTER 12

Teething Troubles for BAe and the 146

The BAe 146 received initially favorable reviews from Dan Air and Air Wisconsin, the first two operators that put the aircraft into service, and was showing a pretty good dispatch rate for a brand-new type (over 95% initially, climbing to 97.8% after six months). Like any new design, the BAe 146 was not without faults, some of which plagued the aircraft for a number of years before they were resolved. Specifically, the power plants were the biggest issue and as a result a BAe acronym began to circulate in a negative fashion, with the definition changed to "Bring Another Engine" or "1-4-sick".

Engine Troubles

The ALF-502's transition from a vertical role as a helicopter power plant to a horizontal role in airline operations had been more challenging than expected. With more than 4,000,000 hours in operations with the military, it was expected that the engine was ready for commercial use. Prior to its application with the BAe 146, it was applied to the Canadair Challenger aircraft. But with a very high shutdown rate and repeated gearbox and bearing failures, the engine was replaced with GE's CF34 on the Challenger, and a lawsuit between Canadair and Avco Lycoming ensued.

Information for Dan Air's and Air Wisconsin's initial operations has been difficult to come by, with early press quoting both carriers as not reporting any reliability or maintainability issues. However, a former BAe rep for the airline reported that Dan Air hated the plane and was only operating it because BAe was reportedly paying for the maintenance. PSA was one of the largest and early operators, and it had been experiencing higher than expected maintenance of the ALF-502 engines. The powerplants required overhauls much earlier than planned, with seals and bearings wearing out faster than anticipated or forecast. Air Wisconsin (and subsequent operators) were told early on not to do full thrust take-offs but ignored the directions and did so anyway. As a result, the engines did not last long and required servicing frequently, after around 2,500 hours of service.

It was understood that many of the issues with the ALF-502 in airline service were due to resonances (vibrations), a result of the higher rotating mass causing vibrations that resulted in extra wear on engine components. Fuel control springs were not reliable, suffering from harmonic vibration and kept breaking, resulting in low take-off power; damper ring failures in the gearbox traced to 'buzzing' bolts that loosened, failed, and ultimately dropped through the engines.

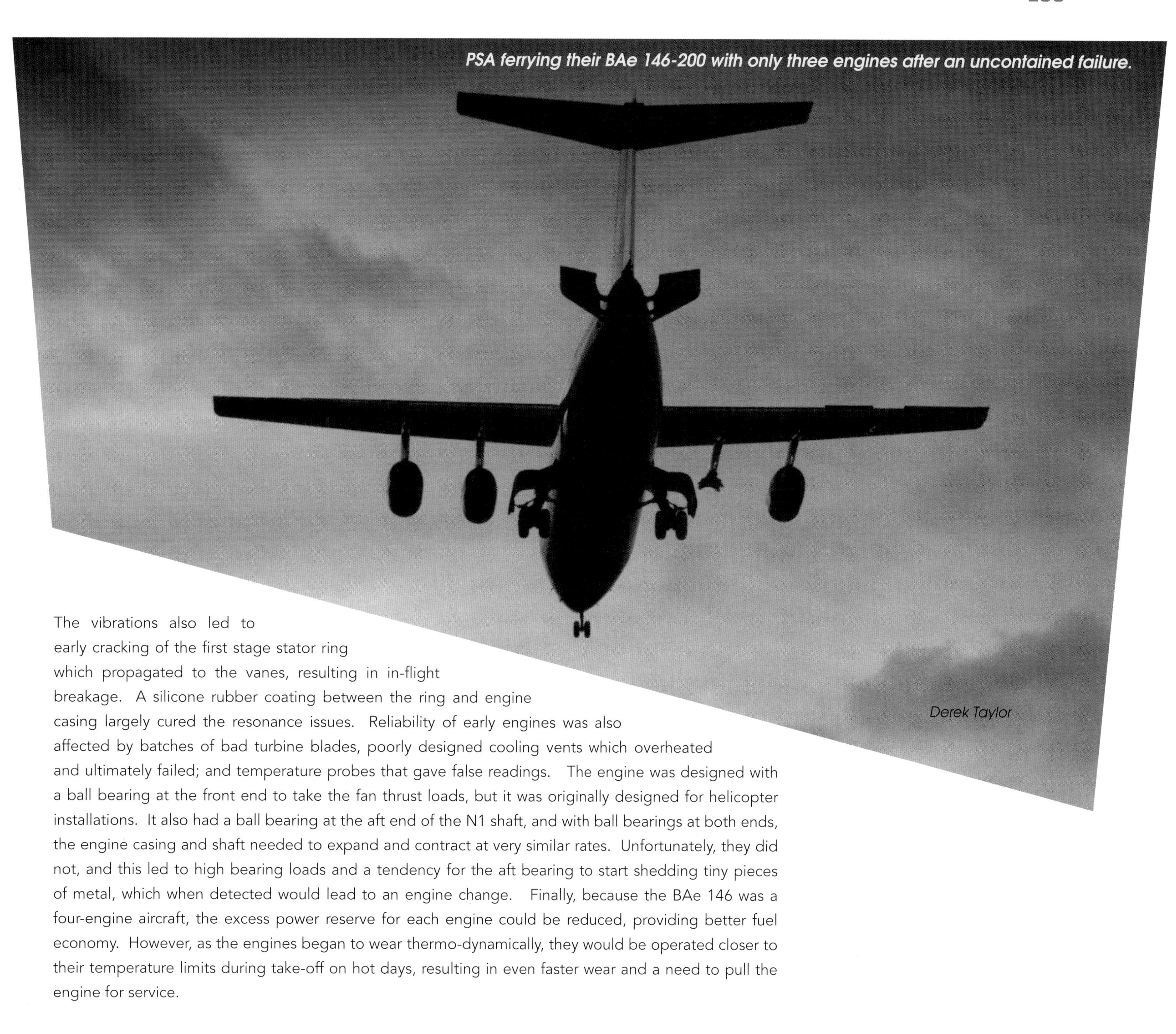

PSA ferrying their BAe 146-200 with only three engines after an uncontained failure.

Derek Taylor

The vibrations also led to early cracking of the first stage stator ring which propagated to the vanes, resulting in in-flight breakage. A silicone rubber coating between the ring and engine casing largely cured the resonance issues. Reliability of early engines was also affected by batches of bad turbine blades, poorly designed cooling vents which overheated and ultimately failed; and temperature probes that gave false readings. The engine was designed with a ball bearing at the front end to take the fan thrust loads, but it was originally designed for helicopter installations. It also had a ball bearing at the aft end of the N1 shaft, and with ball bearings at both ends, the engine casing and shaft needed to expand and contract at very similar rates. Unfortunately, they did not, and this led to high bearing loads and a tendency for the aft bearing to start shedding tiny pieces of metal, which when detected would lead to an engine change. Finally, because the BAe 146 was a four-engine aircraft, the excess power reserve for each engine could be reduced, providing better fuel economy. However, as the engines began to wear thermo-dynamically, they would be operated closer to their temperature limits during take-off on hot days, resulting in even faster wear and a need to pull the engine for service.

Former BAe Chief Executive Officer and Managing Director, Sir Raymond Lygo, joined shortly after PSA began taking deliveries of its large 146 order. In his book 'Collision Course', Sir Raymond mentions he was given sugar-coated reports from management within BAe that PSA was satisfied with the performance, and all was well. This was a common occurrence within BAe's regional jet group which was very hierarchal with managers being told what they wanted to hear rather than what was actually happening. This resulted in no one feeling able to stand up and say, "you're on the wrong track." Frustrated he was not being told the entire story, Sir Raymond flew to San Diego to meet with PSA Chairman, Paul Barkley, to discuss how things were truly working out with the 146. It turns out that the starter was proving to be a major problem in daily operations.

Dan Air and Air Wisconsin observed starter failures (Dan Air 15 in 9,500 engine hours; Air Wisconsin 8 in 9,000 engine hours), and as a result started carrying spare starters to avoid aircraft being stuck away from the maintenance base. Engine starter motors were not completely reliable, but were a part that was listed as requiring less than 30 minutes to change according to the diagrams and manuals provided by BAe,

PSA ferrying their BAe 146-200 with only three engines after an uncontained failure.

Derek Taylor

albeit with specific tools. PSA asked Sir Raymond to demonstrate how it could be done quickly. The clearances to get to the starter were very tight, and accessibility to it was difficult given how far inboard it was from the mechanic's arms. It did not help that there were also four of these starters (one for each engine), which increased the negative impression. Due to how the starters were installed, theoretically the only way to change the oil was to invert the engine or aircraft! Additionally, Plessey increased the frequency of oil changes to every 20 hours for each engine, which was ridiculously short for commercial airline operators. A temporary solution was devised that involved removing the starters to change the oil, then reinstalling them. Clearly this was not a good approach for users of the 146 and brought a very rapid and dark cloud on operations of the aircraft, and the issues were contributing to less than ideal dispatch rates. There was a design flaw with the clutch connecting the motor to the shaft which started the engine high pressure spool rotating. The original Plessey clutch was incompatible with the shaft and kept breaking it. Once a new subcontractor was sourced, modified starters were designed and tested, and then rolled out to airlines and on new build aircraft. All of these problems were eventually addressed with the 502R5 and all 502R7 engines (for the -300).

Spare engines in reserve at the maintenance base for AirCal.

Al Boring

Engine cowling damage from the uncontained failure.

Derek Taylor

Engine teething troubles were widespread amongst early operators, and with its largest customer unwilling to have grounded aircraft, BAe provided two -100s on loan (painted in PSA colors, but with the demonstrator interior in six-abreast seating), and PSA maintained reserve crews at San Francisco and Los Angeles to dispatch the aircraft in the event a 146-200 went 'technical' (flight cancelled due to mechanical issue). BAe provided the aircraft at no charge, while PSA mechanics maintained both loaners. It was reported that someone at PSA coined the term "Bring Another Engine" (after BAe initials) when there was a problem with an engine. It also became apparent that operators were not addressing engine issues while they were installed on the aircraft, contrary to the design philosophy. Instead, airlines were just removing problematic engines to work on later while installing a working powerplant. It was faster, easier, and some of the common issues were not as straight forward to address while an engine was on the aircraft, despite what the BAe literature attested to.

AirCal, even after disclosing how unhappy it was with the engine reliability early on, actually ran out of working spare engines to swap with ones that had gone down. At the time neither Avco-Lycoming nor BAe had any spares as all new engines were being allocated to new build aircraft. This resulted in AirCal having to rebook passengers on other flights, or in some cases having to use a bus to get them from San Diego to Santa Ana. The running joke amongst former AirCal employees was that the airline bussed more passengers than it flew with the 146! This continued into American Airlines acquisition of AirCal, and finally hit a breaking point. During a quarterly meeting with senior executives at American including CEO Bob Crandall, when the time came to discuss the 146, Bob turned to David Banmiller and said, "I've picked up some scuttlebutt that there's been operating problems with the 146." David responded "Yes, the problem is engines. We're out of spare engines, the manufacturer is out of spare engines, I can't get spare engines." Bob responded, "What are you doing about it?" David said, "We're working on it, we're putting as much pressure on both manufacturers as possible."

Bob Crandall was notorious for being a very tough CEO, but he was solutions oriented, and when someone responded, "we're working on it", that just did not fly with Bob. Bob told David "You need to do something right away" and David asked, "what do you suggest?" Bob looked at David firmly and said "I want you to go upstairs, and I want you to call the Chairman of British Aerospace, and tell him you're going to ground the fleet unless we get spare engines right away" (such action is always a very big deal and a black eye for a manufacturer which is certain to grab industry press headlines). Bob finished with "You've got to solve the engine problem." David responded, "well we've tried, but we haven't gotten anywhere." And Bob responded, "try it again." David remembers looking at Bob and said, "What if I can't get him?" Bob replied, "Call the Queen." David succeeded with his call to the Chairman, and the airline got its spare engines. David recognizes that it was the power of American Airlines, the proverbial 800-pound gorilla, not regional carrier AirCal, that likely had the impact. Because who knew, American might have ordered more BAe 146s, right?

The fuselage of the aircraft was pierced with shrapnel from the engine.

On February 27, 1987 PSA experienced the first uncontained engine failure on a BAe 146 (N356PS E2039) during airline operations, with the #3 engine sending shrapnel flying from the turbine into the fuselage and overhead luggage bins. Luckily the flight was not full. The shrapnel was embedded in a suitcase in the overhead bin, some seat backs, and a few bits of it even smoldered on the carpet floor, and yet no one was injured. The failure was caused by overspeed after a reduction gear decoupled following a lubrication fault. In summary, if the fan disconnected from the turbine, a turbine overspeed indicator sensor would detect and shutdown fuel flow immediately. What actually happened was oil that did not cool the bearings down and the overspeed sensors melted, resulting in turbine overspeed and eventually an uncontained engine failure. The oil had been bypassing the filter for over 40 days. The NTSB ruled that the failure of PSA's maintenance personnel to adequately examine the lubrication system after contamination consisting of ferrous/iron chips was detected in the oil was the cause. The BAe technical team that was dispatched from England was surprised upon arrival as they thought the engine had fallen off in flight. What had actually happened was that the aircraft had been ferried back to San Diego on three engines, at low altitude (as the fuselage could not be pressurized), with the defective engine in the lower cargo hold. As a result of the incident, the engine's time between overhauls (TBO) was reduced which increased maintenance costs. USAir cited the 146's per flight hour maintenance cost as the highest in the fleet, which included 30-year-old airframes as well as CATIII and ETOPS aircraft.

All photos on this page courtesy of Derek Taylor

Tape marks the entry point of each piece of shrapnel.

The shrapnel even pierced the overhead bins, landing on the carpet still smoldering.

Leading edge of the wing where the shrapnel pierced the wing.

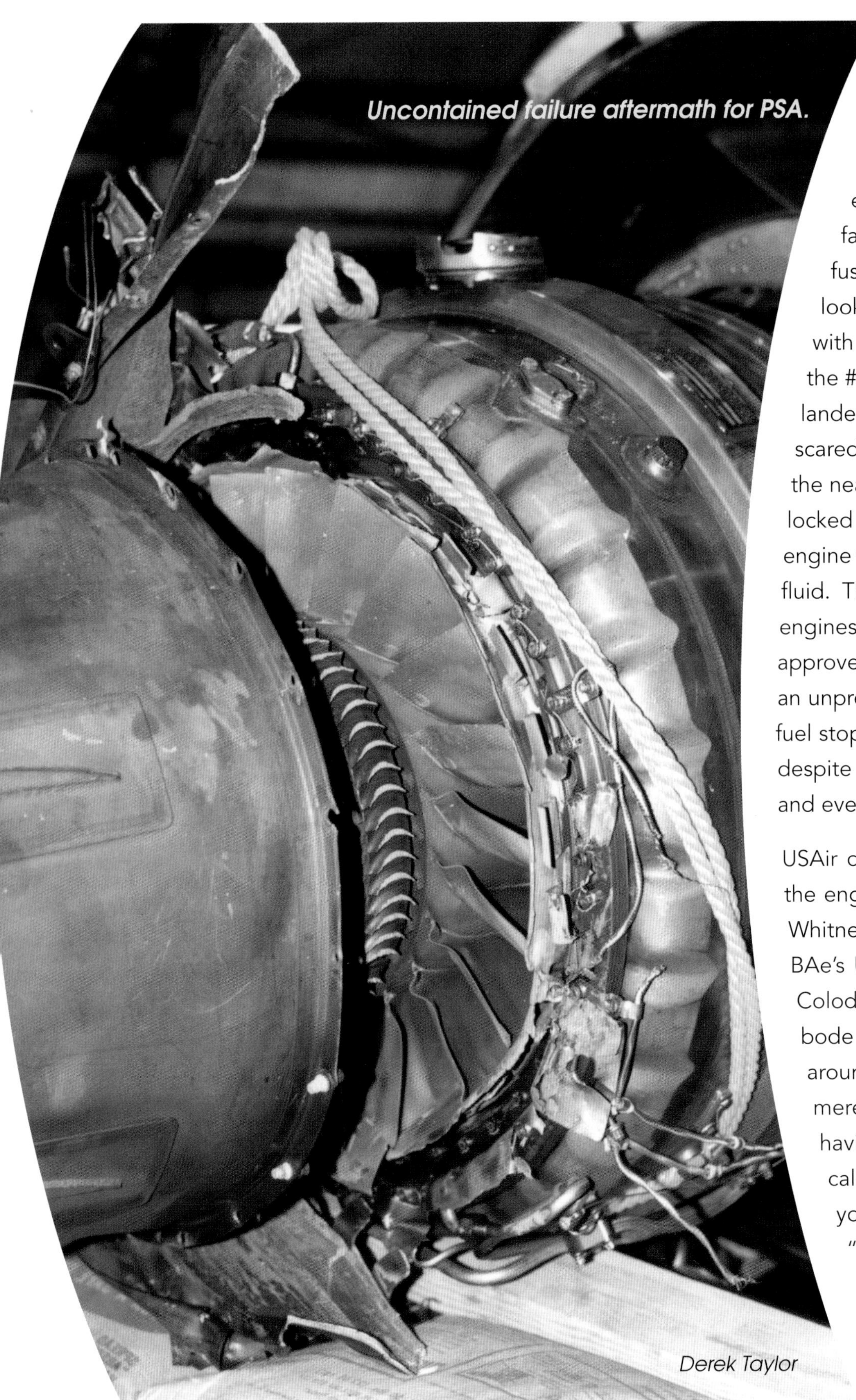

Uncontained failure aftermath for PSA.

Derek Taylor

PSA was not the only airline to experience an uncontained engine failure, or a three-engine ferry. In 1991 BAe 146-200 E2066 XA-RTI flown by Mexican carrier Aviacsa had a massive uncontained failure of the turbine wheel on engine number 4 while cruising at 24,000 feet, 20 minutes after takeoff. The failure threw shrapnel into engine number 3, the pylon, the wing root, the fuselage, and the leading edge of the wing; the damage was substantial. It looked as if someone took a machine gun and sprayed the side of the fuselage with bullets. The damage was so significant that it severed the throttle cable of the #4 engine, and the #3 engine fire shut off cable was also severed. The aircraft landed with thankfully no injuries to the passengers or crew, but the event likely scared everyone on board from flying for quite some time. The aircraft diverted to the nearest airfield (one without airline service), and upon landing the aircraft brakes locked up and the tires left a trail of rubber down the runway. It was alleged that the engine failure resulted from topping off the engine with piston engine oil, an incorrect fluid. The aircraft was ferried to Little Rock Arkansas from Campeche Mexico on three engines by Dan Gurney with the damaged engine in the belly hold. The trip was approved for overland flights only (not venturing out over the Gulf of Mexico), with an unpressurised cabin due to the damage, and with the landing gear down. After a fuel stop in Brownsville, the flaps stuck at 24 degrees. But it was a tough aircraft and despite all the damage it had endured it soldiered on to its final destination for repair and eventually reentered service with Air Atlantic.

USAir cited high operating costs of the 146, specifically related to maintenance of the engines which required higher than normal repairs in comparison to the Pratt & Whitney JT-8Ds installed on its MD-80 aircraft. Sir Raymond became Chairman of BAe's U.S. division shortly after USAir purchased PSA, and began dealing with Ed Colodney, USAir's CEO, regarding the threat to ground the aircraft which would not bode well for British Aerospace. Ed's primary complaint was dispatch was hovering around 98% when it needed to be 99.5%. That may not seem like a big deal, a mere 1.5%, but in the highly competitive airline environment, cancelling a flight and having to reaccommodate up to 85 passengers costs the airline money. Ed even called David Banmiller at AirCal (after it was bought by American) and asked, "are you guys having as much trouble with the 146 as we are?" David responded, "yes."

Ed continued to complain about the 146's high operating cost due to the engines needing so much maintenance, which was resulting in cancelled flights. Sir Raymond arranged a meeting with Avco-Lycoming and its new parent company Textron in New York, at the Textron corporate offices (Avco Lycoming became Textron Lycoming in 1987). In the meeting Ed made his case to the engine supplier, noting all of the issues and operational problems that were occurring. It fell on deaf ears. According to Sir Raymond in his autobiography, Ed Nolan, Chairman of Textron retorted, "You must have known about these airplanes when you bought them, you knew about the engine reliability, don't expect me to pull your chestnuts out of the fire." Mortified is an understatement of the reaction from BAe's Chairman, and in front of a customer no less: a manufacturer clearly admitting it had a poorly performing product, the most crucial one on the aircraft. Sir Raymond began immediately looking for an alternative to the Avco Lycoming engine, and had some fruitful initial talks with Rolls-Royce, with discussions surrounding the RB580 which was based on the successful RB211 design. But in the end, BAe was stuck with the Avco engines for as long as it were committed to a four-engine jetliner. Sir Raymond had these parting words to say: "If ever there was a lesson to be learned in the design of commercial aircraft, it is as Alec Sanson, BAe's then Sales & Marketing Director, said: 'Commercial aircraft hang on the engine – not the other way around'."

Falling in line with Ed Colodny's "Ten Commandments", commandment number 3 "Thou shalt keep an efficient fleet" came to pass. As a result of the high maintenance costs, unreliable engines, and the cutthroat market on the west coast, USAir grounded its entire fleet of BAe 146s on the same day. Former Customer Support Field Rep Derek Taylor said that nearly every 30 minutes a BAe 146 arrived to be parked, noting that two aircraft arrived simultaneously, and flew in formation wingtip-to-wingtip over Mojave before landing. This caused a lot of negative publicity for BAe and made it difficult to sell new build aircraft. It would be nearly four years, with the world economy struggling, before the first aircraft of the fleet left Mojave for good and began a new life with another airline. Six years after the first USAir 146 arrived, the last one left ending the negative image a fleet of fairly new aircraft were bringing to British Aerospace.

It was widely reported most airlines were not doing work with the engines on the wing as BAe intended, but instead were pulling the entire engine off, even in the field, and swapping it with a functional engine (and working on the problematic engine offline). According to mechanic Kevin Govett, the magnesium compressor stage was the Achilles heel of the LF-502 engine – a steel compressor replaced the magnesium compressor in the LF-507, improving performance and reliability. Problems were experienced not only by west coast U.S. airlines, but international carriers as well, as seen in a cartoon by Ansett W.A. (Bring another engine – with a photo of a PSA plane and six engines under each wing). In fact, Textron stated that over a three-month period from August to October 1986, the unscheduled removal rate was 0.256 per 1,000

Like the PSA flight, Aviacsa also experienced shrapnel explosion impacting the wing.

Derek Taylor

hours, three times worse than the rate for the Rolls Royce Spey 555 on the Fokker F.28 which is a two-engine aircraft. Factoring in the BAe 146s four engines, made for a six-time greater rate.

West Coast U.S. operator AirPac purchased one BAe 146-100 aircraft in May of 1983, with plans for more. But after barely two years of 146 operations, the company filed for bankruptcy and BAe repossessed the aircraft. A few years later in September 1989, AirPac sued both British Aerospace and Textron-Lycoming in U.S. Federal Court for $100 million, claiming the aircraft was unreliable and operating costs far exceeded what BAe and Textron promised. The suit alleged that both BAe and Textron knowingly misled AirPac about the reliability and potential profitability of the 146, and deliberately concealed the problems being experienced by other operators, and developmental issues. from AirPac. The $100m amount was derived from estimated lost business due to grounding of the aircraft. AirPac asserted that it relied heavily on the presentations and representations of BAe and Avco Lycoming regarding low maintenance costs, reliability, and other "superior characteristics" of the 146. BAe countersued for $7m for unpaid leasing charges, spares and breach of contract. BAe stipulated that poor management and competition was the result of AirPac's bankruptcy, not its aircraft. Spencer Hall, attorney for AirPac. Said. "A key representation (by BAe) was that the engine was a mature, low risk engine with a proven track record." The ALF-502 was not, however, used on any other commercial jetliner. AirPac submitted claims for over 600 delays or cancellations due to mechanical issues with the aircraft. In a move to protect itself, BAe filed suit against Textron Lycoming aimed at establishing liability if the court decided in favor of AirPac. After two weeks on trial and before the conclusion, BAe settled with AirPac for an undisclosed sum, with the court mediating the settlement.

Once BAe had repossessed the AirPac aircraft, it was sent to AAR aviation services company, where mechanic Kevin Govett began to service it, the first BAe 146 he had worked on in his career. Kevin had heard rumors that AirPac was operating its 146 on three engines because it could not afford to fix the aircraft, and because the engines were consuming so much oil. The cargo holds were in very rough shape from all the seafood that had been carried, the salty ocean air, and the underside of the aircraft was "beat to fire" from the gravel runways it operated from, even with the rough field kit (Kevin mentions the kit was not very well designed back then). After refurbishment, the aircraft provided temporary capacity for PSA when it had an aircraft down.

The Queen's Flight noted the concerns that airlines had with the 146 during initial delivery, but it seemed that the majority of issues were addressed by BAe and Textron-Lycoming prior to ZE700 entering service. Captain Graham Laurie noted that in over 5,000 hours of flying the 146 for the Queen's Flight, he never had to shut an engine down. Early operator TABA said the aftersales relationship with BAe and Avco-Lycoming was poor, and the airline difficulty in securing replacement parts. BAe disputes this by saying the real issue was TABA's refusal to pay the high tariffs the Brazilian government was imposing (100%) on spare parts. Just two months after TABA began flying both of its 146s, one was grounded, cannibalized to keep the other flying. Both were repossessed in 1985.

Quite a few operators later in the life of the BAe 146, like WestAir for example which purchased six BAe 146-200s, insisted on performance guarantees for the aircraft and the engine. This was the result of the under-current in the industry that the engines were problematic, even in the pre-internet days. Former founder Maury Gallagher said that without those guarantees, the 146 would have sunk the commuter airline which was running as United Express out of Fresno. But former BAe commercial executive Peter Connolly said the performance guarantees were not worth the paper they were written on. Peter mentions that even when they were included in a contract, it was in practice impossible to assess if anything was due given the complexity of the issues. Meetings to review any claim could go on for hours without resolution as the parties disputed the true cause of any particular delay. In reality, when things went wrong in a big way, the airline CEO would ring up the most senior person he knew at the manufacturer and give them hell until being bought off with money, more technical support, or a combination of the two. Maury remembers that while the passengers loved the aircraft, and that it had the smoothest landing of any airplane flying then and now (due to trailing arm gear and high wing), at the time the engine was 'junk'.

Gary Ellmer who joined WestAir from Royal West noted that the engine problems resulted in camaraderie between the California airlines. PSA, WestAir, and AirCal (and subsequently Aspen and Air Wisconsin) maintenance crews began freely sharing information and

Aviacsa aircraft that experienced an uncontained failure damaging the fuselage, wing, and engine #4. Note the soldier holding the machine gun to the far left.

The aircraft brakes locked up on landing.

Derek Taylor

operational tips to alleviate the pain the BAe 146 was inducing. AirCal and PSA crew were used to Boeing and McDonnell Douglas aircraft where all that was necessary was 'turning the key on, and it ran'. The BAe 146 suffered from something every day, and it was a referred to as a "maintenance pig" in comparison to other aircraft. AirCal and PSA mechanics grew to dislike working on the 146 because it was a 'pain in the ass' and ended up amplifying the problems they saw with the aircraft as a result. For Air Wisconsin and WestAir, this was their flagship aircraft, and thus the carriers had to live or die by it.

Gary echoed a common theme of other operators, that the 146 was fairly predictable, and you had to get ahead of the problems by fixing them before they happened. The gearbox proved problematic and oil samples were taken constantly. Doing so was not an effective or predictive tool though, and staff spent too much time doing it at BAe's recommendations. But, if something was detected in the oil sample, failure was clearly imminent. At the time it was noted that Textron-Lycoming would point its finger at BAe saying it was the aircraft's problem, while BAe would point the finger at Textron saying it was the engine problem. This was one of the issues that Sir Raymond Lygo highlighted in his memoir, that people were too busy blaming someone else rather than owning up and dealing with the issues at hand.

All photos on this page courtesy of Derek Taylor

BAe and ultimately Honeywell (after acquiring Textron) eventually addressed the majority of issues with the 146. Initially Honeywell and BAe went on a tour to operators and again cautioned them against using full thrust take-offs if they wanted to avoid wear-and-tear on the powerplants. As a result, reliability increased substantially resulting in a doubling of the time between overhaul (TBO) on the engines, which was generally 2,500 hours but increased to 5,000 hours. The LF-507 addressed many of the design and reliability shortcomings, and over time those improvements and parts found their way back into ALF-502 engines, and the aircraft went on to serve for many more years with airlines like Swiss, Cityjet, and Cobham.

All photos on this page courtesy of Derek Taylor

Passenger Cabin Complaints

Early aircraft received less than favorable feedback from passengers when it came to three specific areas: noise, lavatories, and seating.

While the aircraft might have been the world's quietest aircraft for neighborhoods underneath its flight paths, inside it was anything but quiet for passengers. Early operators including PSA all requested redesigned door seals as the rush of air going past created too much white noise. Prior to PSA taking delivery of its first aircraft, the carrier took note of what other operators were reporting and required BAe to work on insulating noise from the cabin interior. The inner flap-track, when flaps were being deployed through the first 18 degrees of travel, created a distinctive howl in the cabin. Further seals were designed to minimize the noise. The lavatories also had very noisy ventilation and suction systems, and Air Wisconsin asked BAe to produce a variable speed motor for the cabin ventilation system because of the noise it caused.

COMPLAINT OR DIFFICULTY REPORT

Report Number 3133
A.T.A. Code 25-40-00

AIRCRAFT
Aircraft Type BAe 146-200
Reg. No. Prod. No.
Operator P.S.A.
Airfield
Date Occurred
Hours Landings
Urgent Normal ans. Req.
Date of Report 31.8.84
Reporters Ref.
Reported By P.S.A.
Reason for Priority if Urgent

COMPONENT
Name TOILET ARRANGEMENT
Part No.
Serial No.
Mod. State
Location in a/c
Hrs Ldgs since o/h /
Total Hrs Ldgs /
Location of Part
Associated Warranty No.
BAe Order No.

Symptom, Cause & Rectification (Photos/Sketches/Rubbings attached):-

The following is a Passenger Comment received by P.S.A.:-

As a frequent flyer, I boarded one of your new BAe 146 "Smiley liners" and when I went into the toilet I found little to smile about. Your British designers either don't know or don't care about Ergonomics or they built the damn thing for British midgets - not for tall Americans. If you happen to be over 5' 5" and want to piss you are in real trouble! Since the side walls of the plane curve in so sharply, you have to twist your head out of the way to get close enough to hit the bowl, then you can't see where you are peeing! If you want to see where you are peeing you have to stand back far enough so that of you are middle-aged with a large prostrate the arc of the stream doesn't make it into the bowl and you wind up peeing all over the seat no matter how hard you try not to! I am happy to pay for a ticket but I'll be damned if I'm going to mop up your toilets.' Solution: Put up large signs all over warning all males over 5' 5" who are over 40 to pee before getting on board or suffer the indignity of having to pee sitting down!

Product Support Note

This same point was raised in Appraisal Report No. 37 dated 12.3.73.

Reply

Mod HCM30033E.F raised to introduce restyled modular toilet S00 P.S.A.

LSC

AIRCRAFT GROUP HATFIELD HERTS

Complaint written to PSA about the bathroom, subsequently passed onto BAe.

Seating was another issue, specifically for PSA. With a six abreast configuration, customers found the cabin to be a tight squeeze, especially for those seated by the window. PSA ordered seating that was too tall which resulted in those next to the window finding their head and shoulders cramped as the fuselage side walls curved into their space. PSA ordered new seats that were approximately 2.5" lower and addressed the issue, and also provided additional seat width. The carrier moved to a five-abreast layout, bringing the 146s into line with its MD-80 aircraft (which passengers favored). The lavatories in the 146 were a tight fit for most passengers, although being a short haul aircraft the design was a bare minimum. On August 31, 1984 PSA forwarded to BAe a complaint sent by a passenger which alleged that anyone over 5'5" was unable to stand up straight while relieving themselves, and that the only way male passengers could save their dignity (instead of sitting down to relieve themselves) was to use the restroom in the airport terminal prior to the flight. BAe's internal memo suggested installing a restyled modular toilet S00 P.S.A from modification HCM30033E.F. Ironically the same memo stated at the bottom of the page "same point was raised in Appraisal Report no. 37 dated March 3, 1973", when it was still in the design phase as the HS146.

Typical bathroom on the BAe 146 which is quite tight.

Other Issues

Both Dan Air and Air Wisconsin experienced an unreliable Thrust Management System (TMS) in the early days. Engineers and pilots were struggling to understand what the issue was and ultimately traced it to unreliable Garrett electrical actuators which form the link between the engines and the TMS. Rather than going into a neutral position when not in use, they had been doing the opposite. Other issues included tolerances that were too liberal, resulting in false warnings and indications. The TMS system was considered unreliable and reportedly the rod never worked correctly. Mechanics allegedly could change the part with their eyes closed because the work had to be done so frequently. The autopilot built by Smiths Aerospace at a cost of $3 million USD was reportedly also a poor performing system (described to the author as a "marginal shit system" by a former BAe Customer Service Rep). BAe reportedly got what it paid for.

Flaps jamming was another issue, which was even experienced on The Queen's Flight aircraft but switching to a grease that had a lower freezing point seemed to 'cure' the problem. Other areas of concern for airlines were windscreens cracking, caused by the windshield heating system, but this too was rectified early on. The heating elements could also be seen in the glass, giving a slight distraction visually. Quality of the cockpit windshields was poor early on, with some having optical distortions depending on the angle at which pilots were looking through it.

Galley redesigns were requested so that a higher level of service could be offered versus what was possible from the earlier facilities, enabling longer segments and a more robust hot food and beverage service for carriers like Ansett Australia and Air Malta. A request was received to redesign the cabin interphone panel because there was not enough differentiation initially between different types of crew calls. PSA reported the early systems made cabin announcements difficult to hear because of static present on the public address system. Air Wisconsin experienced problems with the baggage restraint net system, which was slowing down the quick turnarounds the carrier needed. Air Wisconsin asked BAe to redesign the system to support fast turnarounds. BAe held operators conferences every two years during which operators of the BAe 146 would convene and the manufacturer would address issues that had been submitted. These events also gave customers a chance to discuss amongst themselves what they had learned about operating the aircraft, sharing direct experiences and solutions.

Brakes in the early days started out with steel as the main component, but later they were switched with carbon because it was significantly lighter (about 800 pounds total). But during the early life of carbon brakes, dust was prematurely destroying bearings and axles. Eventually BAe developed not only improved seals, but also adjusted the compound to reduce oxidization. Fans were also employed by operators such as PSA that performed short turnarounds to cool the brakes down quickly. The aircraft's doors were substantial items and were very heavy to operate. What made the doors even more difficult to open were the seals around the sides. Originally lubricant was used to keep the seals from deteriorating, but the result created such a good seal that moving the door became incredibly difficult. It was later determined simple talcum (baby powder) would protect the seals without creating a glue-like seal.

Toxic Fumes

Toxic fumes are an ongoing issue in the airline industry, and the BAe 146 was no exception. It was alleged that the 146 and the Boeing 757 were the worst offending aircraft, creating fumes that resulted illness both temporary and permanent in some airline staff. It's a very sensitive issue but is being included here.

Dan Air was one of the first airlines to report issues relating to smoke in the cabin from oil and acrid smells to the Civil Aviation Authority (CAA). In response, the CAA reportedly claimed that BAe had taken care of the issue. PSA says that another concern was of oil migration from the APU into cabin air system during operation. Cabin crews who were exposed to cabin air for long periods of time complained to managers, and the problem was supposedly rectified by September 1984.

The start-up procedures required learning and finesse to avoid filling the cabin with a carcinogenic smoke and acrid smell related to the temperature controllers. Following procedures outlined by BAe resulted in the dreaded smell with light smoke.

PSA pilots learned that if they managed the temperature controllers manually, and followed a slightly slower procedure, the issue would be avoided altogether. The Number 2 engine bearing seal was determined to have been causing oil smoke in the bleed air feed to leak into the cabin air conditioning system.

In 2011, a documentary film "Broken Wings" was released purporting to document the toxic fumes and their effects on some cabin crews on the BAe 146. It described oil used to lubricate components within the turbine was bled off the engines and contaminated the cabin air supply with toxic fumes inflight. The documentary was produced by Tristan Lorraine and Susan Michaelis of Fact not Fiction Films, both very outspoken pilots who are no longer licensed due to health issues. BAE Systems alleged that both Tristan and Susan were biased, noting that the film was not objective enough for mainstream TV viewing.

In a letter sent to Avco Lycoming on April 22, 1985, J.M. Smith from Mobil Oil noted that Mobil Research and Development Corp (MRDC) had conducted tests relating to odor formation, and determined it was likely to occur when oil was exposed to a condition that resulted in thermal degradation such as leaking past a seal into a compressor (rather than a long-term process such as oxidative degradation). This came in relation to PSA's feedback. Of note was that Mobile stated that Mobil Jet Oil MJO 254 was not approved in the ALF-502, and PSA requested permission from Avco Lycoming to use it, despite using also MJO254 in its MD-80 engines. Avco required additional testing prior to granting approval, and Mobil Oil adjusted the formulation to reduce odor from the burn off of oil. When synthetic oil is heated it breaks down into a number of chemicals, one of which – Tricresyl Phosphate (TCP) – is in itself a lubricant and an anti-wear agent. Side effects of TCP's toxicity is numbness in legs and hands and even temporary paralysis, but long-term exposure could result in permanent damage due to it causing the death of nerve cells. In January of 1986, Ansett cabin crew experienced irritating smoke in the cabin after APU bleed air was used for air conditioning. Once the crew switched to packs 1 and 2 the problem disappeared. BAe issued a directive, Air conditioning ATA document 21-50 (BAE, December 1984) to Ansett, which allegedly provided for workarounds if contamination was present or suspect. In December 1984, BAE produced a revised ATA chapter 21-50 for Ansett relating to air conditioning, providing workarounds if contamination was present or suspect. ATA (Air Transport Association) developed a numbering system for aircraft bulletins, and ATA 21 was the category for Air Conditioning under Aircraft Systems.

In 1991, Dan Air crew operating a -300 experienced an issue with fumes and smells, and the airline wrote directly to BAe/Hatfield with this opening question: "Can Hatfield provide a definitive statement on the medical implications of fumes/smells in the cabin (Dan Air cabin crew have complained of headaches and nausea)?" BAe responded that "SB (Service Bulletin) 21-70-01316A addresses

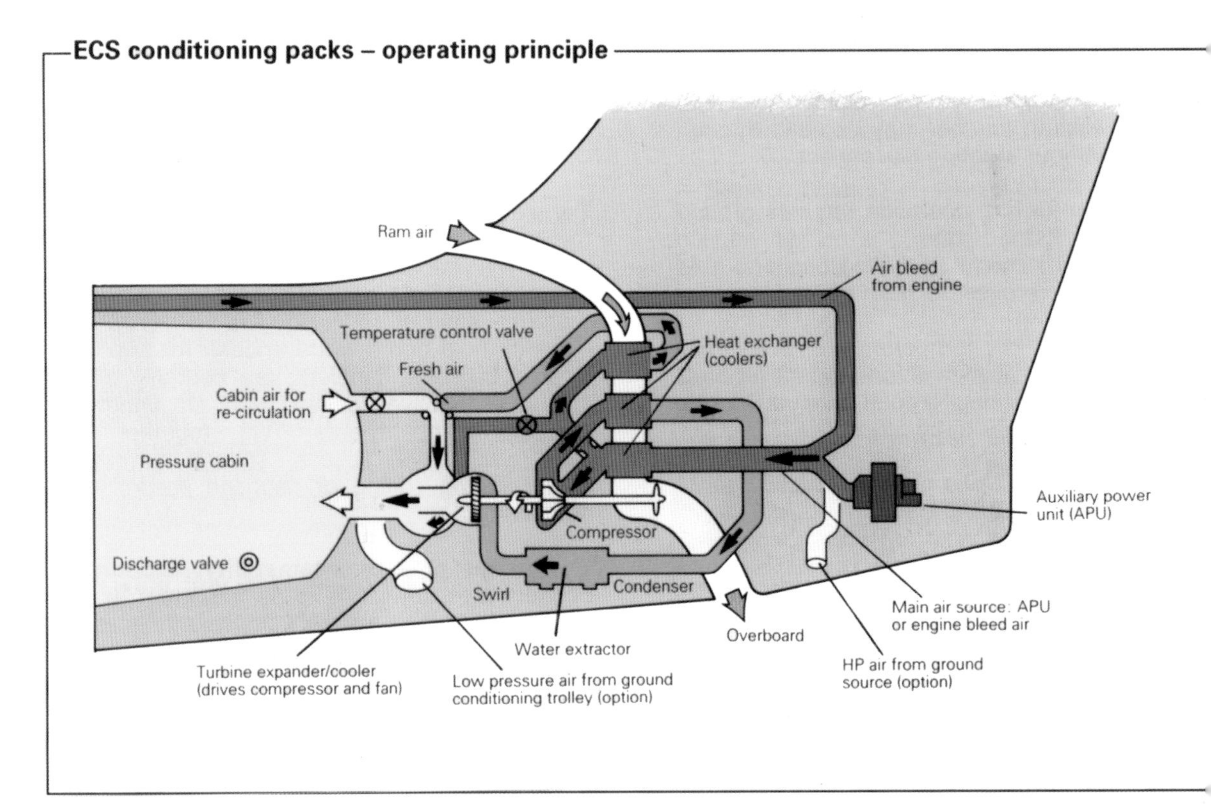

cabin smell problems by installing two coalescers upstream of the E.C.S packs (1 per pack). These prevent oil contamination of the packs. Downstream of the packs in the forward air conditioning bay is a carbon impregnated cloth filter, which cleans the air prior to it entering the cabin. A similar filter albeit smaller is positioned in the hydraulic bay to filter the air supply to the flight deck. With this modification, it is not necessary for catalytic converters in the engine and APU lines." Garrett introduced fixes including reworking of the compressor shaft bellows seal, development of buffered compressor shaft seal, and drawing down of APU gearbox oil pressure.

So, if a seal does fail then the gearbox would tend to draw air in instead of letting oil leak out.

Former BAe test pilot David Thackery acknowledged that the manufacturer was aware of bleed air issues. Dan Air stated in an internal memo that at a BAe 146 operators conference (the fifth to occur) in June 1991, its attendees raised the issue again. The following points were made:

1. It was a serious issue plaguing the -300 series aircraft, and equal in magnitude to technical problems that affected the BAe 146 in early years;

2. Written and verbal complaints had been received from passengers, some directed to the Chairman of the airline.

3. Dan Air was losing passengers once they found out would be traveling on the BAe 146, they transferred to other airlines to avoid the aircraft.

4. Cabin staff complained of headaches, nausea, sore throats, and were referred to the company doctor.

As a result, BAe and Garrett issued directives that the APU air bleeds not be used. Dan Air shot back that this was unacceptable as the aircraft were operating to destinations such as Gibraltar where ground conditioning was essential for passenger comfort. The airline was stuck with the bill for the excessive man-hours used to remove and refit APUs, as well as sealing modifications in the APU bay.

At the same conference, EastWest airlines was equally as vocal regarding the same matter. The carrier went further saying one of its passengers had a very bad reaction resulting from the problem and went on to advise that after take-off (on all flights) captains were to make a PA announcement apologizing for the 'sweaty socks' smell. Shortly after EastWest set up an "BAe 146 odor inquiry committee". But BAE Systems has said these were not true 'fume events' that required the donning of oxygen masks.

In July 1991 Garrett tested the cabin air to determine the cause of complaints from operators at the conference. The firm tested an aircraft that was being turned around for another flight to have a baseline. Later the same evening odor problems started occurring and once the aircraft landed was towed to the hangar for further testing. BAe informed Garrett that air coming from the vents in the cabin is recirculated if the Environmental Control System (ECS) is set to full fresh air. Fresh air enters at the top of the cabin near the light fixtures. Garrett noted that APU s/n 186 was sampled the following morning after it was installed to replace APU s/n 113. The odor was the same, even in full fresh. Air was collected from the cockpit and the main cabin to be sampled and tested. In another test with the same APU, s/n 186, and engine #4 running perfume was poured into the APU inlet. The ECS system was still in recirculate mode. The perfume odor was detected immediately in the cabin and cockpit.

In the end following multiple test samples, objectionable odors were found to be worse in the cockpit than the cabin. It was less noticeable with all engines running and when fresh air was introduced into the cabin. There was a lot of back and forth with BAe and customers such as Ansett and EastWest. Barry Holden of BAe responded to continual complaints from Ansett and EastWest, stating that in "his belief it has always been the APU that is the source of contamination, and this belief is shared by BAe and made clear from the start." He went on to say, "BAe at their own ill afforded expense are introducing a costly filtering system to mask the rubbish coming from another vendors product."

But there were plenty of events, such as the one with VH-EWJ (-300) in August 1992 where after take-off, engine bleed air was switched on and immediately a cabin odor was evident and a flight attendant put on an oxygen mask. APU oil had leaked into the air conditioning system. During another event on September 30, 1993 black smoke was emitted from the cabin air conditioning ducts and passengers even believed a fire had started. Post flight one of the cabin crew coughed up blood, and previous same-day symptoms of fatigue, headaches, lack of concentration were exacerbated. The reports were numerous and could fill a book of their own. The first legal case relating to the BAe 146 was brought by former Ansett flight attendant Alysia Chew in 1995, against Ansett Australia and EastWest.

During 1997, a National Jet Systems (Qantas) VH-NJF was on descent when both pilots became incapacitated when smoke and fume contamination occurred. Not long after on July 23, 1997, pilot Susan Michaelis collapsed after a 146 flight, and following extensive medical testing was deemed unable to fly again. The doctors stated inhalation of oil fumes had caused extensive damage.

However, BAE Systems fought the claim and after six years, and Susan purportedly abandoned her proceedings against the firm, bearing her own costs without any payment made by National Jet Systems nor BAE Systems. Susan was the producer of the documentary "Broken Wings".

AirBC noted that it was having the same issues with oil fumes and odors in the cabin during 1997, with crew reporting identical symptoms: nausea, sore throat, headaches, and burning eyes. One case was even diagnosed as 'neuro-poisoning'. BAe told AirBC that no other operators were experiencing cabin crew complaints relating to air quality until shortly afterward when the firm admitted that it had 127 other reports (AirBC rhetorically said "probably from Ansett"). Reportedly 112 crew out of 200 suffered symptoms of toxic fume poisoning over a period of four months.

Richard Fox of Allied Signal said in an email to Ansett in 1997 that hydrocarbons were still being found in the cabin air of the aircraft, with Tricresyl Phosphate being detected and bleed air contamination monitor during pack burn was four times greater than allowed for engine acceptance at the APU facilities. Suggestions for further changes to operations and maintenance were proposed. An Australian senate inquiry was initiated in 1999, aimed at identifying the source of the odors and health problems. A separate study by Ansett determined that the air failed then new FAA requirements for carbon dioxide, noxious vapors, and contamination, and TCP is often found in the air.

Neils Gomer, a Captain at Malmo, flying at 22,000 feet, reported his first officer was not feeling well and was going to vomit. He donned his oxygen mask, but Neils Gomer started feeling the same way almost immediately afterward. He got his oxygen mask on right away, but was subsequently paralyzed in a slow descent, with about 35 minutes of fuel left. It was determined that during the tear down of an engine, a seal had failed. A lengthy investigation concluded that the pilots were affected by contaminated cabin air but laboratory analysis showed that contaminant residues that were present on the flight deck were still far below occupational health safety guidelines. Nevertheless, the Swedish CAA put in place plans for flight crews to deal with polluted cabin air, or even the suspicion of it, by donning oxygen masks immediately.

The CAA instituted a research program into cabin air quality in 2001 after a number of events had been recorded, including two on UK registered aircraft where the flight crew were partially incapacitated. In 2004, it issued a summary noting that evidence indicated that contamination of the ventilation systems by engine oil fumes was most likely the cause. The report contained photos of a brand new air conditioning duct made by BAe, and a duct that had accumulated 23,000 flight hours, noting that the latter was completely black internally from contamination. Subsequent CAA directives to replace seals on ducts, and from the APU have been ordered. The CAA's work was supported by previous investigations made in Sweden and Australia, as well as discussion with UK AAIB.

Numerous pilots and cabin crew working at airlines around the world experienced these problems. With so many incidents, those affected believe an industry wide cover up was occurring, not just with the 146 aircraft but involving all aircraft types. The British Airline Pilots Association (BALPA) issued a report, concluding with a request that contaminated air be addressed. The CAA issued an AD 002-03-2001 to inspect the aircraft's air conditioning system – engine oil seals, APU and ECS jet pump and air conditioning packs – for signs of oil contamination. The CAA issued subsequent directives, with the FAA doing the same in 2004, requiring repetitive detailed inspections inside of each air conditioning sound-attenuating duct, with corrective action to be taken as necessary. These directives were issued because there was a risk that the inhalation of agents from oil or the breakdown of oil could result in pilots losing control of an aircraft. Contaminated fumes incapacitated the co-pilot a Swiss flight into Zurich in 2005, a prime example of the issue the CAA directives were aimed at solving.

In 2007 Australian senator Kerry O'Brien says he had seen documents that suggest BAe and AlliedSignal paid money to Ansett and EastWest for their silence ($750,000 AUD). This was as liquidated damages in full and final settlement of any and all claims then or in the future which the airlines may have against BAe in respect of oil or other fumes adversely affecting the cabin environment in the aircraft. This was to apply whether under the terms of the warranties contained in the Aircraft Purchase Agreements (APA) or otherwise, or in respect of any failure by BAe to comply with its obligations under the APA in accordance with the terms thereof and any unavailability or unserviceability of the aircraft. BAe was also to provide assistance in respect of Ansett's and/or EWAs legally valid and justifiable claims in respect to costs actually incurred and damages actually sustained. Plus, BAe should be neither a party to nor should it be required to take part in any legal proceedings. BAe's obligation to provide technical data or to join with Ansett or EWA in any settlement discussions was to expire two years after the date of the agreement (September 1993). The agreement was confidential and was not to be disclosed by any party in whole or part without written consent of all parties.

BAE Systems acknowledged that a payment was made to an Australian operator of the BAe 146 (Ansett/EastWest), but it was neither secret nor 'hush money', instead saying it was for the inconvenience and cost of their investigations into cabin air issues, with a standard non-disclosure clause incorporated into the agreement. Even though this took place 21 years ago (1993), opposition groups still used it as the primary exhibit by as key in their evidence of wrong doing, and offered no further exhibits to support their position.

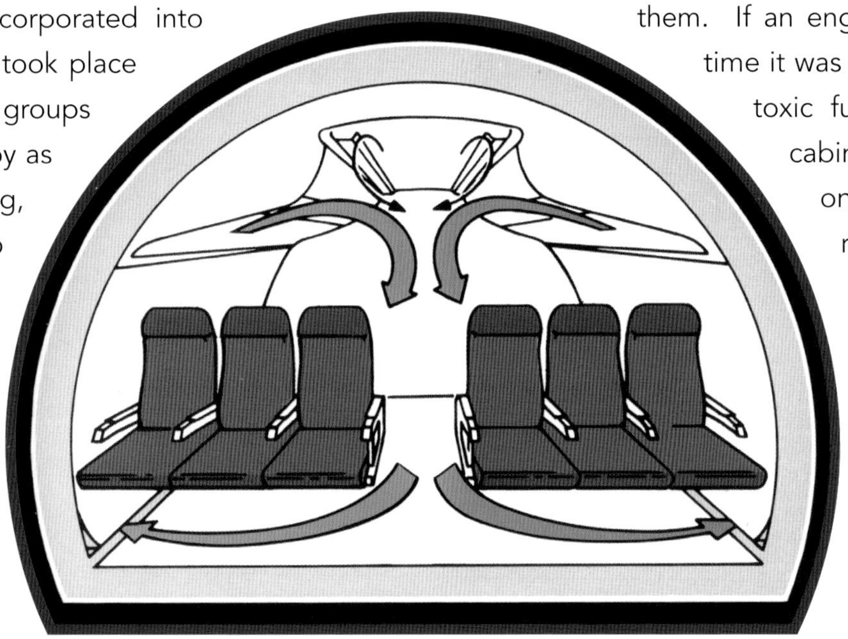

The case relied on reports of TCP being detected in the cabin or flight deck. The airline chose an aircraft and selected Dr. V. Vasak to analyze the cabin air. Vasak reported, "while the contaminants in the cabin can cause discomfort and be considered a nuisance, there is at present no evidence which would support the opinion that the odor would have lasting adverse health effects on the flight crew or passengers."

Diagram showing fresh air enters from above, and is drawn out by passengers feet.

Going further, Allied Signal (AS) had an additional agreement with EWA / Ansett in which it would provide up to $1.235 million in parts and labor credit with regards to operation of APUs on 146 aircraft. The credit would be limited to APUs and parts to convert 85-129(E) APUs to A-129(K) configuration. In exchange, EWA/Ansett agreed to terminate all disputes, differences, and claims. Senator O'Brien felt that BAe and the airlines had covered up the dangers posed by the aircraft to passengers and crews, and there is an assumption that other airlines were compensated for similar issues.

Captain Graham Laurie said that The Queen's Flight aircraft never experienced any of the toxic fume events, but he speculated as to one of the possible causes of it: topping up oils and lubricants too soon after a flight. The Queen's Flight engineers did not top up any oils or lubricants in the aircraft until a minimum of 30 minutes had elapsed after shutdown, to allow proper cooling of components. He surmised that during regular airline operations, hastily done work in between flights could mean the possibility of overfilling was more likely, resulting in spillage of fluids, and subsequently the burn off with fumes entering the cabin. Mechanic Kevin Govett believes that with so many different engineers at an airline touching the engines and aircraft, it becomes difficult to "be in tune" with them. If an engine or APU was leaking oil, by the time it was discovered, it had already produced toxic fumes that could have entered the cabin. There were HEPA filters install onboard the aircraft, but many airlines removed them as they were found to be problematic and did not rectify the problem. In another test with BAe using a Flightline aircraft and they discovered that fumes would appear when the aircraft was in line to take-off, the result of the 146 ingesting fumes from aircraft in front of it (with fumes dropping to zero after take off).

BAE Systems issued an internal, confidential 35-page report regarding the cabin air quality of the BAe 146/Avro RJ aircraft in 2014. It summarized that the various groups, politicians, and selected scientists had taken random events of being symptomatic of a system wide issue. The counter argument was that few aircrew or their unions, nor the aviation industry or traveling public had supported the allegations. It noted bleed air from engines had been used to provide cabin air since the dawn of the jet age, and that fume events were quite rare and usually arose due to failure of a seal or careless maintenance. The seals were by their very nature mechanical devices, and all such devices do eventually fail.

BAE Systems acknowledged in its report that there were approximately ten genuine fume event flights per year that could be traced directly to either engine or APU maintenance issues. With approximately 250 BAe 146/Avro RJ aircraft in service at the time the report was written, using a conservative estimate of 250,000 flights a year would equate to 0.005% of flights being affected, refuting the allegation air contamination was a widespread problem. UK CAA figures from 2011 state that out of a total of 48,000 written complaints in the ten years from January 2001, only 244 (0.5%) were categorized as medical. Of those, the primary issues were pregnancy, infectious diseases, allergies (typically from peanuts), food poisoning, and passengers being scalded by hot coffee or tea.

What the campaign groups never state is the concentrations of the chemicals discovered, because a mass spectrometer is required to detect them, they are always within a very small (and allowable) limit. As a result, opposition groups focus on scientists that don't represent the mainstream views and are outliers in their positions on 'Aerotoxic Syndrome'.

During the first half of 2009 there were 38 occurrences of 'fume events' reported to the CAA and none of them took place on an BAe 146/Avro RJ aircraft. BAE Systems says it has never done anything other than acknowledge the issue exists and to treat it seriously. Another high-profile event on a FlyBe 146 flight from Birmingham to Belfast was not related to the engines at all, but to stronger than specified bleach that had been used to clean the front toilets, resulting in fumes seeping to the cockpit. The investigating team showed some crew were affected quite easily by the bleach fumes, while others were unaffected. BAE Systems claims it takes every fume event seriously and had investigated each event thoroughly. Past investigations over the years have yielded a number of unlikely culprits: a coffee maker left on causing a burning smell; improperly cleaned galleys; electrical failures (leading to a burning smell); even a fish left in an overhead locker resulted in a fume report being submitted. While BAE Systems has taken the position that there are fume events, it stands by its position, supported by the CAA, that they are isolated incidents and not widespread amongst the fleet. Nevertheless, BAE Systems in conjunction with Honeywell and seal suppliers has instigated an improved seal design which has included testing for reliability of both engines and APU's.

A fleet wide replacement has concluded for the engines, and the number of instances of oil seal-related failures has dropped dramatically. Because of the two different APU designs implemented, it took longer than expected to update seals.

The summary report said some people are more susceptible to contaminant issues than others, and that bleed air at altitude was cleaner than at lower altitudes. BAE Systems also went on to note that the biggest and most vocal opponents against Aerotoxic Syndrome and specifically the BAe 146 tended to not be widely supported by the industry, airlines, or pilots and flight attendant unions. This is not to say events have not happened, but it counters suggestions that they happen frequently and that they are a danger to the operators of the aircraft and the flying public. BAE Systems acknowledged such events happen and the source can almost always be traced to failed compressor seals. The irony is that the flight attendants or pilots that claimed a 'life altering' event occurred are isolated, begging the question why other pilots, flight attendants or passengers did not suffer from the same issue.

In the end, the report cited study after study over the years, from aviation to consumer groups, on the BAe 146 and Boeing 757 (the most frequent culprits) as well as other aircraft types, and the results show that incidents are isolated and not tantamount to an industry cover up nor a widespread problem.

Power Rollback Phenomena

In March 1992, an Ansett BAe 146 (VH-JJP) flight experienced an uncommanded power rollback at flight level 310 (31,000 feet), resulting in a progressive loss of power to all four engines. This was due to a sudden and significant rise in outside air temperature. As the aircraft began to descend, normal engine operations were regained at 10,000 feet following numerous attempted restarts. Warmer and more humid conditions than the surrounding air resulted in engine and airframe anti-ice being turned on, placing high bleed air demand on the engines, causing fuel control units to fail to deliver sufficient fuel to the engines and resulting in rollback.

Since 1988, there have been 13 uncommanded thrust reductions (rollbacks) involving the ALF-502. Charter operator Empire had two incidents between 1992 and 1994. American Airlines used to test the anti-ice once a day during a flight to ensure it worked, but noted it was not uncommon for engines to rollback during the test. The United States' Federal Aviation Administration (FAA), Britain's Civil Aviation Authority (CAA), Australia's Civil Aviation Safety Authority (CASA), and Canada's Transport Canada (TC) all issued airworthiness directives applicable to the BAe 146 that provide workarounds if such an event is encountered in flight.

The captain and first officer of Air BC flight 817 were interviewed on May 23, 2000, at the airline's facilities in Vancouver, British Columbia. Both pilots said that they were aware of the engine power "rollback" phenomena, that it was well-known throughout the industry, and that Air BC had trained its pilots and had published an emergency procedures checklist. According to Air BC's BAe 146 Aircraft Operating Manual, "The phenomenon of uncommanded thrust reduction (rollback) is known to occur at altitudes above 26,000 feet in temperature conditions of ISA +9 degrees C. or greater and in the vicinity of thunderstorm activity. Except for descent, flight in icing conditions above 26,000 feet and at an outside air temperature above -40 degrees C. SAT is prohibited within thirty nautical miles of thunderstorms."

C-FBAO was equipped with four Honeywell ALF-502R engines, each rated at 6,970 pounds of thrust. According to the manufacturer, when operated at or above FL280 in certain temperature and moisture conditions, the engines become susceptible to accumulating ice on the supercharger exit guide vanes (EGV). This reduces core engine airflow, resulting in a loss of power. Longtime BAe 146 operator Air Wisconsin experienced something related, when the aircraft in certain conditions would 'bounce' in the air and go through pitch oscillations at cruise.

To protect the engines from ice accumulation, both Honeywell and BAE Systems issued service bulletins that recommended: (1) reducing the length of the core-flow/fan-flow splitter lip (cut-back splitter) "to reduce ice crystal/water ingestion to the core"; (2) insulating the splitter lip baffle "to reduce heat loss"; (3) substituting a heated, single row of 71 EGVs in lieu of the unheated, double row of 88 EGVs in the engine supercharger (compressor) "to prevent ice build-up," and relocating the engine anti-ice air source to the combustor bleed plenum "to reduce system heat loss"; (4) relocating the anti-ice valve (this necessitates the installation of a dished section on the inner surface of the left rear cowling door); and (5) installing insulated plumbing. The engines on the Air BC aircraft, C-FBAO, had not been modified in accordance with these service bulletins.

Running the anti-ice on the engines puts a strain on the engine due to overheating, which could result in the rollback issue, and it was seen by a variety of airlines from AirCal onwards. The FAA issued an airworthiness directives (AD) for ALF-502R5 and ALF-502R3A engines in 1999, that dictated engines that had not been revised to mitigate this issue were confined to altitudes not to exceed 26,000 feet in icing conditions (as the problem had not manifested itself below that altitude). Once engines were modified and the work was verified by the FAA, the restriction was lifted on an aircraft by aircraft basis. Additionally, all four engines had to be modified in accordance with the AD, as the FAA would not [for example] allow two of the engines to be modified and the restriction lifted. Part of the modification included replacing parts from the ALF-502 with updated items used in the LF507. Air Wisconsin went a step further, and after one of the flights experienced bounce and pitch oscillation pulled the flight data recorder (FDR) and put the data into the BAe 146 simulator used for pilot training, and flight deck crew were required to train for such an event.

This problem, while serious in nature, was not commonly encountered.

Ghosts...

While ghosts are not generally a real problem on the BAe 146, there was a reported incident that came from Bolivia associated with a recently delivered aircraft. Aerosur, through BAe service engineer Mike Grout, reported to Hatfield that there was a 'ghost' aboard one of its aircraft and asking how it could be got rid of. Mike responded that there was nothing in the Aircraft Maintenance Manual regarding spirits, and the only airline with experience was Eastern Airlines, which were no longer in business.

Aerosur claimed that following the delivery of E1017, mechanics had reported a ghost on board. The aircraft originally served with Aspen Airways, followed by a very brief stint with Saudi Aramco, but then went back to Aspen and subsequently Aerosur. The reports read that there was a ghost in the rear toilet which often snored at night. The mechanics supposedly were afraid of it, but also said it had not interfered with maintenance of the aircraft until the night of Thursday September 16, 1993.

A Mechanic was sitting in the co-pilots seat updating entries in the logbook under flashlight illumination, and no systems were powered on the aircraft. He had entered through the avionics bay, leaving the door in the open position. All the other exterior doors to the aircraft were shut. The mechanic reported he heard the avionics bay hatch move to a closed position, and frozen in his seat from sheer terror, he heard footsteps come up through the avionics bay and down the passenger cabin aisle. When the footsteps ceased, and the mechanic was sure the ghost was not near him, he quickly exited the aircraft from where he entered, the avionics bay.

All the other mechanics were accounted for immediately afterward, suggesting it was not a prank. The mechanics (except the one who left terrified) claim they were not afraid, yet none were brave enough to walk to the rear toilet of the aircraft in darkness. Mike was facing a dilemma, as the locals were very superstitious, and he had witnessed them smoke out a newly commissioned storeroom with llama dung to chase away evil spirits supposedly living in it. Mike's concern was smoldering llama dung was not a practical solution to exorcise an aircraft.

British Aerospace
Customer Support Dept., Hatfield

SERVICE REPORT

Aircraft Type:	BAe 146-100		
Operator:	Aerosur	Reporters Ref.:	SVC/BOV/GHOST
Location:	Santa Cruz, Bolivia	Date:	20/9/93
Subject:	Supernatural occurances on aircraft E1017.	Service Engineer:	Mike Grout

PLEASE NOTE THAT THE CONTENTS OF THIS REPORT IS CONFIDENTIAL AND SHOULD NOT BE COPIED OR SHOWN TO PERSONS OUTSIDE THE COMPANY WITHOUT FIRST CONTACTING THE SERVICE MANAGER, HATFIELD, HERTS.

On Friday 17th September, I was called to the office of the Maintenance Manager, Sr. Jorge Urioste, and asked to explain what could be done to aircraft CP-2249 (E1017) to rid it of a 'ghost', because, following the event described below, mechanics were refusing to work on this aircraft at night. There is nothing in the Aircraft Maintenance Manual on the subject and the only airline with any experience, although unconfirmed, on this subject is Eastern, who no longer exist. I would like to pass this question onto Avro International Aerospace.

Ever since delivery of aircraft E1017 to AeroSur (their second 146) there have been reports from the mechanics of a 'ghost' in the rear toilet who often snores at night. Mechanics have often been scared by the ghost but until now it's presence has never interfered with the maintenance of the aircraft.

On the night of Thursday 16th September 1993, a mechanic was sitting in the co-pilot's seat of this aircraft updating entries in the log book under flashlight illumination, with no power on the aircraft. He had entered the aircraft through the Avionics Bay and left that bay door in the latched open position, the Flight Deck to Avionics Bay hatch was open, all Passenger and Service Doors were shut. The mechanic reports that he heard the Avionics Bay move to the closed position, frozen to the seat he then heard footsteps come up through the Avionics Bay and walk down the passenger cabin aisle. When the footsteps ceased and the mechanic was sure that the 'ghost' was not near him, the mechanic quickly exited the aircraft the way he had come in.

Of course it is possible that this was a prank by another mechanic, but all mechanics were accounted for immediately following the incident. Also, although some mechanics claim that they are not scared of the 'ghost', there are none brave enough to walk to the rear toilet of E1017 in darkness.

Due to the superstitious nature of the people here, I am concerned as to what action may be taken to rid this aircraft of this 'ghost'. I have previously seen them smoke out a newly commissioned store room with piles of smoldering Llama dung to chase away any evil spirits that may have been living in there, not the kind of practice suitable inside an aircraft, smoldering Llama dung gets awfully hot.

Thanks in anticipation,

Mike

Ghost report from Aerosur written by BAe service rep to Hatfield.

Spare Parts and Customer Service

All aircraft go through a teething process, where generally the bugs get worked out over time through operations, and then they go on to provide a successful service to the airlines they serve. The engines were the most troublesome part of the BAe 146 series, but eventually even these had all the issues resolved to some level which brought sustained reliability. Airlines count every penny spent on maintenance carefully, not to mention that flight cancellations cost carriers dearly. If the BAe 146, and subsequently the ALF-502 engines issues had never been resolved, the aircraft would not have continued to sell, nor would airlines have persevered with its operation. But the 146 had its troublesome issues resolved, and went on to serve operators for years, and in some cases decades.

What added to the challenge was the time it took for BAe to respond to issues discovered by operators. Customer Service reps were often described as 'a pain in the ass' by airline service personnel. Allegedly BAe would often take months to respond to problems reported by customers and supply fixes to address issues. Boeing on the other hand would expect its personnel to respond with a solution no longer than 24 hours after notification. It was alleged that the British had a stubborn relationship with operators, not wanting to deal with reported issues whereas American's had the opposite mentality: "this is broken, let's fix it."

This occurred even in the design phase of the BAe 146 with designer Brian Brasier, who started off doing hydraulic design in the early 1970s and ended up as the chief designer of the aircraft. When PSA was operating the 146 and the aircraft began displaying hydraulic warning lights during landing, the BAe reps at PSA sent difficulty reports back to Hatfield. Instead of offering a solution, the response would come back "the hydraulics worked on the Comet, nothing wrong with them." Tensions boiled over on many points, with the left hand (engineering) not knowing what the right hand (assembly) was doing. One example had Clive Nicholson talking to engineering regarding a problem with changing peri seals, a component that required frequent exchange. Engineering raked Clive over the coals, because as Clive said, "they're never wrong". Clive had to drag engineering down to the assembly hall to show them how parts were installed or changed. While engineering [in one specific example] said, "it only takes two minutes to change this part", they completely overlooked that it took six hours to disassemble the aircraft to get at that part. Then there was a report that water was found in the keel of the aircraft. There are valves that open/close with pressurization that should allow the water to flow out. They rarely worked, but BAe engineering again claimed, "they worked fine on the Comet, nothing wrong here."

It did not help that every aircraft was essentially hand built, with each requiring unique modifications even on the day of delivery. Airlines, pilots, and customer service reps would often develop workarounds, or know how to fix an issue whereas the maintenance manual would have a vague or contrasting fix. PSA mechanic Joe Casper knew instinctively what needed to be fixed and ignored the BAe manual. When BAe began installing ribbon connectors throughout the plane, PSA got an exemption to cut the connector off and hardwire them as they were incredibly troublesome to deal with. These challenges contributed to the aircraft never earning BAe profits. It was alleged that management at Hatfield essentially sealed the planes fate by having the aircraft sold at cost, while planning to 'nail airlines on the cost of spares'. For example, flap computers were made by Dowty. A brand-new example sold to a customer would cost $50,000 USD, while a refurbished computer went for $45,000. But the same spare part from BAe was $73,000. A roll spoiler actuator which was approximately two inches in diameter by seven inches long cost $63,000 USD, while the equivalent part from Boeing was only $8,000 USD. These prices contributed to the excessive maintenance costs of the BAe 146 when compared to other aircraft.

In another example. the APU controller on the BAe 146 was the same as the cabin pressure controller on the Boeing 757. The difference was the codes on the Boeing 757 controller got sorted while the APU controller, despite being the same part, never had its codes sorted and each code could indicate a variety of different issues. These problems combined with high spares costs, slow response times, and internal BAe corporate culture having bought into its own propaganda made for difficult relationships with the customers. Eventually, in the '90s, the culture would change but not before the damage was done to the aircraft's reputation.

Derek Taylor

CHAPTER 13

Marketing and Selling the 146 Aircraft

The effort required to market an all-new aircraft is substantial, with the manufacturer often facing an uphill battle to convince airlines on why they need a new plane. It's a very involved process that takes months and, in some cases, years to accomplish. Airlines are not like consumers who look at a car brochure, take a test drive, and buy. Deep dive analysis reviewing every aspect of how the airplane will operate, what it will cost to maintain, what markets it can serve and how it compares to other aircraft are all considered.

One way to imagine it is if you were to buy a new car, besides a test drive and a sales brochure extolling its virtues, the manufacturer (or dealer) had to present a detailed report showing you the cost to drive it daily to and from work, with every aspect factored in: how far (mileage), estimated speed at various points on the route, the fuel you are likely to burn (and the cost per gallon), the estimated cost to change the oil and other maintenance items, your monthly payments broken down including insurance and interest, and finally the residual value of the vehicle after five or ten years. It might surprise you how much each mile you drive in your car costs you. Airlines require these analysis reports in order to purchase aircraft.

The process often starts with initial aircraft data and sales material being sent to an airline. Sometimes the potential customer requests this information, and in other cases an BAe salesman calls to present the case for the aircraft. Follow up meetings occur during which a manufacturer will not only bring artwork and even models of the aircraft in the prospective airline's colours, but will also present a very detailed analysis of the airline's route network featuring the new aircraft. The best way to sell a new aircraft is to explain the financial benefits to the customer, from acquisition and amortization, to maintenance, fuel burn, as well as cost to operate it on routes, per hour, per seat, and so on. This information is necessary for the prospective operator to properly evaluate how the aircraft will fit in with its current operation, including assessing all costs associated with a new aircraft and how it compares not just to its existing aircraft but to competitive aircraft as well. In the end, the question a would-be operator needs to answer is whether it can make more money with the proposed aircraft than any other alternative? The process does not always happen in this order, and each manufacturer and sales team will approach every airline differently. Large, well-established airlines tend to operate slowly and are committee driven, whereas a smaller or independent carriers often move much faster.

Before a new aircraft actually flies, most of the information presented to airlines is based on calculated assumptions. Only when the first aircraft has been built and flown does the data change from assumptions to reality. This provides the manufacturer with an opportunity to correct any performance targets that are not being met. For example, an aircraft might be forecast to be a certain weight from calculated assumptions, but actual data might have the aircraft heavier (or lighter) than expected. That data will affect the performance parameters which will naturally affect the bottom line in operating costs.

During a sales tour in 1988 (North and South America), BAe took the 146 that was destined for Air Wisconsin (operating as a United Express affiliate) and demonstrated to Pan Am in consideration for a domestic express service fleet replacement.

From there, a manufacturer will make design changes as needed in order to try and hit the promised performance targets calculated assumptions. Occasionally, an aircraft will outperform the pre-flight calculations. As former BAE Systems salesman Steve Doughty notes, it's about "matching what you have to sell with what people need to buy", and results in hunting for the right customer. For example, you would not try to sell a short haul aircraft to a transatlantic carrier. In other words, there's no point talking to operators that do not have the need. Instead Steve mentions he would comb through directories of the world's airlines and review what they were currently operating, and which had aircraft in the category he was trying to sell.

Once the production line is running and aircraft are delivered to airlines, the manufacturer will revise the sales material and detailed analysis given to prospective customers based on in-service performance. For airlines that are looking at a sizable initial order, the manufacturer may fly the prototype to the customer's primary hub or maintenance base for evaluation and review. Early production aircraft are often engaged in sales tours both to committed customers and prospects, to further bolster interest and generate further sales.

Potential airline customers are also able to get a sense of the interior of the aircraft before it had even been built. At Hatfield, where final delivery of finished BAe 146 aircraft would occur, airline customers could view a full-size mockup of the interior, including a full seating layout that highlighted the fabric and colors available for the interior (assuming a customer did not want to supply their own).

Sales Process

"Hey buddy, wanna buy some jets? Low mileage, I'll give you a good price."

Thankfully, that's not how jets are sold (usually – there's always one guy), but the process of buying a jet can range from very simple, to a long drawn out process with absolutely no guarantee of a sale. At the time BAe began pitching the 146 to airlines, its primary competitors were Fokker offering the similarly F.28, Boeing (with the 737-200) and McDonnell Douglas (with the DC-9), while other manufacturers only had smaller turboprop aircraft available.

The reality was the Boeing 737-200 series (plus the forthcoming -300) and the McDonnell Douglas DC-9/MD-80 120+ seat series were much larger, and the 65-70 seat Fokker F-28 somewhat smaller. That left a gap for the BAe 146 series to fill, assuming that gap constituted a real market demand. The BAe 146 slotted in between all other aircraft available at the time, being a 70-100 seater geared for short haul markets. Manufacturers produce sales brochures for their aircraft, features, and other informative aspects. BAe produced a number over the years dealing with the aircraft, the engines, ease of maintenance, improvements, and variants. These were not tailored to a specific airline or routes, but just provided general information; a take-away for airlines, the media and the industry generally. The majority of initial brochures, including the artwork featured inside, were designed by Tom Sheppard.

The process begins when either an airline puts an inquiry to a manufacturer relating to a specific need, or a

Many airlines were targeted by the sales team, including this retouched image featuring Greece based Oympic Airlines.

manufacturer sees an opportunity for their aircraft to replace an older or less profitable member of a fleet, and a sales campaign gets underway. Unlike buying a car, when a customer generally deals with one or two people (the salesman and maybe the finance department), selling multi-million dollar aircraft along with spare parts is a very labor-intensive process. Regardless of whoever initiates the process, it usually entails the manufacturer studying the route structure of the airline to determining where an aircraft would fit in.

BAe supplied airlines with data analysts to determine how to extract every last penny of revenue from the aircraft. They continually worked with airlines and studied their current and proposed routes, considering all of the costs involved to determine how customers could best and most lucratively extract every bit of profit from a new fleet. The data included the number of flights, frequencies, average passenger load factors, estimated fuel burn, cost of spare parts, training, fuel costs and more. It is a way to sell aeroplanes; to show the prospective customers airlines how to make every last cent from its operation.

The manufacturer uses this data in a presentations, showing prospects all of the upsides and benefits and explaining why the aircraft is needed. Sales reps show up with artists depictions, retouched photographs, customised brochures tailored to the customer, and even large-scale models emblazoned with the airlines livery. They use every sales 'tool' to help convince an airline they need the aircraft, hopefully a lot of them. Some opportunities involve a series of meetings over a short period of time, while others entail a long drawn out process with no guarantee of a sale. The airline's staff are often invited to the factory to take a tour, demo flights are arranged and include having the prospective airline's pilots fly the aircraft, while a visit to a mock-up is used to help determine interior configurations, materials, and to select options that are available through the manufacturer or third-party companies.

But even when the time that an airline is leaning towards signing on the dotted line come, the sales process is far from over. For airlines nearing a purchase decision, demonstration flights are often required before sealing the deal. Seeing the aircraft at the manufacturer's premises or performing at a trade show (i.e. Farnborough) is rarely enough to persuade a prospect to commit millions of dollars to an order and the manufacturer may have to fly the aircraft to the requisite airline's country to prove to airline executives that the it is the right one. Occasionally politics play into the purchase decision. According to former members of BAe's sales and marketing team, selling to legacy versus small airlines can be a very different experience. Legacy and large airlines are very corporate, incredibly slow, and steeped in committees that make decisions. They have a formal process and there is little deviation from it. With smaller airlines, the sales team is generally dealing directly with the founder or CEO and get to know them, their business, their needs, and the relationship is more personal. Additionally, they are typically not slow, corporate nor driven by committee; the CEO is the decision maker, and thus closing a sale is generally a faster process.

It's not just the airline that the salesmen have to sell to – that is just half of their job. Selling internally within the manufacturing organization outlining why the deal should be done is the other half of the process – why the price proposal should be offered, which delivery positions can be provided, and all of the terms of the deal must be mutually accepted. In the early days of the 146 between 1983 and 1990, many deals did not have the proper oversight, disclosure or understanding internally within BAe, thus leading the firm to the edge of bankruptcy in 1992. During that period the pressure to do the deal too often overcame any reservations about its financial viability or the true state of the customer.

Former BAe salesman Barry Lloyd detailed (in his book 'Wings for Sale') how a sales campaign with South American airline Varig Brasil almost resulted in a large purchase of 146 aircraft that were to be used on the "air bridge" route from the downtown Santos Dumont Airport (SDU) in Rio de Janeiro, to Sao Paulo. The airfield in Rio has a short runway, measuring a mere 4,341 feet and limiting it to turboprops at the time. The 146 was the ideal candidate to fulfil Varig's requirements thanks to its STOL capabilities, but it faced competition from Boeing's 737-300 and the Fokker 100. The Fokker 100 was eliminated from the competition quickly because it could not operate out of SDU with any kind of a load. Boeing's 737-300 as it was then could not either, but the US manufacturer promised an performance upgrade to enable it to function as a short field performer.

It turned out the Queen's Flight BAe 146 was going to be carrying Prince Philip to Belize for a World Wildlife Fund event in 1988, and was then planned to continue onto Brazil to transport Princess Anne who would arrive a few days later. BAe requested permission from the RAF, and with the Queen's blessing, it was agreed her plane could be used in the sales campaign. Once the aircraft arrived in Brazil, a carefully orchestrated series of demonstration flights took placed (The Queen's aircraft had limited seating as it was in VIP configuration), along with presentations to many local airline officials. Even competing

Just some of the covers for the many sales and marketing brochures produced over the years for the BAe 146 and Avro series.

Aerosur did end up operating the BAe 146-100 aircraft after a successful sales campaign.

Brazilian airline VASP was given demonstration flights. At the conclusion of the visit BAe rolled out another tool in the war chest of its sales department: wining-and-dining the clients with a formal banquet for all the requisite personnel, especially the senior management.

Varig provided additional feedback after further presentations and demonstration flights, which resulted in the sales department coordinating with all the data analysts, engineering, and pilots to deliver an even more finite analysis (this time with strict seating, bulkheads, catering, weight limits, etc.), and BAe then presented the new figures to the airline. Boeing was doing the same thing for their 737-300 aircraft, but it became clear that the 737-300 would not be able to land at SDU in certain weather conditions such as heavy rain. Nevertheless, Boeing promised a fix for the 737. After more back and forth involving further refinements to the proposals, 18 months later it seemed almost certain Varig would place an order with BAe for the 146.

Politics – a necessary evil

This is when the last, final, and in some cases most important tool in the war chest of a manufacturer's sales organization can come into play: politics. It seemed that Boeing was informed it were not going to get a deal for its aircraft, and so the US manufacturer, starting with the local sales agent in Brazil, took its plight to Washington looking for a diplomatic solution. There had been a recent trade mission from Washington that had expressed interest in buying a substantial amount of Brazilian goods, ranging from fresh fruit to natural resources and other material such as steel The contingent from Washington was not too keen on buying from Brazil if the Brazilians were not going to reciprocate and instead elected to buy British: no sale to Boeing, no U.S. purchases of Brazilian goods. As a result, Varig acquired 737-300s, but years later confirmed Boeing still had not modified the aircraft to allow it to land at SDU in all conditions. Even the leaders of the Brazilian pilots' union opposed operating the 737-300 out of SDU, citing slim safety margins, while Varig ended up with a larger and more expensive aircraft that it could not fully utilize for the task at hand.

As related elsewhere BAe, in its hubris, later helped establish Air Brasil, a new competing carrier for the Airbridge,. Unfortunately, it was a short-lived commercial disaster and proved the point that if anyone not rowing with the political tide in places like Brazil was going to end up stranded high and dry.

Politics also played a hand in preventing Iran Air from being possibly sold or leased up to 50 BAe 146 aircraft in 1991. The United States of America blocked the sale of the aircraft because of the U.S. content (more than the 20% maximum) and the trade sanctions that were in place against Iran. BAe lost far more from the blocked campaign than the U.S. supplier AVCO-Lycoming. In direct contrast the U.S. approved the export of 12 Fokker 100s to Iran, but that aircraft contains far less U.S. content than the 146.

BAe looked at ways to get around the U.S. ban and briefly considering wet-leasing BAe 146 aircraft to Iran Air. A wet-lease would have involved BAe providing the aircraft (which it would have retained ownership of) along with the operating crews, and handle all maintenance. In this case, Iranian carrier Bon Air would have been the recipient, and BAe's leasing operation Trident Aviation would have managed the process. However, the arrangement did not come to pass because the U.S. did not make a distinction between sales and leases. It would be more than 15 years before Iranian carrier Taban Air took delivery of a used BAe 146-300 from Bulgarian airline Hemus Air. Mahan Air uses a fleet of 18 BAe 146-200, 300s and Avro RJ85/100s today. The U.S. Department of Treasury has detailed that if these embargoed aircraft (along with non-BAe aircraft) land at airports outside Iran, those countries may be sanctioned too. It's clear the matter has not been resolved.

Aircraft demonstration boarding pass (AeroSur).

With more competition in the RJ space, BAE advertised in aviation trade journals touting the cabin width over other regional jets.

How the RJ70 makes the competition feel uncomfortable.

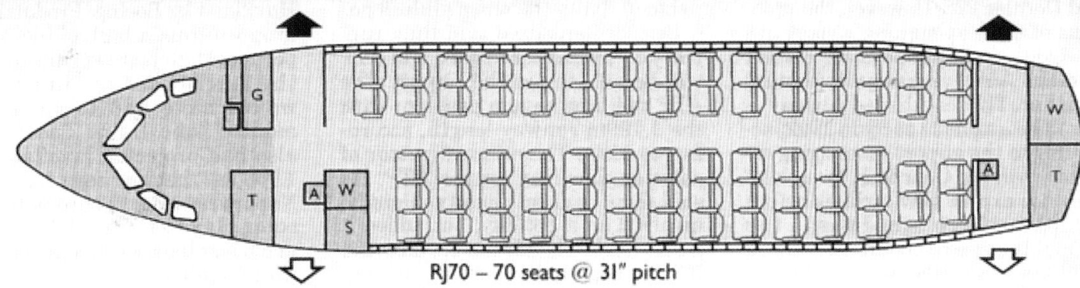

The standard 5-abreast interior comes complete with two toilets, a choice of one or two galleys, wardrobe space, and more carry on stowage per passenger than any of its rivals.

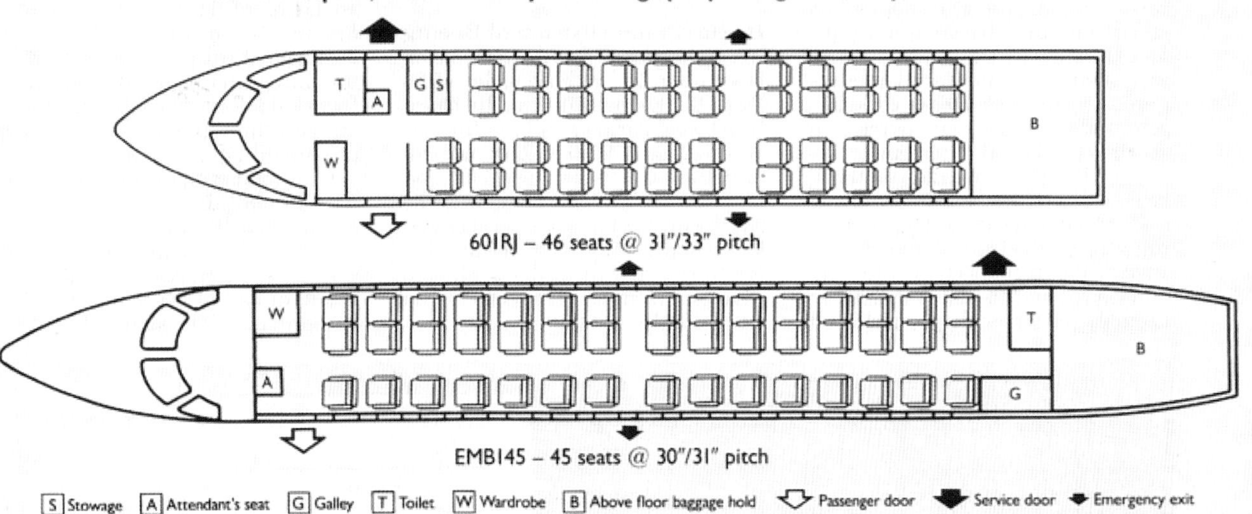

The British Aerospace RJ70 regional jetliner provides a standard of comfort and amenities that no other regional jet can match.

Designed and built to real airliner standards its wide cabin offers a quiet and spacious environment that allows your regional operation to satisfy the most demanding passenger.

Unique among regional jets, the RJ70's flexibility gives you a clear marketing advantage. The standard business class 5-abreast seating can be adapted to a mixed class layout as the market demands.

The addition of a genuine 4-abreast first class section means that high yield passengers can travel in extra comfort and style.

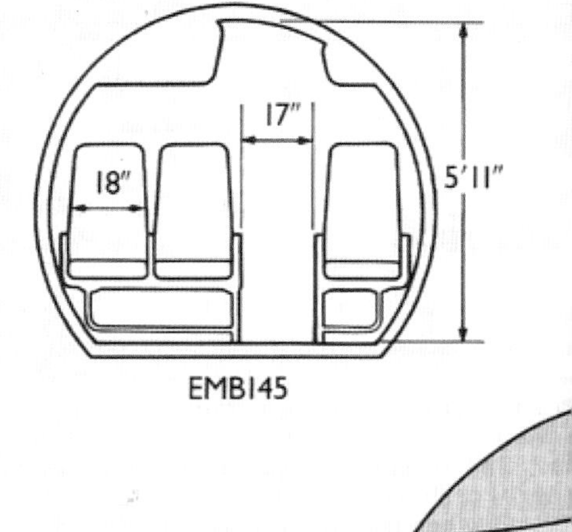

19"

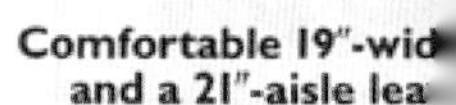

BUSINESS EXPRESS

Comfortable in all airport environments too, it's the only regional jet able to use an airbridge, with integral airstairs also available a an option.

Turnrounds are made quick and easy wit twin passenger and service doors. Whilst the waist-high underfloor holds ease the moveme of baggage and cargo.

But it's not just very accommodating. It'

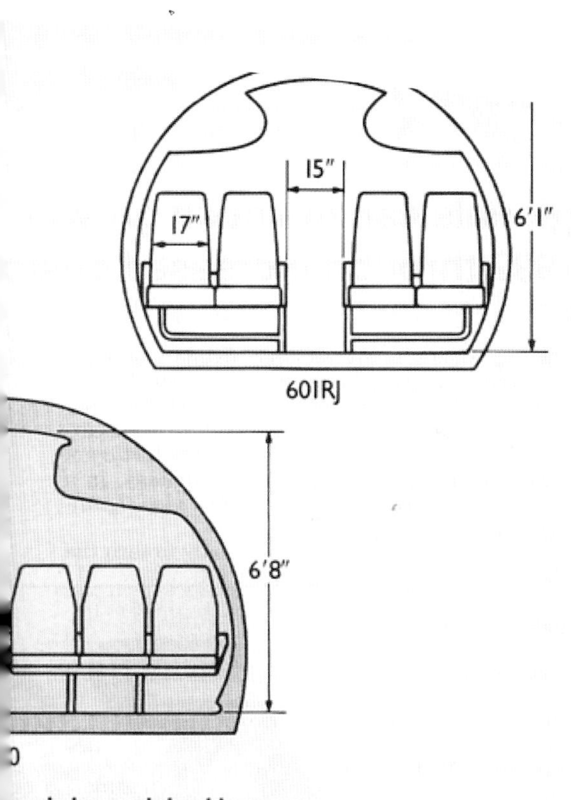

mple leg- and shoulder-room,
npetition feeling squeezed.

also very economical, offering the lowest total operating cost per seat of any regional jet.

All things considered, with the RJ70 you stand to make a comfortable profit.

UNDERGROUND POLITICS - "Agents"

One of the less talked about aspect of aircraft deals is the use of agents employed by BAe (and all the other manufacturers), to assist the sales team in selling their aircraft. At the time of the 146's re-launch BAe had agents in most countries, some appointed for specific products, others representing the company's whole product line. Some had represented BAe or one of its predecessor companies for many years and did little more than facilitate the salesman's visit by booking accommodation, arranging appointments and accompanying the visit. With the improvement in communications there was less need for these services, so this kind of representation died away. Others who worked on a company wide basis could also be problematic because they may have been originally appointed to pursue military deals and had little knowledge or interest in the civil market. Getting rid of them could sometimes be difficult because they often had good contacts at senior levels within BAe who might support their retention, regardless of what the 146 team wanted. Such people were then little more than a tax on the product whose commission had to be allowed for in any price quoted, but who were going to add very little to the marketing campaign.

In general manufacturers felt a need for agents in countries where the state played some part in determining or approving the equipment selection of the national carrier, and in the 1980's many airlines were still either state owned or heavily influenced by state bodies. There was a widespread belief that trying to pursue a sales campaign, excluding operators in Western Europe, North America or Australia/New Zealand, without an agent was going to leave the company at a disadvantage to its competitors. In nearly all such cases an agent was employed not only to guide the company through the often Levantine bureaucratic structure of such countries, but primarily to pay off whoever needed to be paid to win the deal. Bribery was never talked of as such and was often specifically prohibited in the agreements signed with the agent but that did not alter the reality. At the time BAe and its competitors were not breaking any laws in their own countries or anywhere else but were outsourcing the task to someone else who was, in most cases, breaking the law in their own country. Politics is an expensive business so politicians in such countries were always attempting to get a piece of the action on any government contract, personal greed was also never far away and with such examples at the top it was unsurprising that people lower down were also looking for their slice of the pie.

Agents came in all shapes and sizes but it was an activity that attracted more than its fair share of fantasists, misfits, con-men, chancers and all round incompetents. In many countries certain companies or individuals would tend to be favored by British companies and would be recommended by the British Embassy, while others would work more with, say, French companies. It was as important for the agent to have some understanding of his client's culture and constraints as it was to understand those of the customer. If there was a opportunity brewing BAe would be approached by a variety of people claiming that they could secure the deal, and it then became a case of trying to sort the wheat out from the chaff.

There was a tendency to seek someone who had the best contacts at the highest level in the naive belief that if the top man was on side then he would just say buy this plane and the deal would happen. Senior managers of the company were very susceptible to this kind of approach. There were occasions when we were told to appoint person 'x', at some exorbitant rate, because he had taken a Director of BAe in to meet the President of country 'y' and the deal had been agreed. Disaster almost invariably followed because that approach was akin to placing all your money on one number and a spin of the roulette wheel. There were so many variables involved that could not be predicted or covered; the President may have made the same promise to a competitor. or it might turn out that the key man was the Prime Minister, or the decision might get thwarted by people at a lower level who did not like being dictated to. Complex issues like this were never solved by simplistic one-shot solutions.

Derek Taylor

The paradox was that if the system concerned was corrupt then how could you trust anyone to do what they said they were going to do when they were going to benefit whichever product was selected. In that sense it became a zero-sum game where each competitor had this aspect covered and the competition could thus revert to a more conventional analysis with the airline probably choosing the aircraft it favored for perfectly rational and appropriate reasons. To overcome this and make a difference would require an exceptional person. What we were looking for was someone who could take you into the sewer system, guide you through the innumerable passages and then when you came out the other end you were both smelling of roses! Such people did exist, but they were a rare commodity and finding one was like hitting the primary vein in a mine. There were certain deals done on the 146/Avro where we essentially had the whole process rigged from top to bottom, all through the airline and then into the approving ministry bureaucracy and wherever else was necessary. Some agents were overt and would accompany you to meetings and the relationship would be public and acknowledged by all, even if there was an onshore agreement for the local taxman and an offshore one where most of the money would go. Others were covert where the person concerned would be known only to a small number of BAe personnel and meetings were clandestine and often outside the country concerned. Even in such situations events often conspired to cause problems, not everyone in a network was reliable and of course opponents would be doing their own scheming so the sales campaign remained a roller-coaster ride to the very end.

Derek Taylor

Identifying, managing and controlling agents was a difficult job and it was all too easy to stumble into pitfalls. Some people became too close to an agent and ended up being manipulated by them for their own ends; after all such people were experts. With so much money sloshing around in such an unaccountable and unfathomable way there was some splash back and there were cases when BAe people almost certainly took kickbacks. For example, the agent would flag up a plausible excuse, maybe even a true one, stating political party 'x' will support the deal but they want $200K USD, senior BAe contact supports the case due to the deal being at a critical stage, and the company grudgingly agrees to an exceptional one off payment of $100K USD, with $10k going to the political party for which agent gets all the credit, BAe man gets given $20k by agent (having been told by agent that they are sharing 50/50!), and the agent keeps $70k. Not hard to work out who the real beneficiary is but at all times only the agent knows the true position. It may seem perverse to say it in such morally questionable circumstances, but such a relationship only works if the agent and the salesmen maintain their integrity.

Sales personnel sometimes had to spend long periods in country pursuing a deal. Sometimes I felt like I was visiting Colonel Kurtz in the film Apocalypse Now when I went to see them. A few had become intoxicated by the pleasures of being a young man with an expense account in a place like Manila and had completely lost sight of what they were trying to achieve, being seduced by incompetent agents into believing the deal was just about to happen; anything as long as they could carry on enjoying the local nightlife. Others had created a labyrinth of petty informants and agents and thought they were controlling some great network when in fact they were merely surrounded by deadbeats trading low level gossip designed to tell them what they wanted to hear.

In fact what I was witnessing was the end of an era. The Foreign Corrupt Practices Act of 1977 had knocked U.S. companies out of the game and they were not happy to see their foreign competitors benefiting at their expense. Pressures tightened on other western companies and gradually the most egregious examples were squeezed out, but it was only after further well documented scandals involving BAe amongst others, that other western governments and the companies themselves called time on the kind of activities documented here. However, I have little doubt that even today non-western companies in the aviation business are treading the same paths we did and certainly making the same mistakes.

Derek Taylor

Flight through mountainous terrain in India.

Derek Taylor

[AUTHORS NOTE: this was an anonymous, independent and unverified contribution from "Caractacus" – a former BAe employee.]

AIRLINE INVESTMENT

Years later in 1990, Air Brasil, another Brazilian airline was formed. Founded in 1987, it was as speculative venture with three investors getting it off the ground: Lider Taxi Aereo S.A., the country's largest air taxi operator with JAA (Jose Afonso Assumpsao) who started the airline at the helm; TNT-Sava, a Brazillian entity associated with TNT Australia; and BAe, which was interested in using the 146 to operate as a sales tool for South America. BAe was asked to take an equity position to show 'good faith' and began what was known as "Integration Financing". The arrangement did not last long with TNT-Sava pulling out leaving only JAA and BAe, with BAe being asked to contribute more money to the airline. In the end BAe funded the airline operations for a short time before it became clear that the manufacturer would be pumping in money indefinitely while JAA would reap whatever rewards he could. From changing terms, mysterious partners that never invested, and refusal to sign subleases of the aircraft, BAe took a $1.6m charge against the airline it was hoping would bring more 146 sales once it was operating the Air Bridge routes.

Air Brasil was not the only airline British Aerospace invested in with the hope of securing more 146 sales. Much earlier BAe had put money into U.S. carrier Pacific Express, which was founded in 1981 as a discount carrier and became one of the first airlines to commit to the 146 with an order for six. BAe provided seed money for the startup, guaranteed financing on the BAC 1-11 aircraft that were to serve as interim capacity until the 146's were delivered, and purchased stock and debentures in the airline at a public offering (amounting to nearly $1.9m dollars). Pacific Express had shut down by 1984 after posting losses year over year, and not a single BAe 146 was delivered to the carrier. Former airline Discovery was yet another business in which BAe invested nearly $2m and it ended up operating five aircraft but was shut down by the U.S. Government after five months for violating foreign investment laws.

Investing in airlines, overwhelmingly start-ups, that would go on to buy your aircraft was a tactic not limited to BAe. Many manufacturers have tried such methods to place their aircraft. Airbus in its early (and struggling) days wanted to make a sale and break into the U.S. market so badly, it reportedly offered Eastern Airlines four whitetails (built aircraft with no buyer) on a six-month no cost lease. McDonnell Douglas offered American Airlines (AA), which wanted no new aircraft, such favorable terms on MD-80s that the carrier was able take advantage of a 'month to month' lease (a five-year term, but AA could cancel at any time with just a minimal penalty). Aircraft manufactures have always been creative in finding ways to sell or place aircraft.

The first South American operator of the BAe 146 was TABA which, in another sales tactic to move aircraft, was given incredibly favorable terms in 1983 to enable BAe to place aircraft. A deal was reached to sell the airline two BAe 146-100 aircraft for $40m USD, with zero down payment and a payment plan over a ten-year period. Instead, a payment of $2m for spare parts was required. Both aircraft were returned in 1985 when the spares and tariff issue could not be resolved with the Brazilian government.

A more positive story was that of Malmo Aviation, with which BAe had placed a used 146-200 on a pay as you go short term lease in the late 1980's. BAe wanted to see the 146 operating into Stockholm's downtown Bromma Airport which had long been threatened with closure, and where commercial operations had dwindled to a few domestic turboprop services. The strategic prize would be a sizeable order for new aircraft as a result. With significant assistance from BAe, Malmo Aviation was able to overcome all environmental and operating obstacles to commence scheduled services to Bromma. Passenger numbers grew quickly, and BAe supplied further 146's on lease, but the airline continued to struggle primarily because it was under capitalised. Matters came to a head in 1991 when BAe faced the stark choice of extending credit or seeing the airline go bankrupt. After a swift analysis it decided to loan the carrier $500,000 and that kept it going long enough for new investors to come in and rescue the business. Subsequently Malmo became the longest serving operator of the 146 and ultimately flew a large fleet of used Avro aircraft that AMO placed with it. In purely financial terms it was a success and BAe earned a good return, but the manufacturer never placed a new aircraft with the airline, so the original strategic goal was never realised.

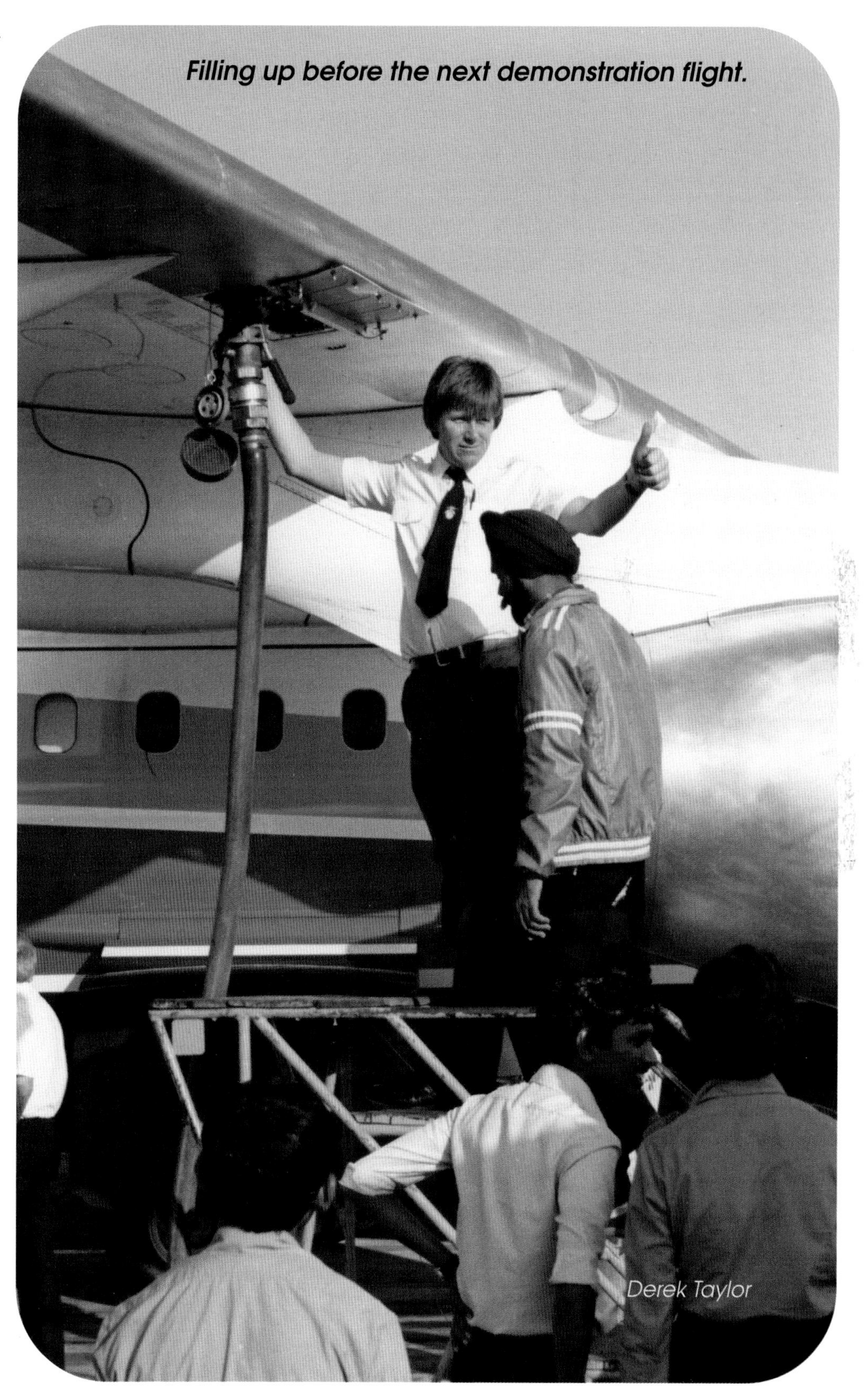

Filling up before the next demonstration flight.

Derek Taylor

There are certainly occasions when such risky strategies pay off, but correct targeting is essential, and this was where BAe went awry. If start-up carriers are struggling to raise cash or established carriers are facing difficulties, it is generally because investors question their viability. In such circumstances manufacturers like BAe effectively become a banker of last resort.

HRH Prince Philip flew the aircraft (under supervision) between cities during the Far East Tour in Japan.

HRH Prince Philip Duke of Edinburgh touring the BAe 146 for the first time during a visit to Japan.

At Narita Airport, Japan with British Airways and Pan Am Boeing 747s in the background.

This Cover was flown on BAe 146 (E1005) G-SCHH from Hatfield on October 24th 1982, on a sales tour of the Far East.
Captain - Mr. P.A. Sedgwick
Crew - Mr. M.S. Hopgood
Mr. J.V. Creswell

The cover that was carried to each country and stamped accordingly.

Back of the tickets that are issued to both passengers and employees.

Visiting Mount Cook, New Zealand.

The team that accompanied the aircraft on the Far East Tour.

All photos courtesy of Derek Taylor.

The short-term gain is essentially all to the carrier, with the manufacturer simply obtaining a dubious PR bounce in return for taking all the financial risk up front. As the instances above show, BAe rarely saw much if any return for its investments.

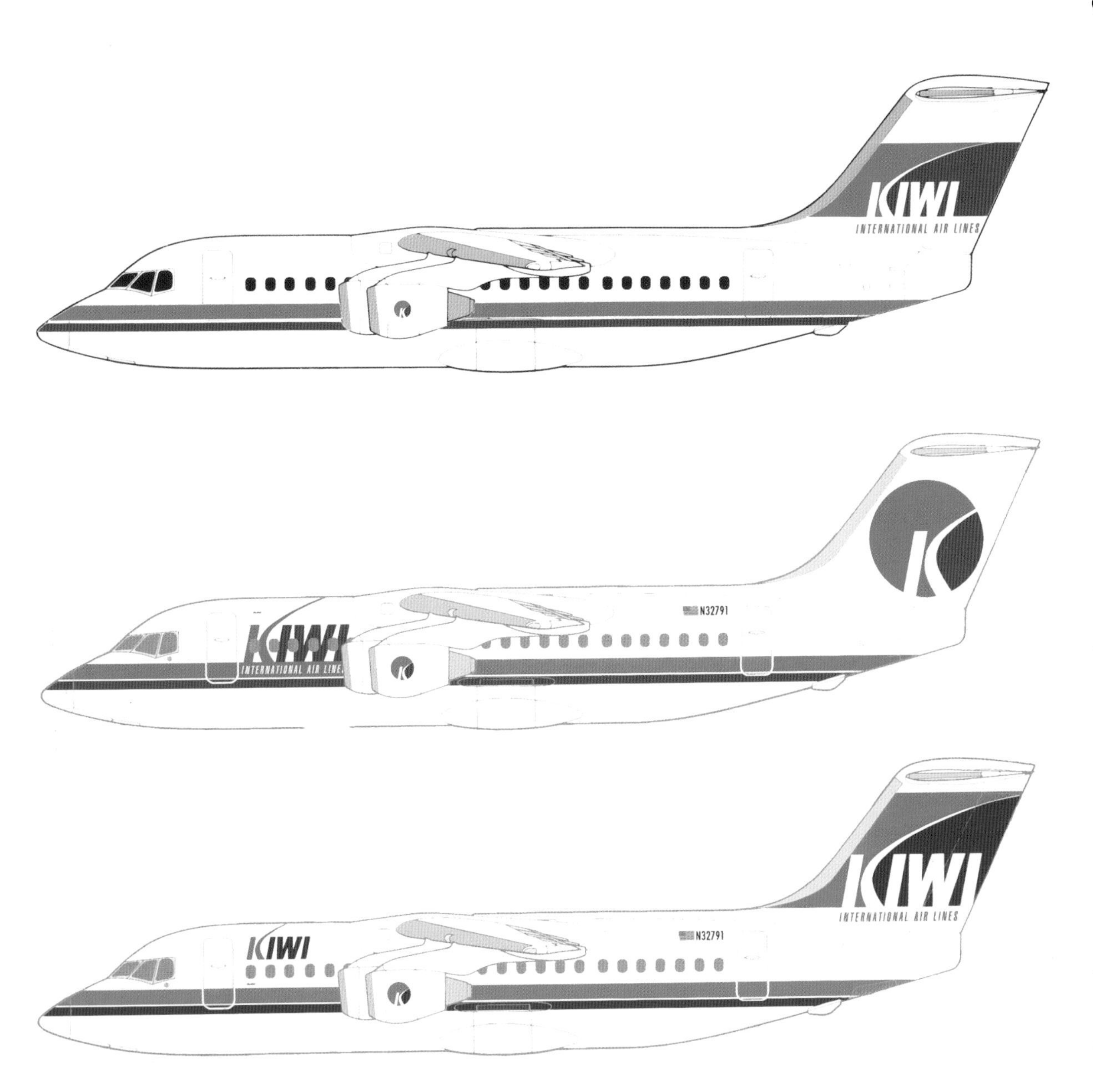

On the other end of the spectrum you had start-up carrier "Kiwi" which was formed by Bob Iverson, a former Eastern Airlines pilot. From his book "When Kiwis Flew: The Diary of a Mad Airline Entrepreneur", Bob detailed how the airline was incredibly close to buying Avro RJ aircraft (instead of Boeing 727s) when the airline got going. Avro's U.S. office arranged to pick the key individuals up in a demonstration Avro RJ70 and fly them to Virginia where BAE's North American headquarters were located. Avro was pitching the RJ70 as the right aircraft, but Bob said they were more interested in the RJ100, and if they could certify one with overwater equipment to allow flights to Bermuda from Newark, as well as the Bahamas. BAE agreed if they placed an order for twenty (20) aircraft.

Bob told BAE that integrating 20 planes in 18 months was a big pill to swallow, and they wanted Avro to share the financial burden especially because BAE had no North American customer for the Avro line of aircraft. Kiwi would agree to take the aircraft if Avro would arrange financing for the aircraft, wave cash down payments, and compensate Kiwi $1m per aircraft sold to other carriers in the United States as a 'commission', including the initial 20 Kiwi would take. If the deal was consummated, it would result in $20 million being paid direct to Kiwi at the contract signing, plus any other sales to other carriers should they occur. Meryl Hershman who headed BAE's Sales and Contracts for North America said she'd present the deal to Woodford. A few days later, models of Avro RJ70s in a proposed livery arrived to Bob's office.

The irony of this proposed deal was it went directly against what BAE (Avro) planned when it restructured the company, setting up AMO, and Avro only looking to deal with currently operating airlines that had financing in place and a strong balance sheet. Kiwi was neither at the time, and to hear that there was an agreement in principle for a no cash deal, financed arranged by BAE, with a lump sum cash payment at signing was surprising given how recent the financial wounds of former sales campaigns brought BAE to its knees. In the end, other principles in the company scoffed at the idea of operating smaller aircraft, viewing it as beneath the airline, and internally through political maneuvering killed the deal off.

Aircraft sales are generally valued in U.S. dollars because the U.S. was the dominant force in aircraft manufacturing between the 1960s and the 1980s. As a result, other manufacturers marketed their aircraft in USD as well. But the hidden beast of burden was currency fluctuations, which could wipe out profits on an aircraft overnight. It affected not just the sale of aircraft, but the value of parts sourced from outside of the U.K. as well (remember, 40% of the 146 was built in the U.S.), as prices could change significantly overnight. When the exchange rate dropped (e.g. $1.20 to the Pound) BAe was in a bad position financially, but when the exchange increased (e.g. $1.80 to the Pound) BAe was in the opposite situation.

This is neither a complete nor comprehensive guide for sales of BAe 146, but aims to give the reader an overview of the different facets of selling aircraft into airline. All the hurdles the manufacturer needs to overcome, as well as some of the creative ways manufacturers have used to try to place aircraft.

BAe was hopeful that its relationship with PSA would net them an order for the now launched BAe 146-300 from USAir.

Models presented during sales campaigns to respective airlines.

CHAPTER 14

Introducing the Statesman – The VIP BAe 146

The 146 seemed to be the perfect candidate for a VIP transport, and BAe thought it knew how to position the aircraft: large and spacious enough to accommodate a flexible passenger load; quiet and unassuming for unobtrusive airport operations; exempt from curfew; ability to land on short airfields (including unpaved); integrated APU and air stairs making it self-sufficient; and finally four engines for the utmost in safety which gave it an advantage over twin-engine aircraft. BAe believed the 146, from the series 100 to the 300, could provide flexibility no matter the mission at hand. And with fewer passengers (and subsequently weight), the range could be extended to approximately 1,800 nautical miles including reserves for a 150nm diversion and a 30-minute hold when the pannier tanks were installed. What did not improve was the speed of the aircraft as it was still limited to a maximum of Mach .70 (high speed cruise), nor would it have transatlantic range.

BAe heavily marketed the 146 and Avro series as VIP aircraft, referring to it as the "Statesman". It was aimed at foreign (and domestic) governments and the firm built a mockup of a VIP interior at its factory. In later years, BAe added the QC variant to the Statesman brochure, showing artist renderings of a -200 aircraft with the forward half being a VIP interior, and the rear carrying either a full-size SUV (loaded via an impractical ad-hoc ramp) horses, or anything a customer felt they needed to travel with. The compartments were separated by a 9G bulkhead, and there were no shortage of ideas as to how the aircraft could be configured.

This chapter is neither a complete history of concepts nor of all the actual aircraft that served in VIP roles but a sampling of intriguing airframes.

Mockup of an Statesman interior.

The Queen's Flight and No. 32 (The Royal) Squadron

After the successful tests of the two BAe 146-100 the Royal Air Force purchased in 1983, the UK government approved the RAF purchase of two BAe 146-100s for The Queen's Flight, with delivery in 1986. The Queen's Flight consisted of 180 officers and men of the Royal Air Force.

Construction of the first aircraft began at Hatfield on January 30, 1984 with work on the second starting on June 9. On November 23, 1984 the aircraft left Hatfield and were flown to British Aerospace's Chester factory for completion of the interior fit. The 146 replaced two of the three Andover CC.2s which were then over 22 years old, which were slower and noisier. The 146 was selected because it was British designed and built, flew twice as fast, had twice the range of the Andover, and was certainly quieter to travel in. The series 100 was chosen because it had the best short field performance and would enable travel to pretty much any airfield in the United Kingdom. On November 3, 1983, the Queen herself reviewed one of the development aircraft which was flown to RAF Marham (near the Queen's Norfolk home at Sandringham) with a mock-up Royal interior installed for approval. Unlike the cabins of VVIP aircraft owned by billionaires and monarchs of the Middle East, in the mockup BAe showed to the Queen, there would be no gold plated anything. The only extravagance was an extra-large changing room.

The Chairman and Directors of
British Aerospace PLC
request the pleasure of
the company of

---

to attend the hand-over ceremony
of the first BAe 146 to
The Queen's Flight of
the Royal Air Force
at
British Aerospace, Civil Aircraft Division
Hatfield, on
Wednesday April 23rd 1986
at 2.30 pm

R.S.V.P.
D. Dorman
Public Relations Manager
British Aerospace PLC
Civil Aircraft Division
Hatfield
Hertfordshire AL10 9TL

The livery and materials were carefully chosen and the Queen was involved in the design and selection throughout, starting with fabrics in three shades of her favorite blue. BAe proposed a very lavish interior with gold taps and brown and beige interior, but HRH turned down those suggestions. Inside, blue and beige colors were chosen to provide for a relaxing feel. The exterior cheatline was added in order to downplay the 'frumpy' appearance of the 146. Initially the wings and horizontal stabilizers were not painted red, and the Queen's Flight logos were displayed on a white background under the cockpit and at the rear of the aircraft within the blue cheatline. About a month after delivery at RAF Benson, the wings and horizontal stabilizers were painted red because the aircraft looked a bit too understated. To avoid it being obscured or damaged by the doors rubbing against them, The Queen's Flight emblem was moved opposite to where the doors opened, and had the same blue background as the cheatline.

The first of two 146-100s was handed over on April 23, 1986 at a ceremony at British Aerospace in Hatfield. Air Vice Marshal John Severne, Captain of The Queen's Flight and Mr. Bowes Lyon, Lord Lieutenant of Hertfordshire accepted the first (ZE700) for Royal Family transport. There was an official hand-over of the logbook, along with an oil painting commissioned by Sir Austin Pearce. After the acceptance there was a brief flying display, accompanied with celebratory champagne and strawberry tea for guests while music from the Welwyn Garden

City Band played. The delivery marked the 50-year anniversary of The King's Flight in July 1936.

Prince Charles and Prince Philip were both expected to pilot the 146, as they had previously flown the Andover. Neither would be required to obtain type certification, as they would always fly under the instruction of a senior pilot. ZE701 was delivered just over two months later on July 9, 1986, with both Prince Philip and Prince Charles receiving a conversion type rating shortly after, allowing them to fly under the supervision of an RAF pilot.

The Queens Flight BAe 146's were not for exclusive use of just the Queen herself, but were also available for other members of the Royal Family, the Prime Minister, the Chief of Staff and other visiting heads of state. Flights within British airspace carrying the top six members of the Royal Family are conducted in air traffic control lanes known as Purple Airways, and Royal Low-Level Corridors (for the Royal helicopters) are established separating the aircraft from other air traffic.

Invitation for the hand-over ceremony for the Queens Flight.

The aircraft were configured very differently than one would imagine a head of state aircraft to be configured. Normally the front and centre of the cabins were allocated for VIPs, and the rear would be configured with economy seating for household staff. On other RAF aircraft, seating for the crew was in the front of the cabin behind the entrance door and galley. The Queen preferred to be in the rear of the aircraft in order to be able to wave goodbye to everyone on the ground without the views being obscured by the overhanging engines. In the centre was the seating for household staff such as HRH's traveling secretary, security detail, and others as needed. This area was configured in a 3-2 economy seat layout and included a small desk and workstation area (originally it contained a typewriter and fax/printer). The Royal Cabin in the rear of the cabin featured a 2-1 'business class'-style layout with small desks/tables and seating for six persons. The aircraft was configured to carry approximately 20 passengers plus a crew of seven: two pilots, a Navigator, two cabin attendants, a crew Chief, and a Policeman, all Royal Air Force personnel. The navigator was eventually dropped after GPS was fitted onboard. There were also three technical stewards that travelled with the aircraft, each handled engine, airframe and electrical systems in case an issue developed in a remote location.

Even with the unique layout, the aircraft were certified as civil public transports and the RAF adopted civil maintenance procedures to speed up their introduction. The Royal Flight operated with three key words as its mantra: Safety, Comfort, and Timing (in that order). An on-time arrival was + or – 5 seconds, and that was achieved on more than 90% of occasions. The MOD insisted on having Smiths eight-day clocks fitted at a time when the rest of the industry was switching to Davtron Electronic digital devices. On the day of when the aircraft left, an RAF airman noticed the clock was not working and complained about the issue. The fitter at Hatfield leaned forward and wound it up for him. Needless to say, there was no further complaining about the clocks not working.

Handover ceremony for the first Queen's Flight BAe 146-100.

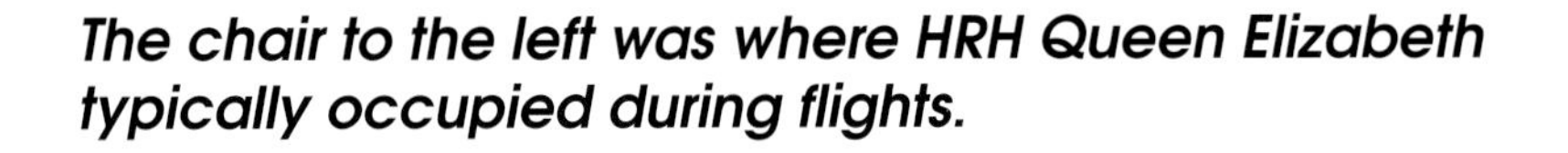

The chair to the left was where HRH Queen Elizabeth typically occupied during flights.

The couch directly opposite of where HRH Queen Elizabeth would occupy during flights.

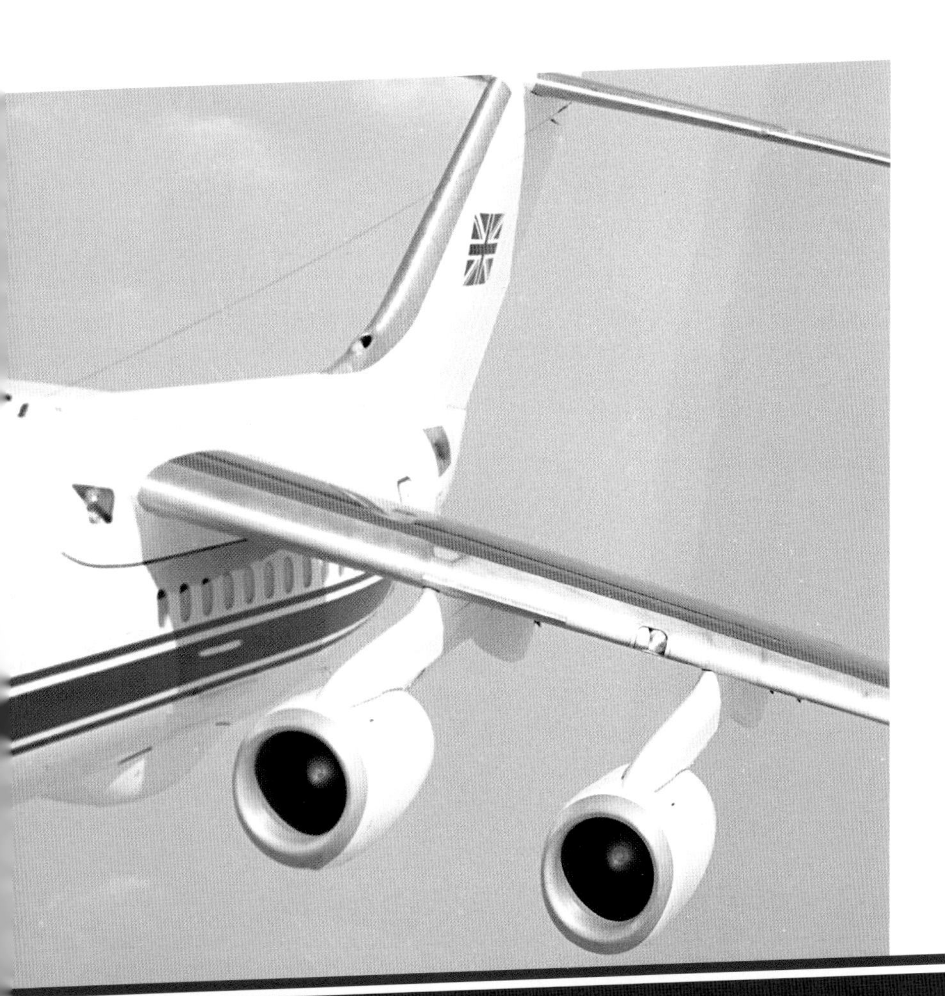

Press Event after the handover ceremony.

The forward cabin occupied by pilots, engineers, and other members of the RAF team.

The 146's first Royal Flight carried the Duke and Duchess of York to the Azores on their Honeymoon. The 146 would be taken to the limits of its range during the lengthy trip. RAF engineers playfully added the words "Just Married" to the inside of the rear tail brake which appeared on both petals when they were opened, along with decals for bells and horseshoes. Neither the Duke and Duchess nor the Captain of the Queens Flight knew about this, even though the engineers tipped off the press to go to Heathrow during the departure where the petal brakes were opened for everyone to see. Had the Captain of the Queens Flight known, he would have been incredibly angry. He ended up chuckling about it four days later when the ground staff were removing the decals.

By August of 1987, the first defensive aids were added to E1021 and E1029, both being fitted with LORAL Matador Infrared Countermeasures Systems (IRCM). Two rear facing pods contained a 4KW incandescent lamp that was modulated by a computer with various codes to match any incoming missile threat should there be one. These devices were installed after a potential threat from the IRA was brought to the RAF's attention. Initially tested by Sqn Leader David Gale, this system consisted of two pod structures mounted on the aft fuselage. There was an on/off switch on the flight deck as well as a selector for the missile type or threat expected in a given area. The modifications did not have any range or fuel burn penalty.

The Royal Flight 146s were used to transport the Royal Family of course, but also dignitaries and other members of the government as needed. The "batting order" (order of importance) for securing the Royal Flight aircraft was the six senior Royal members, followed by the six junior Royal members, then the Prime Minister. While the aircraft unfortunately never had transatlantic range, this did not stop the RAF (and members of royalty) from taking the aircraft long distances, even if this required multiple stops or travel over several days. The aircraft have traveled as far as Brazil not only on Royal duties, but as a sales tools for British Aerospace, and have also visited African and Asian countries. They have even landed on unpaved or poorly paved fields on occasion, which necessitated tire pressures being reduced to avoid ruptures, until low pressure tires were fitted. Her Majesty the Queen travelled as far east as China in the 146 during a tour in late 1986. By the late 1980s, the Queen's Flight was averaging around 1,200 flights per year. When missions took them to distant locations such as North America, South America, or Africa, the aircraft would be ferried there while the Royal(s) travelled commercially to the destination, and then joined up with the 146.

When the Royals needed to use the aircraft, the household staff would coordinate all logistical planning with the RAF. The lower rear hold of the aircraft was used for all Royal baggage to ensure proper weight distribution. The front hold was used to carry a spares pack, except on a tours like the far east tour HRH The Queen embarked on, when two spares packs were carried. The rear APU saw very minimal use because the Royal's boarded from the rear of the aircraft. It was only used at remote locations when no ground power available. During all flights where any member of the Royal family was aboard, the Captain would perform both the take-off and landing. When the flights were not transporting members of the Royal family, the pilot in the right seat would perform all the flying. The Captain would also test co-pilots with a variety of scenarios on these flights, including what they would do if they lost an engine, or what airfield should be selected in case of emergency. The answer was not always the closest one, but was based on a number of factors such as the easiest approach including glide, etc.

Sqn. Leader Graham Laurie noted that traveling with the Royal's dogs occasionally presented some challenges or humorous situations as the dogs had a mischievous side. In one example when all of the Queen's Corgi's were aboard, Sqn. Leader Laurie remembers he felt something 'soft and squishy'and it turned out to be a sleeping Corgi that had snuck in behind and under his feet directly in front of the rudder pedals while at cruising altitude. On the same flight, a Corgi went missing, and they spent some time trying to locate it after the aircraft had landed. The dog was found unharmed behind a service cart that contained food and was trying to find a way in to get a nibble. Then there was the occasion the black Labrador dogs of Anne, Princess Royal, rolled around in wet grass before departing on the RAF BAe 146. After the flight, the entire crew reportedly spent over a half hour with sticky tape trying to remove all the hair the dogs left, not to mention opening all the doors to air out the 'wet dog' smell.

The last remaining Queens Flight Andover turboprop was retained until January 1991, when it was determined the 146 offered greater reliability for the task at hand and the Foreign Office wanted another aircraft on call. Thus, a third 146, ZE702, was ordered. The final aircraft had further improvements including increased sound proofing, and carbon brakes to replace the steel examples installed on ZE700 and ZE701. On January 14, 1991, ZE702 was delivered to The Queens Flight, and it was also flown into LCY to mark the opening of the extended runway.

High Altitude Operations (by former Sqn Ldr David Gale)

In the summer of 1989, there was a requirement to operate a Queen's Flight BAe146 into La Paz (El Alto International Airport) and Potosi (Captain Nicolas Rojas Airport) in Bolivia. At that time BAe had only flight tested and certified the airplane to operate into airports up to a maximum of 8,000 feet elevation. Because the airport at La Paz sits at 13,325ft, the trip required new takeoff and landing performance data as well as modifications to the oxygen system – after all, no one wanted the cabin oxygen masks to automatically drop out of the ceiling as the airplane made its approach!

The oxygen system was modified to include a switch which prevented the masks from dropping during a high altitude landing, as well as avoiding nuisance 'CABIN HI ALT' warnings on the flight deck. In addition, using performance data from flight testing below 8,000ft, the data for higher altitudes was interpolated by BAe using a degradation for each one thousand feet above 8,000 which ensured the performance data erred on the side of safety.

As is normal for the more complex royal flights, a proving flight without the Royal party was accomplished to iron out any safety and security issues. The flight into Potosi Airport was even more challenging than the arrival at La Paz, because Potosi was around 13,000ft elevation but also had a short runway and was surrounded by even higher terrain in all quadrants. Performance calculations showed that the 146 could safely land and depart from Potosi as long as the airplane carried minimum crew, fuel and baggage. This was rather unusual because the airplane normally carried around a ton of spare parts when flying overseas on extended tours. There would be no backup airplane and no quick and easy way of acquiring essential spare parts such as main and nose wheel tires, essential avionics components, etc.

The arrival into La Paz on the proving flight was uneventful even though the indicated air speed (IAS) during landing of around 125 knots was actually equivalent to around 155 knots (True Air Speed) due to the thin air. La Paz had more than 13,000ft of runway available, unlike Potosi which was only 9,000ft long. The latter might seem like plenty, especially for an airplane like the BAe146, but in actuality it was quite short considering the poorer acceleration and reduced climb performance resulting from the thin air, and with high ground all around. Nevertheless, the 146 was up to the task and safely landed at Potosi.

Unfortunately, there was no suitable Ground Power Unit (GPU) available for engine start. The GPU that was provided was able to keep the airplane lights on but little else and it didn't generate enough power to assist with the engine start. So, the airplane's auxiliary power unit (APU) was used without an assist from the GPU. The engine start took more than twice as long as usual and the APU suffered from a few compressor stalls during the sequence. However, the first engine was successfully started without exceeding any limitations and subsequent starts on the other three engines were also successful.

Because of the challenging issues presented during the proving flight, it was decided to modify the operation into Potosi for the Royal visit. The 146 arrived at Potosi and the number four engine was kept running during the royal deplaning procedure to ensure subsequent successful engine starts. Once the passengers had left the airport the airplane was flown to Sucre (Juana Azurduy de Padilla International Airport), just a 10 minutes flight away. The airport at Sucre was 'only' a 9,500ft elevation and it had a runway length of 9,400ft with no significant higher ground in the vicinity. The 146 returned to Potosi later in the day to pick up the Royal party and an engine-running embarkation was completed before it returned to Sucre. It was all downhill for the remainder of a very successful royal tour!

The Royal Family

Prince Charles, as well as Prince Philip, often occupied the cockpit and flew the BAe 146 (but never with the Queen aboard), always in the left seat with the supervising pilot acting more like an instructor sitting in the right seat. Prince Charles did however fly the aircraft when his sons or the Princess of Wales was onboard. However, an incident occurred during an approach into Glenegedale Airport on the island of Islay on June 29, 1994, with Prince Charles at the controls alongside supervising pilot Sqn. Ldr. Graham Laurie (plus 11 passengers). The aircraft touched down very late on the runway (reportedly on its nosewheel), with a 12-knot tail wind and reportedly 32 knots over the reference landing speed. The official report into the incident stated that wheel brakes were applied before full activation of the anti-skid protection systems, causing the main wheels to lock up and resulting in the tires blowing out, the nose gear collapsing, and the aircraft running off the end of the runway into a ditch. The cost of repairing the damage to the aircraft amounted to around $1.5m.

The Prince of Wales told the press "It was not quite a crash… We went off the end of the runway, it's not something I recommend happening all the time, but unfortunately it did." St James Palace announced that The Prince of Wales had given up piloting royal flights following the incident.

ZE700

For the Duke and Duchess of York who departed London on their honeymoon.

Safety card for the Queen's Flight.

Visiting Chitral, Pakistan September 1991.
Graham Laurie

Princess Diana.

Departing the ceremony as the Captain leads a cheer.

HRH Queen Elizabeth accepting a gold plated model of the BAe 146.

The RAF determined that Prince Charles was not to be blamed because he was regarded as a passenger (invited to fly the aircraft), even though he held the RAF rank of Group Captain and was type rated on the aircraft. The investigation placed blame largely on the Captain, Squadron Leader Laurie, for failing to intervene when performance limitations were exceeded in the final stages of the flight. The navigator was found negligent for failing to advise the captain of the tailwind and subsequent inaccurate approach parameters. Squadron Leader Laurie was not grounded, but he would no longer be allowed to act as an instructing or supervising pilot.

The Queens Flight disbanded on 31 March 1995, with the aircraft moving from their home base at RAF Benson to RAF Northholt. The very next day, April 1, 1995, 32 Squadron became the 32 (The Royal Squadron), Royal Air Force, but with civilian engineers. As a result, only the flight crew and general staff remained, and approximately 150 Royal Air Force staff assigned to the Queens Flight went back to regular duties in other capacities. The purpose of the transition was to make the aircraft available to all VIPs when military transport was not available or required. The role of 32 (The Royal Squadron) was first and foremost military use, but the initial year of it being run mainly by civilian staff was a rough start as there was not the precision of RAF-guided maintenance. This had repercussions that would manifest themselves in a near disaster on what was thankfully a training flight on November 6, 1997.

Maintenance was performed on all four engines, and during a training flight two days later, the crew saw oil depletion from all four engines. This required two engines to be shut down, and prompted an emergency landing into Stansted airport. The cause of the incident was magnetic chip-detector plugs (MCDP) had been installed without oil seals (O-rings) in all four engines. The civilian contractor in this case, FRA SERCO, performed routine maintenance on the 146 engines at RAF Northolt air station. There were supposed to be 12 civilian contractors working on the aircraft during the night shift, but due to a staff shortage only nine were available. One was a supervisor who formerly served with the military as an airframe technician, but he had no engine maintenance training. There were very documented actions performed and they did not include getting samples of the engine oil within 15 minutes of shutdown, nor was an MCDP change kit used. The supervisor did not consult the aircraft maintenance manual, which had strict requirements for the engine servicing. After completing the work, engine ground runs were supposed to be undertaken, but these did not take place.

Subsequently the supervisor who completed the work asked a technician to sign the card as if he did the work, and the supervisor who actually performed the work signed off on it as the supervisor validating the work. Two days later, the training flight departed at 3:10pm, and no oil was seen leaking from the engines during start up or taxi. At 3:25pm as the aircraft was climbing through 5,000 feet, the pilots noticed that the oil gauges for engines number 2, 3 and 4 were showing empty, and engine 1 was showing less than ¼ full. The crew immediately turned around to head back to Northolt when the low oil pressure warning light for the number 3 engine illuminated. Two minutes later the crew declared an emergency, shut down the engine and received immediate clearance to land at Stansted. On final approach the oil low-pressure lights for engines 2 and 4 illuminated, and with thrust on those reduced to idle number 1 was at full power.

After the aircraft touched down and had been parked, the captain watched as the crew chief opened the engine cowls (which were covered in oil) and oil spilled to the ground. The MCDPs were removed and found to have no seals. The incident occurred because of poor planning and failure to follow explicit maintenance procedures and instructions. Sqn Leader Graham Laurie noted that RAF procedures would never have permitted work to occur on more than half the engines, specifically to avoid such incidents, which could be tragic. In September 2018, BAE Systems won a contract to support the RAF fleet of 146 aircraft and will carry out all future maintenance.

In 1998, the aircraft's countermeasures were enhanced. The revised system used a series of four strategically placed Directional Infrared Counter Measures Systems (DIRCM). These external sensors are on the left side ahead of engine nacelle #2; one underneath the fuselage behind the nose gear; and two rearward facing units behind each of the aft gear doors which include steerable laser lamps. Each one contains an ultra-violet infrared sensor and the set up provides complete coverage for all areas of the aircraft.

Should the system be triggered, the CPU located on the flight deck (ironically in a 'black box') would give the pilot an audible warning it had been activated. Pilots would relinquish control of the aircraft so the system could do its job of protecting the aircraft. The system will then shine an infrared beam at the incoming threat in order to disorient its tracking system. The result, in theory, is that the missile begins to change direction radically resulting in it losing the ability to home in on the aircraft and plummet to the ground after failing to find its target. Reportedly the system results in a 3% fuel burn penalty, but that's a small price to pay for safety. Fortunately the system has never yet had to defend an aircraft against a threat. The RAF continues to upgrade the RAF 146 aircraft with the latest DIRCM systems as the aircraft have begun operating in the Middle East where threats are considered higher.

Until 1997 The Queen's Flight BAe 146s had only carried one deceased person, Major Hugh Charles Lyttelton Lindsay, who died in an avalanche while accompanying Prince Charles and Princess Diana on skiing holiday in Switzerland. The third aircraft, ZE702, had the unfortunate task of carrying the body of Diana, former Princess of Wales, after her tragic death in Paris, France in the early hours of August 31, 1997. Squadron Leader Laurie learned of the accident from his son, who had arrived home quite late (just after 2am) and had woken up his father with the news of the car crash. It was not clear at that point if the Princess had passed away, and Squadron Leader Laurie was already scheduled to pick up Prince Charles and Prince Harry from Aberdeen later that day and fly them to RAF Lyneham. At this juncture it became clear to that the schedule already in place would likely be scrapped. When he arrived at Northolt at 6am he learned from the press Princess Diana had died. The household staff began to ring him at the office at 7am, and he learned Prince Charles wished to fly to Paris at lunchtime. There was suspicion that a crew may need to be sent to retrieve the body even though royal protocol called for a four-day pause to be enforced to allow plans for management of public crowds to be implemented, and services to be scheduled at the church in RAF Benson. The tragedy hit home hard because a training event took place just the previous week, which rehearsed procedures for a member of the Royal family perishing while abroad. Squadron Leader Laurie's schedule was originally set for a departure to Aberdeen at 10:30am, but at 10:00am he was told that Diana's body might in fact be repatriated that evening. A change of aircraft from ZE700 to ZE702 was necessary as the latter's rear cargo belly was fitted with a flat floor and a rolling platform with ball bearings that would make moving the casket in and out easier, as a result of the recent training session.

ZE702 took off from Northolt bound for RAF Wittering where Diana's sisters, Lady Jane Fellowes and Lady Sarah McCorquadale, were collected after which the flight continued on to Aberdeen to pick up Charles, the Prince of Wales. At the same time, an RAF Lockheed C-130 Hercules aircraft was dispatched to Paris with a lead-lined coffin. With Prince Charles and Diana's sisters aboard, ZE702 headed to Vélizy – Villacoublay Air Base southwest of Paris. It touched down around 3:00pm, and after arrival both Diana's sisters and Prince Charles visited the hospital to collect the body. They returned with Diana, and a 45-minute holy ceremony with a priest, choir and Guard of Honor was performed at the airport prior to the flight back to Britain. At the conclusion of the brief service, Prince Charles gave the crew start clearance for the return trip.

Piloted by RAF Squadron Leader Graham Laurie, ZE702 took off into a setting sun. French and English air traffic controllers offered a perfect direct route and almost all of them asked the crew to pass on their condolences to Prince Charles. En-route, Sqn. Ldr Laurie switched communication frequencies to a military UHF band to avoid any aircraft spotters, who monitored civil channels, from listening in to messages from ZE702. When Sqd Ldr Laurie took off from RAF Northolt that morning there were only 47 persons at the base. Upon his return, the number had swelled to more than 600. As ZE702 was at about 100 feet and passing over the highway at the end of the runway, he noticed that it was jammed full of cars and everyone was standing outside watching the aircraft arrive. ZE702 touched down at RAF Northolt at 6:51pm on Sunday, August 31, 1997. Once it was clear of the runway, it was briefly parked behind the air traffic control tower with its engines idling, hidden from public view, while two engineers released the straps holding the coffin in place. With that task complete, the aircraft was powered back up and taxied around the apron to a spot where hundreds of press and TV crews were waiting and watching every move being made.

At 7:00PM exactly and to the second, the hydraulics were shut down first (the noisiest), then the engines were powered down, and the rear lower hold door was opened. The Queen's Colour Squadron acted as pallbearers to the nearby hearse, and her coffin was draped in the Royal standard as she was moved from the belly of the aircraft to a temporary resting place. The guard inside the hold told the Colour Squadron to hold the coffin until he gave permission to lift it, ensuring it would not hit the top of the hold door. The Colour Squadron had no time to rehearse and performed a flawless transfer. Even though Diana had lost her royal status after her divorce from Charles, her children retained their royal status and she was afforded the same treatment any royal family member would have received, still referred to by the media as the Princess of Wales.

BUCKINGHAM PALACE

Mrs Alexandra McCreery MVO
Archivist and Librarian to The Duke of Edinburgh

7th November, 2018

Dear Mr Wiklem

The Duke of Edinburgh has asked me to write and thank you for your letter in which you ask about His Royal Highness's association with the BAe 146 from 1986 to 1997.

I am afraid it will not be possible for Prince Philip to give you an interview, however he has asked me to send you a copy of a page from his Flying Log showing his conversion training, amongst other flights he made in 1986.

With best wishes

Alexandra McCreery

Mr Brian Wiklem

Year 1986		AIRCRAFT				
Month	Date	Type and Mark	No	Captain or 1st Pilot	Co-pilot 2nd Pilot Pupil or Crew	DUTY (including number of day or night landings as 1st Pilot or Dual)
—	—	—	—	—	—	— Totals brought forward
JUL	04	ANDOVER	XS 793	SELF S/L WILLIAMS	S/L ANDERSON	LHR → NEWCASTLE
JUL	07	ANDOVER	XS 793	S/L WILLIAMS SELF	S/L ANDERSON	NEWCASTLE → LHR
JUL	14	BAe 146	ZE 701	S/L WILLIAMS SELF	S/L ANDERSON	CONVERSION TRAINING
JUL	14	BAe 146	ZE 701	S/L WILLIAMS SELF	S/L ANDERSON	MARHAM → GATWICK
JUL	17	BAe 146	ZE 701	S/L WILLIAMS SELF	S/L ANDERSON	LHR → BLACKPOOL
JUL	21	BAe 146	ZE 701	S/L WILLIAMS SELF	S/L ANDERSON	SALMESBURY → BENSON
JUL	21	BAe 146	ZE 701	S/L WILLIAMS SELF	S/L ANDERSON	CONVERSION TRAINING
JUL	21	BAe 146	ZE 701	S/L WILLIAMS SELF	S/L ANDERSON	BENSON → LHR
JUL	23	BAe 146	ZE 701	S/L WILLIAMS SELF	S/L ANDERSON	LHR → EDINBURGH
JUL	25	ANDOVER	XS 790	S/L WILLIAMS SELF	S/L ANDERSON	EDINBURGH → WICK
JUL	25	ANDOVER	XS 790	S/L WILLIAMS SELF	S/L ANDERSON	DOUNREAY → EDINBURGH
JUL	30	BAe 146	ZE 700	S/L WILLIAMS SELF	S/L ANDERSON	INVERNESS → LOSSIEMOUTH
JUL	30	BAe 146	ZE 700	S/L WILLIAMS SELF	S/L ANDERSON	LOSSIEMOUTH → EDINBURGH
AUG	02	ANDOVER	XS 790	S/L ROWE SELF	F/L STANTON	BLACKPOOL → LHR
AUG	06	ANDOVER	XS 790	S/L ROWE SELF	F/L STANTON	SOUTHAMPTON → BLACKPOOL
		ANDOVER	XS 790	S/L ROWE SELF	F/L GUTTERIDGE	CARLISLE → LHR
		BAe 146	ZE 700	S/L WILLIAMS SELF	S/L ANDERSON	LHR → ABERDEEN
		BAe 146	ZE 700	S/L LAURIE SELF	F/L STANTON	ABERDEEN → FARNBOROUGH
		BAe 146	ZE 700	S/L LAURIE SELF	F/L STANTON	FARNBOROUGH → EDINBURGH
		BAe 146	ZE 700	S/L WILLIAMS SELF	S/L ANDERSON	LHR → PERUGIA
		Ae 146	ZE 700	S/L WILLIAMS SELF	S/L ANDERSON	PERUGIA → LUXOR
		Ae 146	ZE 700	S/L WILLIAMS SELF	S/L ANDERSON	LUXOR → SEEB
		Ae 146	ZE 700	S/L WILLIAMS SELF	S/L ANDERSON	SEEB → MADRAS
		Ae 146	ZE 700	S/L WILLIAMS SELF	S/L ANDERSON	MADRAS → PHUKET
						Totals carried forward

A page from Prince Philip's flight log, showing BAe 146 conversion training (July 14, 1986).

Unpaved airfield operations in Jijiga, Ethiopia.

Graham Laurie

Keeping ZE701 clean while on tour in Africa on an unpaved field. Note the anti-missile countermeasure system next to the ladder behind the rear door.

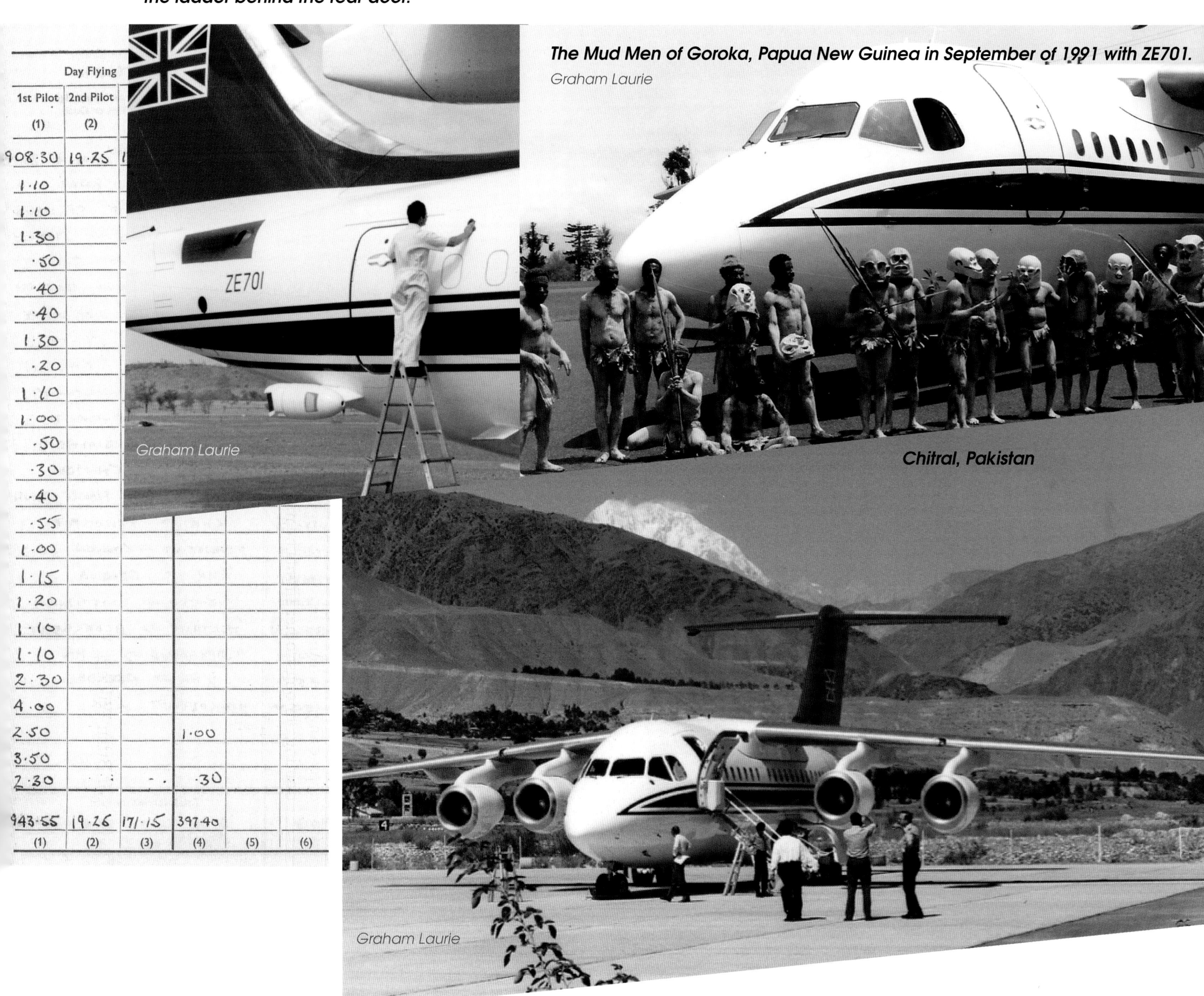

Day Flying					
1st Pilot	2nd Pilot				
(1)	(2)				
908·30	19·25				
1·10					
1·10					
1·30					
·50					
·40					
·40					
1·30					
·20					
1·10					
1·00					
·50					
·30					
·40					
·55					
1·00					
1·15					
1·20					
1·10					
1·10					
2·30					
4·00					
2·50			1·00		
3·50					
2·30			·30		
943·55	19·26	171·15	397·40		
(1)	(2)	(3)	(4)	(5)	(6)

Graham Laurie

The Mud Men of Goroka, Papua New Guinea in September of 1991 with ZE701.

Graham Laurie

Chitral, Pakistan

Graham Laurie

ROYAL AIR FORCE

Paul Seymour

Twenty minutes later, ZE702 departed RAF Northolt for Aberdeen so that Prince Charles could return to Balmoral with his children, Prince William and Prince Harry. When the aircraft landed at Aberdeen, Prince Charles came up to the flight deck to say goodbye as he usually does. Much to Squadron Leader Laurie's surprise, he and Charles talked for nearly five minutes. But Squadron Leader Laurie was so overwhelmed emotionally with the day, he cannot recall what they talked about! He surmises he said what anyone would have said in that situation. Captain Laurie overnighted with the crew in Aberdeen, and the next day at lunchtime he departed to RAF Northolt with some of the royal's staff. That evening he went to Kensington Palace to lay some flowers, and went back again on Thursday to sign the book of condolence. On September 6, 1997, Squadron Leader Laurie flew Her Majesty the Queen and The Queen Mother to Aberdeen after the funeral. The entire week was a daze, but upon personal reflection, Captain Laurie realized he had flown Princess Diana over 200 times, and was honored to fly her on her final flight home.

ZE700, ZE701, and ZE702 served the RAF, both in the role as transport to the Royal Family, as well as military transports. Squadron Leader Laurie reflected that other than some minor issues with the 146s, it was not only a reliable aircraft but a joy to fly. ZE700 and ZE701 remain with the Royal Air Force to this day, but ZE702 was decommissioned in 2001 due to budgetary cuts and because it had not received the military countermeasure modifications installed on the other two aircraft. ZE702 eventually joined Airfast Indonesia as a VIP aircraft, but having the interior updated to white and gold, including a bed. It now carries the registration PK-TNV and flies for TransNusa Air Services.

In 2004, the Ministry of Defense required the livery carried by the 146s to be updated to a lower key, airline-like livery. In an effort to make the aircraft less of a target of terrorist aspirations, the red vertical and horizontal colors were dispatched. The electronic countermeasures were also updated again in 2008 with a system called Large Aircraft Infrared Countermeasures (LAIRCM), with missile warning sensors located on each side of the forward fuselage. The squadron also began modernizing its fleet and acquired two additional BAe 146 aircraft in 2011, this time series -200 QCs, to provide tactical airlift. Today the RAF operates a fleet of four BAe 146s, two series -100s CC.MK2 and two series -200QC CC.MK3 aircraft with countermeasures installed in 2011. As recently as August 2019, the MOD has been hinting it may be time to retire ZE700 and ZE701.

pelita air
G-6-239

Pelita Air Services

Pelita Air Services took delivery of series -200 PK-PJP, the 50th BAe 146 built, on behalf of the Indonesian government on June 27, 1986. This aircraft also had pannier tanks fitted for extended range operations. Pelita Air Service was the aviation division of the Indonesia state oil company Pertamina, but in practice PK-PJP was a presidential aircraft used to fly the President around Indonesia. Pelita Air Service took delivery of a replacement aircraft, an RJ85 ordered in 1993. The aircraft was flown 'green' (empty) to be outfitted by Marshall Aerospace at Cambridge, UK. Interior concepts developed by Design Research Associates in Warwick used a specified color scheme, which was blue and black with gold trim. Outfitted with soft leather interior, dark blues, and pile carpeting in conjunction with strategically placed mirrors that gave the cabin a much larger feeling. It also had no overhead bins.

The forward VIP area had a 'club' like arrangement with each side featuring four seats (two and two facing each other) along with a redwood burr veneer table complete with gold trimmed cup holders and Pelita's seahorse logo. All metal surfaces such as belt buckles, sink fixtures, ashtrays, vents, and more were gold plated aluminum. The headrests were designed to accommodate the elaborate and traditional hair styles of Indonesian women on ceremonial occasions. The center section of the aircraft featured two swiveling and fully reclining sleeperette seats (which could rotate through 360 degrees). The forward cabin had an entertainment system complete with laser disc video and a surround sound system connected to a 20" television. The rear of the aircraft featured one row of first-class seats (in a two-two configuration) followed by five rows of six-abreast economy seats with 32" pitch. A large rear galley provided meal services.

Marshall of Cambridge

Air Zimbabwe

BAe 146-200 Z-WPD (E2065) "Jongwe" was delivered on November 25, 1987 and was intended for dual use as both a VIP and an airline aircraft. It was rumored that in order to win the order, BAe borrowed eight rows of seats and furniture from a local hotel to give the aircraft a luxurious look in the front of the cabin. President Mugabe inspected it and was pleased with what he saw but declined to join the crew for a test flight and following a brief aerial demonstration immediately afterward (with the furniture strapped down), he decided to purchase the 146. It was painted in Air Zimbabwe colors and although it was a government aircraft, it was shared with the airline.

The front of the aircraft could reportedly be converted to a VIP layout within 40 minutes, the amount of time it took officials to get from Harare city to the airport. The aircraft was divided into three compartments with the front of aircraft with eight luxury seats (2+2) and a folding table in between. The second compartment featured two sofas with three seats each and a table in front of them facing each other. The rear of the cabin was set up for 40 coach passengers in typical airline layout. The interior even featured a television on a pneumatic shock system (no flat panels then) to provide inflight entertainment, which could be stowed away into a compartment when not in use. The 146 was sold with the provision that four Hawker Hunter jet fighters and two spare Avon engines for the Air Force were included in the deal. It was believed Mugabe wasn't interested in the 146, instead after the fighter aircraft. A second aircraft (non-VIP, A2-ABD) was leased by Air Zimbabwe, and served for two years. After suffering from lack of capital to maintain both aircraft (resulting in lack of spares), they were withdrawn from use in 2000 following 13 years of service (two years for A2-ABD).

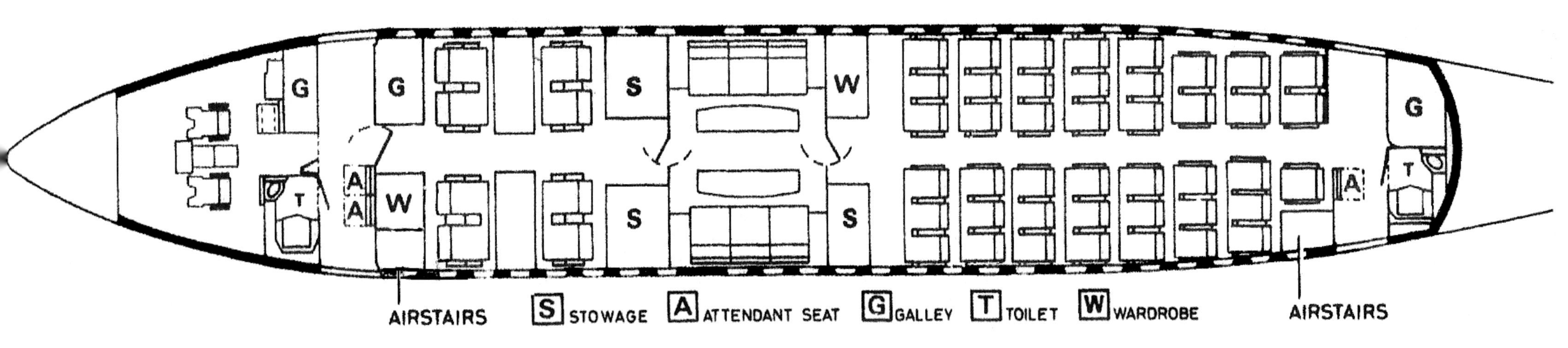

The center of the aircraft featured two sofa's facing each other with folding tables in front. This section could seat six.

Abu Dhabi Amiri Flight

The Abu Dhabi Government purchased three aircraft over the years, starting with BAe 146-100 A6-SHK which was delivered on December, 20 1988. A second aircraft, Avro RJ70 A6-RJK (later reregistered A6-LIW) was added to the fleet of the newly named Royal Jet (Amiri Flight) in October, 2005. RJ100 A6-AAB was the third and final aircraft to the (again newly named) Abu Dhabi Presidential Flight. Jet Aviation AG of Switzerland fitted out the series -100 A6-SHK with an adjustable interior that could accommodate 10 to 18 passengers in a VIP configuration, with the rear of the cabin configured for 29 to 44 seats in a business class set up. The aircraft served for nearly 11 years with the Abu Dhabi government.

Boutsen Aviation

Boutsen Aviation

E1267 former UAE Amiri Flight RJ70 was sold to Bernie Ecclestone who used the aircraft as a private transport for three years before listing it for sale.

In May 1999, maintenance personnel at Crossair servicing the aircraft discovered corrosion in the wing tanks. It was determined to be uneconomic to repair and after a mere 3,251 hours of flight time, the fuselage was flown in an Antonov An-124 to the UK where it was broken up by Air Salvage International. A6-LIW was configured with six plush seats in the front of the cabin with tables, while the mid-cabin had four business class reclining seats, and the rear had a row of five economy seats for the staff. The bathrooms featured gold plated sinks and faucets. It was rumored that this aircraft served as a transport for the Dubai Royal Family and their falcons, where falconry is a very popular (and controversial) sport.

Air Salvage International - Mark Gregory

The forward cabin of the RJ70, with seating for six with tables for working or dining.

Boutsen Aviation

The rear of the cabin featured sleeper seats for eight persons, and a single row of 5 economy seats for staff.

Boutsen Aviation

The aircraft including the bathroom was equipped with gold plated fixtures.

Boutsen Aviation

F1

Formula One's Bernie Ecclestone acquired a second hand BAe 146-100 in 1998. Fitted with 52 economy seats and eight first class seats, the aircraft was used as a corporate shuttle for F1 personnel. An early build airframe (E1006, dating back to 1983) delivered to Dan Air initially, and flown by Malmo (1994) and Cityjet until 1997, the aircraft was registered G-OFOA and painted in a stunning and aggressive livery with "F1" on the tail. A second aircraft, G-OFOM (MSN E1144), was delivered in March 2000 and carried the same livery but on a white painted airframe and lacking the designation on the tail. G-OFOA was retired in 2017 as a replacement Avro RJ70, M-STRY (MSN E1267), was added to the fleet. The RJ70 was formerly VIP transport for the Abu Dhabi Amiri flight before being delivered in 2015 to F1. In 2018, M-STRY was put up for sale by Boutson Aviation. It was still configured in the same layout as it was originally delivered to the Abu Dhabi Amiri flight.

E1144 , also operated by F1 carries the interior of former operator P.T.N.A.C.

Montex Drilling (Moncrief Oil)

Aircraft E1068 made its first flight on December 20, 1986, approximately one month before heading to China, where it would serve in airline operations with both China Eastern Airlines and CAAC for nearly seven years. Upon being returned to BAe and following a brief stint at Carib Express, it served as a corporate shuttle for BAe for approximately a year, transporting employees between company facilities, before it went to a new owner, Moncrief Oil based in Dallas, Texas. The majority of Moncrief Oil's oil drilling operations took the principles across the state of Texas but as the company grew, so did its operations and production, expanding into the nearby states of Colorado and Wyoming. With Texas being the largest state, and operations in nearby states, getting staff to its sites quickly and without the hassle of scheduled airline services necessitated a private aircraft. Moncrief operated a BAC One-Eleven (N114M) between 1974 and 1997 but due to it age as well as high maintenance and fuel costs, it was time for a new aircraft.

Captain Jim Skinner contacted British Aerospace's office in Washington, inquiring about performance specs for operations out of KGUC (Gunnison-Crested Butte airport in Colorado), located at 7,658 feet altitude. Upon receiving the results, the boss was sold and with the aircraft having four engines, safety was the number one reason the 146 was chosen. British Aerospace demoed a series 200 to Moncrief Oil, noting that the series 100 was only 'negligibly' smaller. The series 100 BAe was then operating was selected, a deal done, and conversion began with a delivery to

Kevin Govett

Using the airstairs during a visit to Ft. Lauderdale, Florida.

Jim Skinner

Kevin Govett performed most of the servicing of N114M. The aircraft is being supported by jack stands.

Moncrief on October 25, 1997 after which the aircraft was ferried to Innotech Aviation, Montreal for the VIP conversion. N114M was ultimately configured to carry 25 passengers comfortably, and up to 1,000 pounds of luggage per passenger.

The pilots began simulator training at Reflectone (now Pan Am Flight Academy) in Virginia, and went through additional training at Woodford, United Kingdom before obtaining a type rating shortly after. In later years, check rides were performed not in the simulator, but in the actual aircraft. The aircraft flew approximately 150 hours per year, with a maximum of 300 hours in some years. Most flights were to the sites of company operations in Texas as well as neighboring states. Occasionally it visited destinations in California and the East Coast too, all of which were within the range of the aircraft. N114M was not fitted with the Pannier tanks to extend its range, as not only were they not available (very few customers took them up), but there would have been a significant expense to install them, and they added another layer of maintenance.

Windshield cracked during flight due to a faulty defrost heating element.

Jim Skinner

Captain Skinner loved flying the 146 and was quick to note that with exception of the early years of operation, it was a joy to fly. Prior to mechanic Kevin Govett joining Montex Oil, the first few years after service entry were a bit rough. The engines and all the quirks were not ironed out, meaning dispatch reliability was not where it should have been. Jim encountered a windscreen that experienced an electrical arc that resulted in the windshield completely cracking inflight at cruise altitude. The PPG glass held up as expected, but the aircraft had to be grounded until a new one could be installed. Jim also had to make a flaps up landing, as there was a 'quirkiness' in the flap system that Kevin discovered. If the pilots did not move the flap lever to the desired setting in one motion, but instead stopped at each gate setting before moving the lever again, the flaps would often seize. He ended up finding a solution and designed a modification to prevent the problem from happening again. He also noted that it was necessary to pay extra attention when maintaining the flap system with lubricants, and do so frequently to avoid any jams.

Kevin started working at Montex about three years after N114M joined the organization. He noted that if he did not stay on top or ahead of maintenance, the 146 would "eat your lunch", especially affecting anyone not familiar with the aircraft. Having worked on them, including former AirCal planes at AAR, Kevin knew all the quirks. He was responsible for all upkeep on N114M, and generally maintained the aircraft as the only mechanic. For bigger jobs or heavy maintenance, additional staff were contracted to assist.

Kevin Govett

N114 featured a very plush interior including a separate dining area.

Kevin felt BAe's low utilization maintenance program was too 'light' and as a result, and just did not make sense. He redesigned and submitted a revised program of calendar inspections (one-month, three-month, 12 month, then yearly afterwards), which was approved by the FAA.

The aircraft had been used as a shuttle at BAe, and Kevin was in "shock and awe" that the manufacturer had flown it in the condition it was in, which included air cycle machine leaks. By the fifth year of ownership, all the bugs had been sorted out, and the aircraft served its expected missions perfectly. He really liked the 146 and described it in his light Texas accent as "like putting on an old shirt, the cockpit just makes sense as well as the way the aircraft was put together." It was easy to maintain and work on. In periods when the aircraft was not flown (i.e. every 30 days), the crew would take it up for a technical flight. If a scheduled trip was planned, it was prepared and inspected ahead of time to ensure smooth operations and dispatch reliability. Kevin did not travel with the aircraft, but if there was an issue it would be discussed over the phone and if it was serious enough he would go to the location and address the problem.

Pat Doyle

Pat Doyle, former Air Wisconsin pilot who ferried N114M down to Chile.

Kevin noted that the outboard engines cabling was chafing due to the cowlings rubbing against them. He redesigned how the cables were run, and subsequently BAe reportedly adopted his modifications and relayed them to other operators. One interesting item to note is Kevin compared the maintenance cost of the Gulfstream 4 [executive jet] to the 146 and determined that the 146 was substantially cheaper. In order to avoid any engine issues, Kevin ensured that all the LF-502s had the upgraded LF-507 parts. In later years, it was cheaper and easier to just buy new engines instead of looking for parts to service the existing ones. It also became harder to find spares as the years went on, and Prestwick was no longer manufacturing many of the parts that would keep the aircraft going. He had to creatively source parts all over the world.

Mr. Moncrief was slightly disappointed with the -100 versus the -200 he toured, noting that his seat was right next to the #3 engine, whereas on the -200 it would not have been. N114M was the only VIP operated BAe 146 aircraft in the United States, or even North America. It, like the previous BAC One-Eleven, carrier the name "Lucky Liz", after the principal's mother. Her name was put on an oil lease that ended up generating the bulk of their wealth. After nearly 17 years in service with Montex, the 146 was retired and subsequently sold to Aerovias DAP by Tronos. They will retain its VIP configuration and it will be used for charter flights.

Handover ceremony for Bahrain Defense Force aircraft E2390.

Bahrain Amiri Flight

In May of 2001, the Emirate of Bahrain ordered a single Avro RJ85 configured for VIP use. This RJ85 was one of the last five on the production line prior to the switchover to the RJX variant. It joined the Amiri Flight which also operated a Boeing 747SP, Boeing 727-200, and two Gulfstream aircraft. Registered A9C-BDF (E2390), the aircraft was delivered on November 10, 2001 and was joined by a second RJ85, A9C-HWR, in 2005. A third aircraft, RJ100 A9C-AWL (E3386), was also added to the fleet in 2008.

Forward cabin of the Bahrain Defense Force Aircraft.

България
Team of Bulgaria
swissport
BAE SYSTEMS
Asset Managment

Paul Seymour

Design Q aircraft

Design Q, based in Worcestershire, United Kingdom is a multi-faceted design studio with a background primarily in automotive and aviation, from corporate branding to full exterior and interior design. Founded by current CEO Howard Guy, the team assembled at Design Q have strong passion for the work they produce. Having produced a number of corporate interiors for a variety of jet aircraft, Design Q landed its first corporate client for the British Aerospace jet in 2008. Hemus Air (now Bulgaria Air) of Bulgaria approached the firm to design a VIP interior for its single Avro RJ70. Hemus Air wanted high end but modern styling for a VIP aircraft that would be used by both the head of the airline as well as chartered out to heads of state and sports teams. The interior was outfitted with luxuriously appointed club seating featuring light colored Parchment leather framed with rich dark Burr Walnut veneers.

When a passenger first entered the aircraft, they encountered a bar complete with stemware which implied "this is going to be an enjoyable flight". Entering the split cabin, travellers would encounter two groups of four seats with tables for working or dining. Immediately aft were two chairs facing a table, with a large screen TV in between for entertainment. Across the aisle was a large couch for up to four facing the television. Passing through the partition accessed a separate cabin with twelve more seats.

Design Q, at the request of BAE Systems, produced a series of concepts for the Avro RJ as a VIP aircraft with very extravagant styling. They were given free rein to come up with innovative designs. The first in a series of four aircraft concepts to be unveiled, and the most controversial, was the ABJ Explorer. BAE Systems had received a QC aircraft back off lease and was going to remove the cargo door by welding it shut. Design Q said "no!" It had a VIP design that would work very well with the rear door and was by far the wildest of those it came up with, pun intended. Matching the Avro RJ's ability to land on unpaved or gravel airstrips, the idea behind the Explorer One and Four was that of a 'Safari' aircraft, the ultimate "go anywhere" flying machine. Using a QC or modified QT aircraft, the idea was that the aircraft would land where there was animal life, the rear main deck cargo door would open, and a viewing platform (with removable railing) would emerge and slide out from under the main deck floor. The two chairs in the rear of the aircraft would then swivel to face outwards. A second set of portable chairs and table could also be placed on the viewing deck.

On the starboard side of the aircraft the forward lower hold cargo area would be converted into crew sleeping quarters, while an awning could be erected over a makeshift patio to provide a sitting area. The rest of the aircraft was set up with a forward sitting/eating area (like the Hemus Air aircraft), with four separate cabins in which couches and chairs could be converted into queen sized beds. In the center of the aircraft was a large shared bathroom. A revised secondary interior concept (Explorer Four) turned the front of the aircraft into a large lounge area (with dual couches and chairs), and the center of the aircraft under the wing into a six-seat dining area.

Shown at the May 2010 EBACE show, the Explorer certainly baffled attendees at how 'out there' the design was. Critics and enthusiasts found it hard to fathom, with some interpreting the observation deck as being deployed while in flight as well! Needless to say, Howard received some pretty strong hate mail from

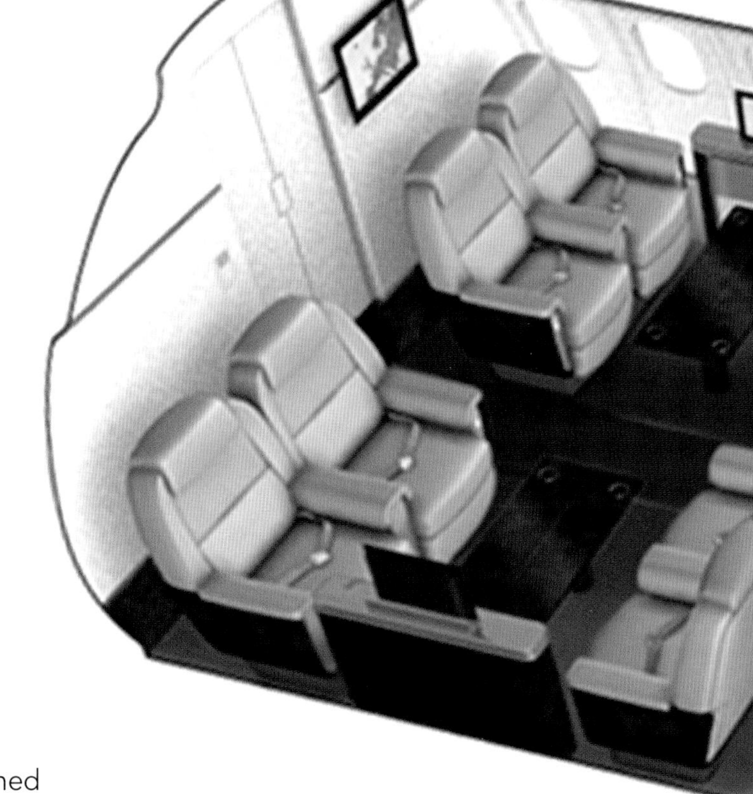

those that just did not "get it". But in another turn of events, there was a lot of very serious interest from potential customers in the Middle East to acquire and convert the aircraft into a fully functioning Explorer. What happened? "Lack of [aircraft] range killed off the Explorer," Howard explained.

Shortly afterward, a second design was unveiled at the October 2010 NBAA (National Business Aviation Association) show in Atlanta, Georgia. This, known as the "Fusion", went in the opposite direction to the Explorer. It was aimed at the charter market for sports team, corporate VIPs, celebrities, even high-end hotels shuttling VIP customers around. It was meant to generate the response "wow, this is going to be a lot of fun" the moment a passenger entered the cabin, seeing the lounge-like seating and lighting, and the sit-at bar to the right. The front of the cabin featured two VIP bathrooms, followed by a large lounge with two couches and chairs, a small intimate dining area on the starboard side, and the bar on the port side under the wings. At the rear of the aircraft was a second sitting area with a larger dining table able to accommodate six guests. The rear furnishings were designed 'love seat' style.

The third design, the ABJ Elegante was unveiled at the May 2011 European Business Aviation Convention and Exhibition (EBACE). While still portraying a modern style, it was aimed specifically at the Middle Eastern market for VIP aircraft. It featured contrasting black and white coloured leather, white patterned cabin linings, and liberal use of wood trim and golden fixtures. It also had forward and rear lounges separated by a central meeting or dining area, and a large bathroom at the very rear of the cabin. The final design was unveiled not by BAE Systems, but by new owner Falko which had purchased BAE Systems aircraft and spares support assets in 2011, at the NBAA show. The ABJ Q was a reference to the fictional "Q Branch" from the James Bond series of films noted Gary Doy, Director of Design Q. The aircraft had a number of design elements in hidden compartments that were activated by finger print recognition. The most notable accent was the 'tunnel' with LED accent lighting that also cleverly hid the wing box. Designed to mimic the gun barrel viewers see at the beginning of a James Bond film (a visual trademark), the lighting colors could be changed and there were also full light panels that could be turned on to mimick daylight. Unfortunately, none of the designs were ever selected by ABJ customers, but if this author ever wins the lottery the ABJ Q will see the light of day.

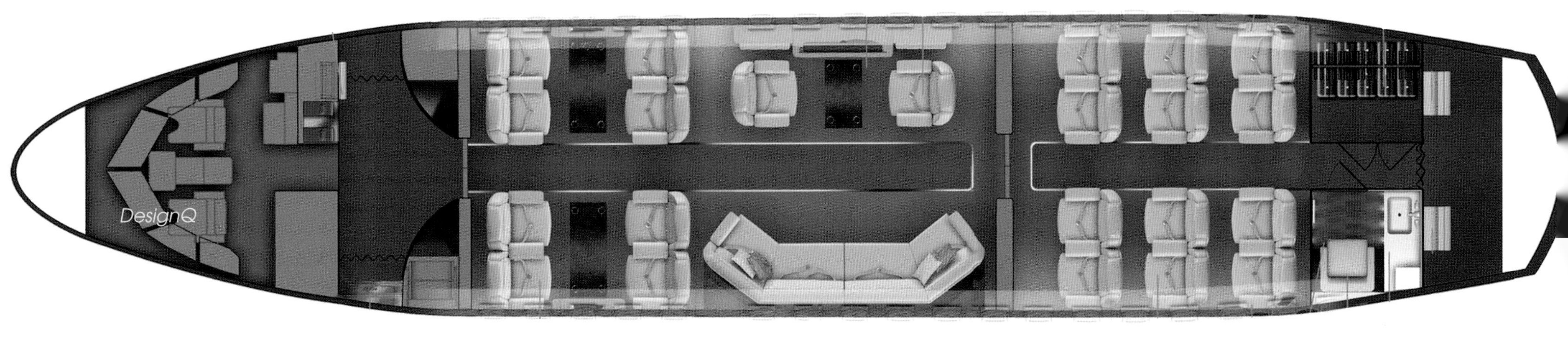

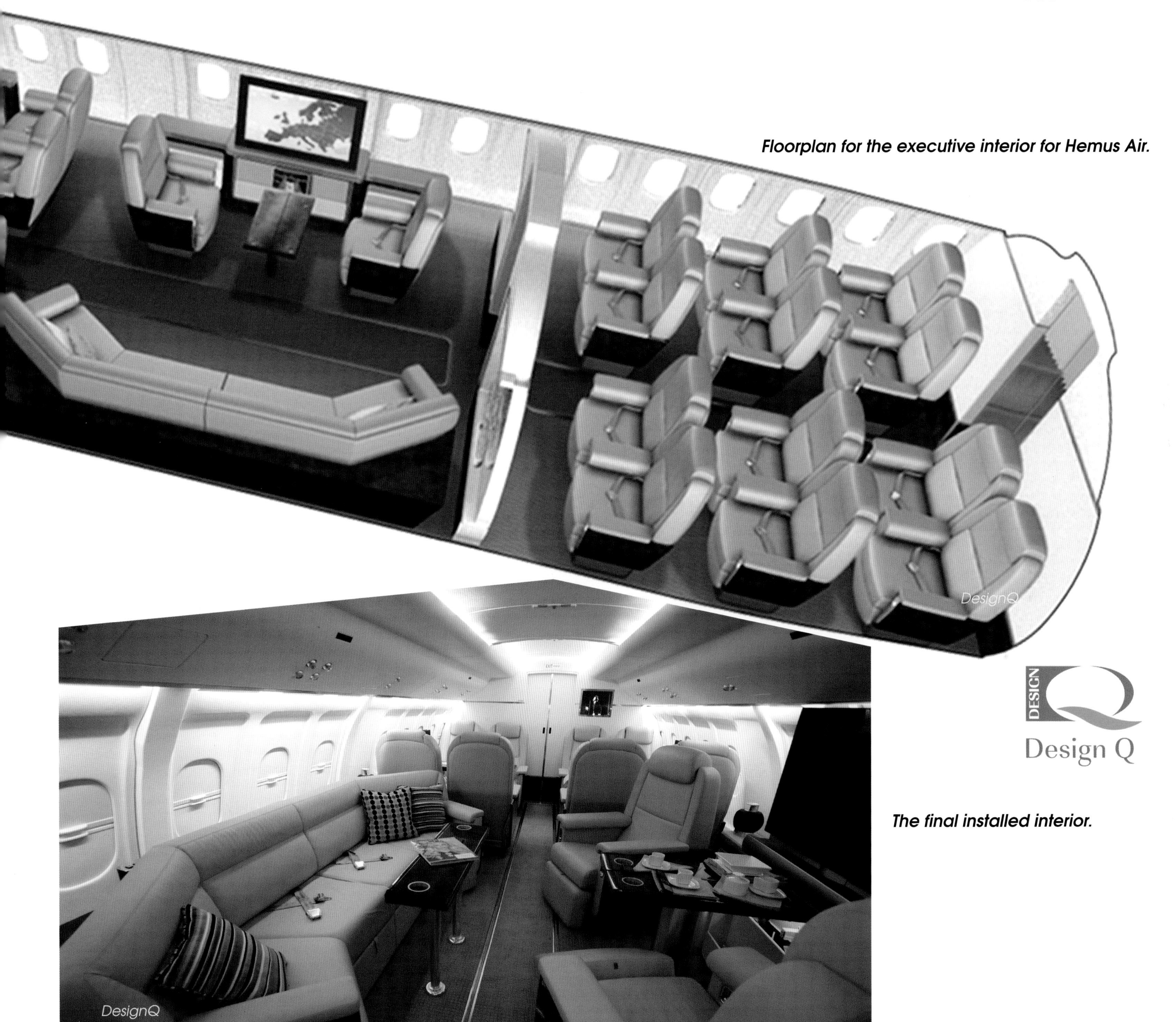

Floorplan for the executive interior for Hemus Air.

The final installed interior.

The controversial Explorer concept based on the Avro RJ100.

The Explorer lounge with the observation deck deployed.

DesignQ

The Fusion concept.

The Elegante bedroom concept.

The "Q" Concept with the gun barrel ceiling hiding the wing buldge.

British Aerospace

BAE Systems operated two aircraft for its internal Corporate Air Travel program (CAT). It used G-TBAE (E2018), a BAe 146-200 that served with a number of airlines, between 2003 and 2013 and BAe 146-100 G-BLRA which joined in 1997 following service with Aspen Air and Tristar Airlines. Neither aircraft, unlike the examples referred to earlier in this chapter, had luxurious VIP interior configurations. Instead, the series 100 had 66 seats and the series 200 had 97 seats. These aircraft were used to shuttle employees to different BAe manufacturing sites or partner locations, and both have since been scrapped.

One of the aircraft British Aerospace shuttled employees and executives between locations.

In later years when the updated Avro RJ series came online, a more extreme version of the Statesman was marketed as the "RJ Sportsman". It had an interior in line with 1990s trends which was designed to 'Mulliner Park Ward' standards of quality. Featuring the QC door and the ability to carry horses (and the staff to attend to them with seating for six in the rear), all-terrain vehicles (ATV), 'virtual reality stations' and watercraft, the designs pushed the "play hard" envelope. Missing from this iteration was any representation of carrying

Uzbekistan's first Avro RJ, E2312 RJ85, which was configured in VIP configuration. Note the vertical stabilizer leading edge has been painted.

full size passenger vehicles – perhaps BAe realized this was far from practical. But in nearly every iteration there was seating for six household staff (in economy type seats), separated from the main cabin. The 146 and subsequent Avro RJ did serve a variety of roles in the VIP market, and a few still continue to operate in an executive capacity today.

The large executive jet market demands two things above all else: speed and range. Neither were attributes of the 146 so inevitably they type was going to be restricted to a niche market where its strengths of airfield performance and interior space were the prevailing factors. BAe never put great effort into this market, perhaps recognising that it was distinctly limited. A more dedicated commitment might have yielded a few more sales but that would have been it.

VIP interior of E2312 in the forward compartment.

CHAPTER 15

Paper Airplanes

Every manufacturer investigates variants or spin-off versions of their in-production aircraft. Each is looking for ways to build upon the platform and leverage it for new markets, without the substantial R&D required for new aircraft development. Often these are design exercises, involving no formal plans to come to market. They are used to gauge customer interest in a derivative that may fit into an existing route structure that the current aircraft cannot support. Why invest substantial sums of money for new aircraft when you can leverage the current aircraft further? Since its launch, BAe had always planned for variants of the 146 and there has been no shortage of design exercises over the life of the type. This chapter explores just a few of the many generated over the years.

Prop-Fan

The high mounted wing provided the more than one option for power plants. In 1984, airlines were not abandoning turboprops for jets (yet), and BAe announced in that it was studying both pusher and tractor propellers. Tractor propellers power the majority of prop aircraft currently, where the propeller 'pulls' the aircraft forward. Pusher-props are mounted behind the wing and of course 'push' the aircraft forward. There were designs for both quad (four) propeller systems as well as twin prop variants. Additionally, geared and ungeared open rotors as well as ducted rotors were considered. The advantages of a geared contra-rotating propfan were improved economics and higher power versus a standard turboprop. The powerplant would in theory have higher rotor propulsive efficiency, and a lighter and more efficient gearbox with less system complexity. With a tractor pull powerplant, there were more efficiencies to be gained: fewer blades per disc; less intake blockage; elimination of the need to contour the wing to remove slipstream swirl; the possibility of lowering tip speed to reduce source noise without loss of propeller efficiency. Additionally, cruise speed would also be considerably higher than most props, with an average of MACH .75, and the ungeared propfan having higher power giving true cruise speed of MACH .80. These were forecasts for typical short-medium range airliners, whereas regional aircraft in the 80-120 seat range would be looking to a MACH .65-.70 cruise on a smaller powerplant with either open or ducted fans.

Pratt & Whitney had three engines under development, all candidates for installation on the BAe 146. The STS742 powerplant had a dual contra-rotating prop-fan, each fan having six blades. Pratt & Whitney was putting more effort into a pusher version (STS743) than the tractor type open rotor (STS742), which would need to be scaled down for the BAe 146. Allison had the 578-B2 turboprop, which BAe was considering in design exercises for the BAe 146-300. Rolls-Royce was also working on the RB529-04, a ducted high bypass contra-rotating fan.

Two variations of the BAe 146: the aircraft above features unducted turboprops, while the aircraft to the left features ducted prop-fans.

By 1986, BAe unveiled a scale model with a twin-prop power plant, each having two geared counter-rotating props delivering a cruise speed of Mach 0.65. The logic that the 146 airframe could be kept competitive well into the 1990s by stretching it (up to 120 seats) and re-engining it was presented by Alan Blythe, Project Engineer for BAe Civil Division to the International Council of Aeronautical Sciences Congress. BAe was focusing on derivatives of both engines and airframes to keep development costs to a minimum and cost of ownership low for airlines, especially those that might operate both the jet and prop-fan version of the 146.

Unswept contra-props using a turbo-shaft engine could add up to 45% range to the 146, with direct operating costs up to 6.5% less than the ALF-502 powered aircraft. BAe even generated design proposals with the airframe having a lower wing and rear fuselage mounted engines as an alternative.

Outside of some limited testing on a static aircraft wing and on structural specimen 201 in July of 1987, BAe never built nor marketed a prop-fan version of the BAe 146. With fuel prices continuing to fall (instead of increase), and the increase in cabin noise, history shows manufacturers abandoned the idea of open-rotor engines for commercial airline aircraft and by the end of 1988 BAe had dropped the idea of a prop-fan version of the 146.

Ducted-Fan

At the Paris Air Show in 1987, BAe showed a twin-fan version of the 146-300, with ducted fans nearly twice the size of the ALF-502 engines. This aircraft, which did not get beyond the concept phase, was pitched for airline service as well as for military missions where clandestine operations were needed. One of the engines in development and under consideration was the Pratt & Whitney STS749 ducted prop. The aircraft was forecast to have an increase of 350nm in range compared to earlier versions, as well as a substantial drop in fuel burn. The BAe 146-200 was the 100% baseline, the -300 was at the 89% mark with the ducted fan coming in at 68%.

146NRA – New Regional Aircraft Twin

October of 1989 brought on an internal British Aerospace evaluation involving moving the then current BAe 146 from four-engines to a two-engine turbine design. BAe also sought out risk sharing partners to be a part of the project. Evaluations of new power plants such as the Rolls-Royce Tay 620, 650, 670, and the 690; BMW/Rolls-Royce BR715 and 720; IAE V2500; and the Pratt & Whitney RTF180 were undertaken. British Aerospace also sought input from airlines currently operating the BAe 146. Even a temporary internal designation of '126' was used for the new 146NRA aircraft, a throwback to Hawker Siddeley nomenclature.

By late 1990 the 146 Twin was beginning to evolve into a more complete and robust aircraft, and was expected to go head-to-head against the higher density and larger Fokker 100 derivative being designed (Fokker 130). Northwest Airlines offered a potentially large order for a twin engine aircraft, pushing BAe to pursue the designs. The manufacturer had plans to unveil its 146 Twin at a press conference on December 4, 1990, but the event was abruptly cancelled at the last moment. It was rumored that the cancellation was due to BAe shifting from offering a single choice of the CFM56-3B-2 for the new 146 Twin, to offering multiple choices which included a version of the Rolls Royce Tay 690 power plant. The 146NRA was initially scheduled for an entry into service around 1994-1995. Deutsche Aerospace (DASA) originally sought to participate in a joint development with BAe, but the UK manufacturer baulked at DASA's insistence on leading the program. All of this was occurring while BAe was testing the new LF-507 on a -300 demonstrator in preparation to also offer an updated BAe 146 line.

Shortly after the cancellation of the press conference, BAe announced it was looking for risk sharing partners to share the estimated $1 Billion USD it estimated the 146 NRA was going to cost to bring to market. It also began to release details of the 146NRA including a new wing with a 25-degree sweep, higher aspect ratio (allowing for a cruise speed of Mach 0.82 @ 39,000 feet), and twin engines. The advantages of the existing 146 – STOL and very quiet performance – were traded on the 146NRA in favor of faster cruise speed and engine commonality with larger airliners.

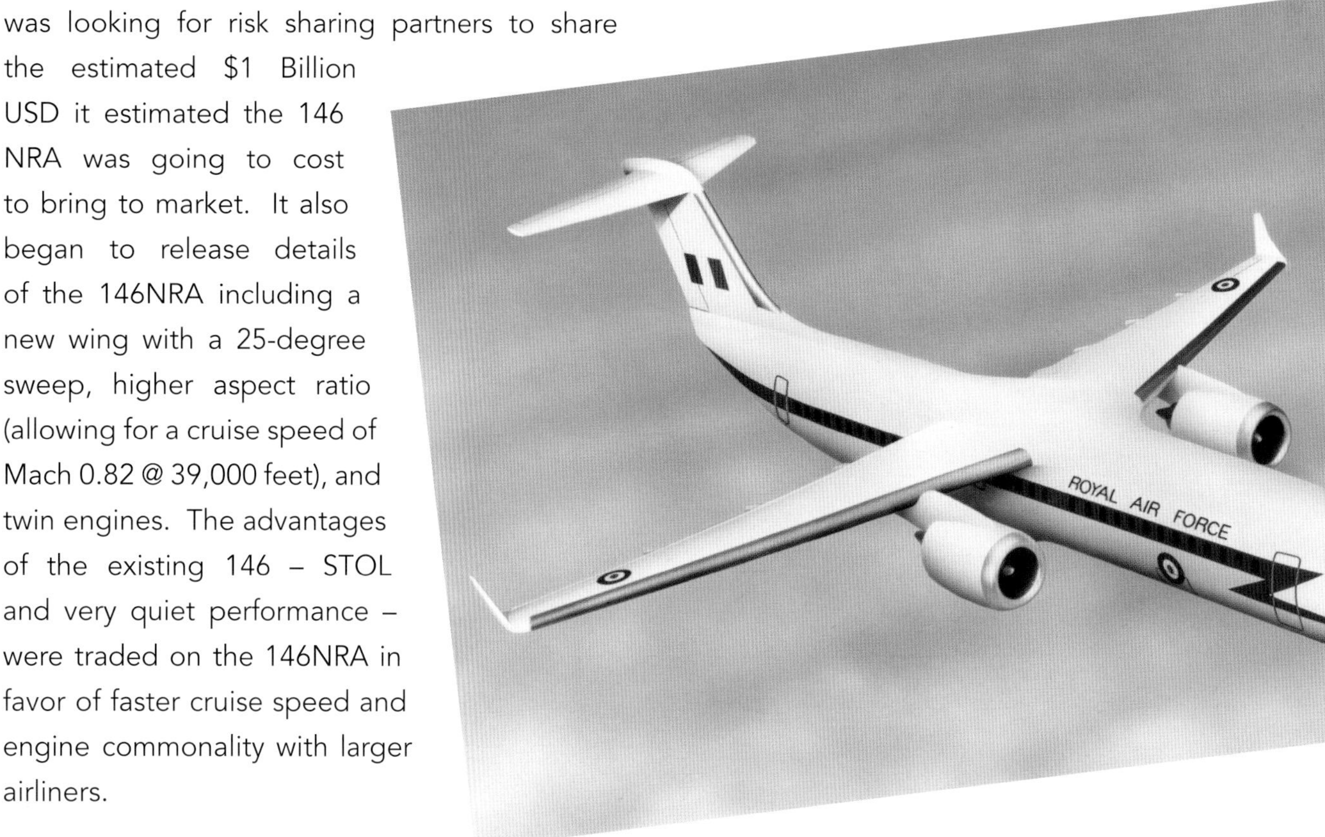

This was discussed in the press during 1990, but in April 1991, BAe released a brochure detailing an aircraft known as the BAe 146-NRA with the following features:

• 100-130 seat category (with capacity for 110-136 five-abreast at 29" pitch)

• Mid-cabin exits

• Engines to be determined, but CFM56-F5 were mentioned, as were V2500 derivatives (with BR715, MTU RTF180 and CFM M123 studies proposed) – each having 18,000-21,000 lbs. thrust

• Fully digital flight deck with fly-by-wire controls

• Fits into an airline's fleet above the BAe 146, but below a full-size Boeing 737 or Airbus A320 aircraft, offering commonality with the 146

• Entry into service of 1996 (with first flight in 1995)

• Fuselage and cockpit based on existing 146 airframe

• All new wing with winglets at 106.2 feet span, higher aspect 25-degree sweep for faster cruise at Mach 0.82 or long-range cruise at Mach 0.77. Utilizing the Airbus A320 wing was also under consideration.

• Average range of 1,750nm with up potential for 2,500nm (based on payload and fuel capacity options)

• Elimination of the tail brake, and integration of thrust reversers on engines.

• Comparisons with Boeing 737-500

• Conventional yoke with optional side stick controllers

• Use of composite materials for leading edges, horizontal stabilizers, engine nacelles, and fuselage fairings

From this summary, it seemed like the 146NRA was well defined, but that is far from the truth. BAe had indications of what it was planning in the way of the 146NRA, the design was still changing and would continue to do so for the next two years. To muddy the waters further, BAe said that if Airbus went ahead with the proposed 130-seat Airbus A319 the 146NRA would likely be dropped.

Dr. Tupolev visiting Hatfield during the early RJX twin engine studies.

In mid-1991 it became public knowledge that BAe had held talks with the Tupolev Design Bureau regarding the possibility of the Russian firm becoming a risk sharing partner on the 146NRA. The discussions were initiated by a BAe board member Ivan Yates and everything from design work to fabrication and some manufacturing was on the table, while Tupolev had the most interest in the 146NRA's wing design. During January 1992 an internal summary detailing the possibility of using the Tu-334 wing (in a high mounted configuration) was released, while an evaluation of the Lotarev D-436 engine deemed it to offer unsatisfactory thrust for various flight scenarios. The D-436 was subsequently employed on the Antonov An-148, a high wing regional jetliner that looks similar to the BAe 146. Design work was carried out by Tupolev in Moscow and the

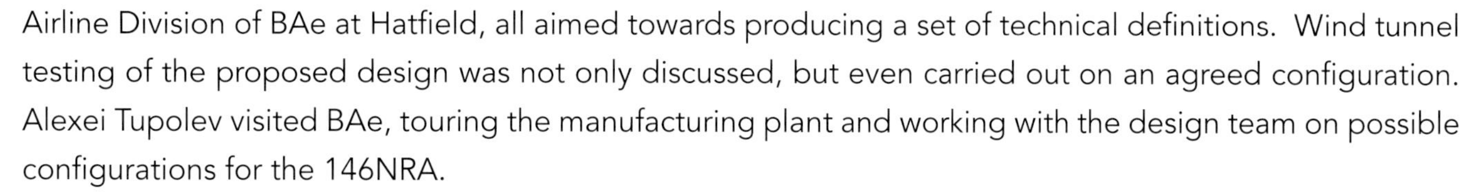

Airline Division of BAe at Hatfield, all aimed towards producing a set of technical definitions. Wind tunnel testing of the proposed design was not only discussed, but even carried out on an agreed configuration. Alexei Tupolev visited BAe, touring the manufacturing plant and working with the design team on possible configurations for the 146NRA.

While this work was going on with Tupolev, a twin Tay powered variant of the 146-100 with 70 seats was being pitched to North American regional airlines and associated majors. This variant was proposed by Guinness Peat Aviation, and was received enthusiastically by the Airline Division management. Unfortunately, the economic performance of the aircraft was very poor and did not support a business case. But it spurred discussions over utilization of the larger CFM56 engine with airlines that had a poor view of the 146 because of its four engines. BAe marketing and sales then began approaching airlines with the proposed CFM56 146NRA configuration to gauge interest. The manufacturer was not just tied to the 146NRA as it then stood, which was internally coded as the "152-120" and was a high wing aircraft with wing mounted engines. It also began pursuing evaluations for other configurations such as the 153-120 (low wing, wing mounted engines) and the 153-120RE (low wing, rear mounted engines). The 120 designation was for the seat count of approximately 120 seats, part of a family with the -96 which was a 96-seat version (152-96). These configurations were achieved with five abreast seating at a 32-inch pitch as standard and were based on the existing BAe 146 cross section of 140". However, a 148" cross section was being explored as well, which would have led to an all-new aircraft.

Dr. Tupolev sitting down on a VIP equipped BAe 146.

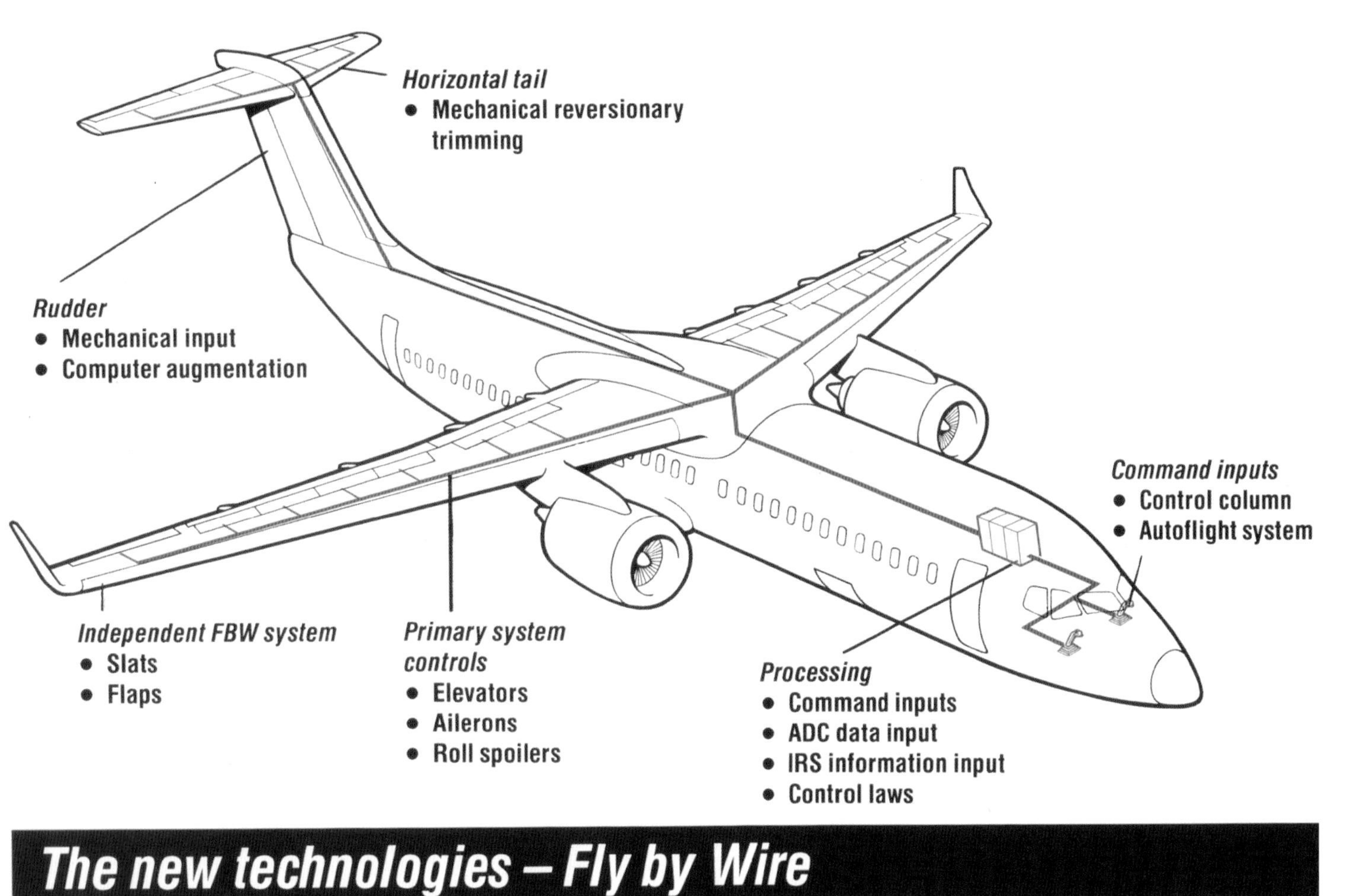

The new technologies – Fly by Wire

The aircraft was assessed on the basis of 'Airbus' technology by BAe (digital flight deck and use of composites). Airlines responded that its speeds and operating altitudes needed to be increased relative to the 146 while its systems technology, particularly automatic and flying surface controls, needed to move to a standard that would be more appropriate for this type of aircraft. The study of the 146NRA had used the A310/A330/A340 as examples, establishing recommendations for a fly-by-wire flight deck. Internally, BAe was grappling with limited experience with such technology, and questioning whether it was affordable.

The manufacturer used the BMW BR715 engine as an example, and determined through simulated testing that a low wing and wing mounted engines provided the best fuel economy over 1,800nm and 2,400nm ranges, with the rear engine variant having the poorest fuel economy. Over a simulated range of 2,400nm, the 146 derivative (152-120) had a fuel burn of 25,950 lbs /4,325 gallons (Mach 0.79 cruise at 39,000 ft. ISA). The low wing version (153-120) burned 25,650 lbs/4,275 gallons, while the rear engine 153RE-120 burned 26,750 lbs/4,458 gallons. Further testing showed that the P153 could "hold its own" against other aircraft including the A319, but was considerably better than the 737-300 and -500/MD-87/F130.

Despite internal numbers showing a low wing design was superior to the high wing design (for the mission envisioned: faster and longer range with more passengers), BAe went with the high wing 146NRA derivative as the baseline in order to avoid the overwhelming expense of developing an all new aircraft. In the end, this derivative showed only approximately a 5% improvement in direct operating costs. Airlines were not impressed, even given potential engine commonality with the rest of their fleet and thus the 146NRA was abandoned.

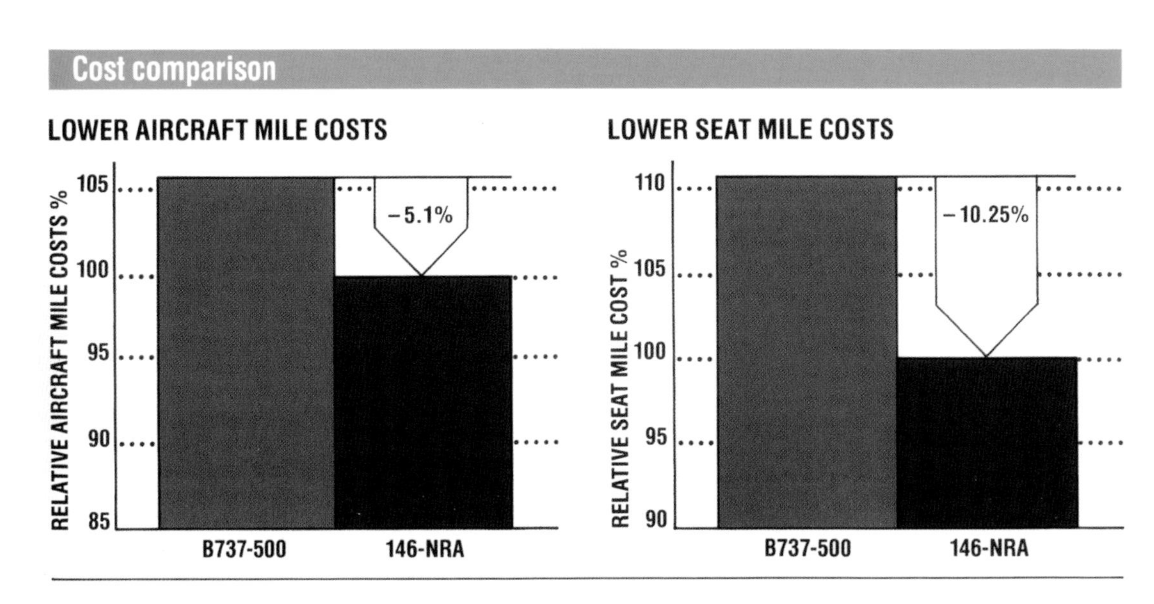

RJX Twin

In mid-1992, BAe began promoting technical descriptions for the new twin-engined 146, the RJX (again), to potential airline customers. The aircraft was offered with a MTOW of up to 129,000 pounds with a payload of up to 35,500 pounds while carrying up to 32,500 pounds of fuel (5,416 gallons). The wing would have a gross area of 1148 sq. feet with a sweep of up to 25 degrees and a 4-degree anhedral. Cruise speed was at Mach 0.82 at up to 41,000 feet and it could carry 119 passengers at five abreast and a 32" pitch (up to 139 seats at six abreast, 33" pitch. Engines were now the 19,660 lbs CFM56-F5/C5 with a 5.0:1 bypass, or the 20,100 lbs BMW/Rolls Royce BR715-55 with a 4.7:1 bypass.

The flight deck would continue with a fully digital system with two LCD displays per pilot, and digital flight guidance system. Control columns would no longer be offered, and instead side stick controllers connected to a fully digital fly-by-wire system, similar to Airbus flight decks. In a change from earlier iterations, the fuselage would feature either a tail cone, or optional petal air brake as per the 146. The wing root was substantial with more bulk to it to allow for the larger engines and fuel capacity. Whilst the 146 did not have leading edge slats, the RJX incorporated them. Everything associated with the BAe 146 that the manufacturer promoted as offering an advantage over other aircraft (no thrust reversers, no leading edge slats, four engine) was now in the firing line, with the RJX offering all of these systems.

Rear engine concept for the twin engine RJX.

Twin engine low wing mounted concept. Ironically, this version had a lower estimated fuel burn versus the high wing mounted version.

The general construction of the RJX was similar to the 146, with the fuselage being the same (except length/wing root) including all the subsystems such as cabin comfort, galleys and cargo areas. It was clear from the documentation that the BMW/Rolls powered aircraft had a range and payload advantage over the CFM56 variant, but the CFM56 was more widely used by airlines. Fleet commonality was not insignificant when it came to powerplants, as it would in theory reduce the spare parts an airline required to hold when compared to maintaining different engine types.

In the end, there was never a twin-engine BAe 146 (or derivative aircraft) that went into production, neither with jet engines nor prop-fans, and all work on the concepts ceased once the programme moved to Woodford.

Ironically, in 2004 Antonov launched a twin-engine regional jetliner called the AN-148 (and a higher capacity AN-158 version). It featured a bigger wing, like the renditions of the BAe RJX, and had two larger engines. It too also sported unpaved field performance, as well as a fly-by-wire system. Even the wing root bore some resemblance to the proposed designs for the 146NRA, while the fuselage interior diameter of 10.25 ft versus the BAe 146 at 10.50 ft made for a very similar-looking aircraft.

Payload range - BMW/Rolls-Royce BR715-55

Payload range - Standard

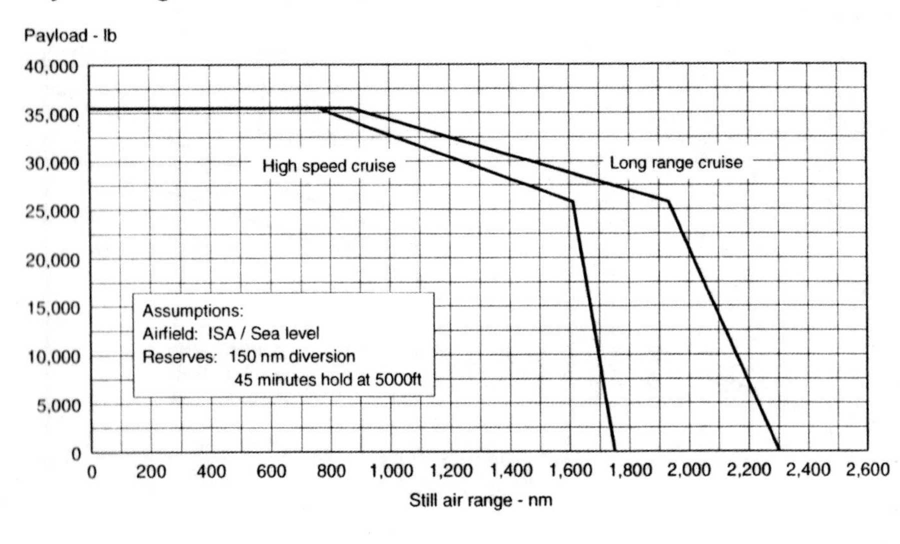

Payload range - Developed

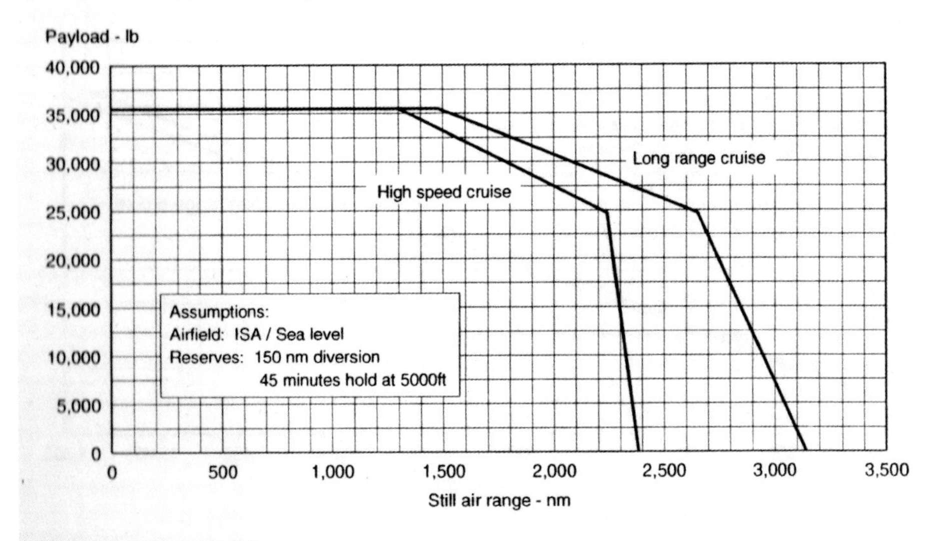

RJX

Payload range estimates for the BMW/Rolls-Royce BR715-55 on the twin engine RJX proposal. This is the same engine used on the Boeing 717 aircraft.

Seating layouts

High density
139 seats , 6 abreast, 33inch pitch

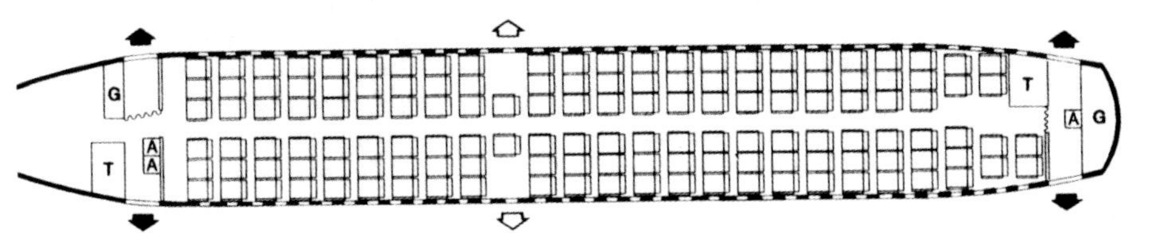

Standard
119 seats , 5 abreast, 32inch pitch

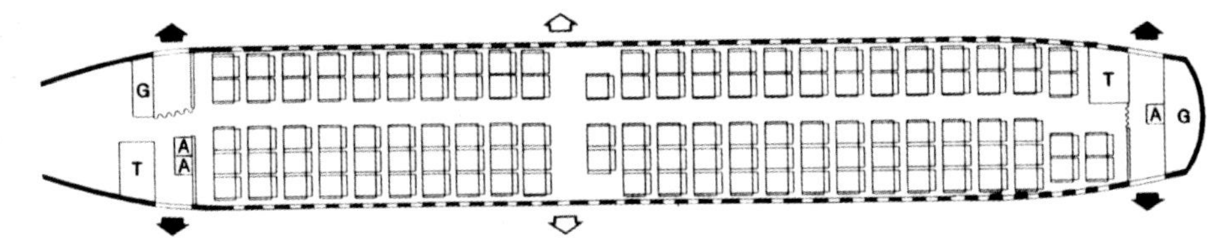

Typical customised mixed class
96 seats, 5 abreast, 32inch pitch
plus
12 seats, 4 abreast, 38inch pitch

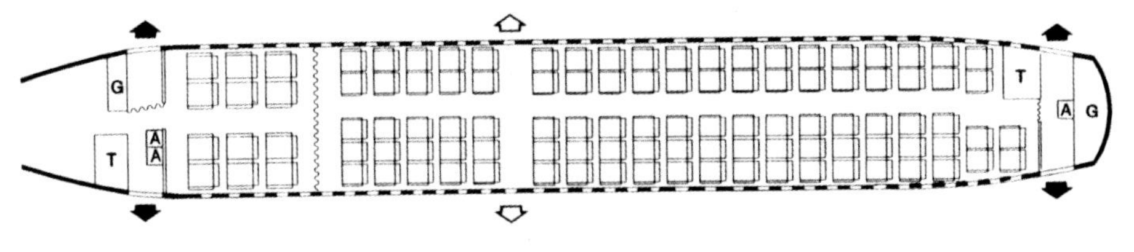

A Attendant's seat G Galley T Toilet W Wardrobe
Passenger/service door Emergency exit

RJX

Seating layout proposals for the baseline fuselage.

The An-148-100 had seating for up to 85 passengers in one class, while the 158 accommodated up to 99 passengers in a single-class layout. The specifications for the An-148 reads like the RJX series, almost right down to the fuel capacity of 26,570 lbs. as well as range which was reported up to 2,600 nautical miles. Antonov has secured orders and built approximately 37 aircraft before production ceased, with operators limited to Soviet airlines as well as Communist state airlines Air Koryo (North Korea) and Cubana (Cuba). Cubana has since grounded the aircraft due to lack of spare parts.

There were many other 'paper aeroplanes' based on the BAe 146, including the proposed rear loading ramp version. But such designs remain on paper for a reason: no perceived commercial viability or orders from customers.

Concepts for military variants with larger rear opening cargo doors.

One of a dozen Military variant concept proposals (this one for the Navy).

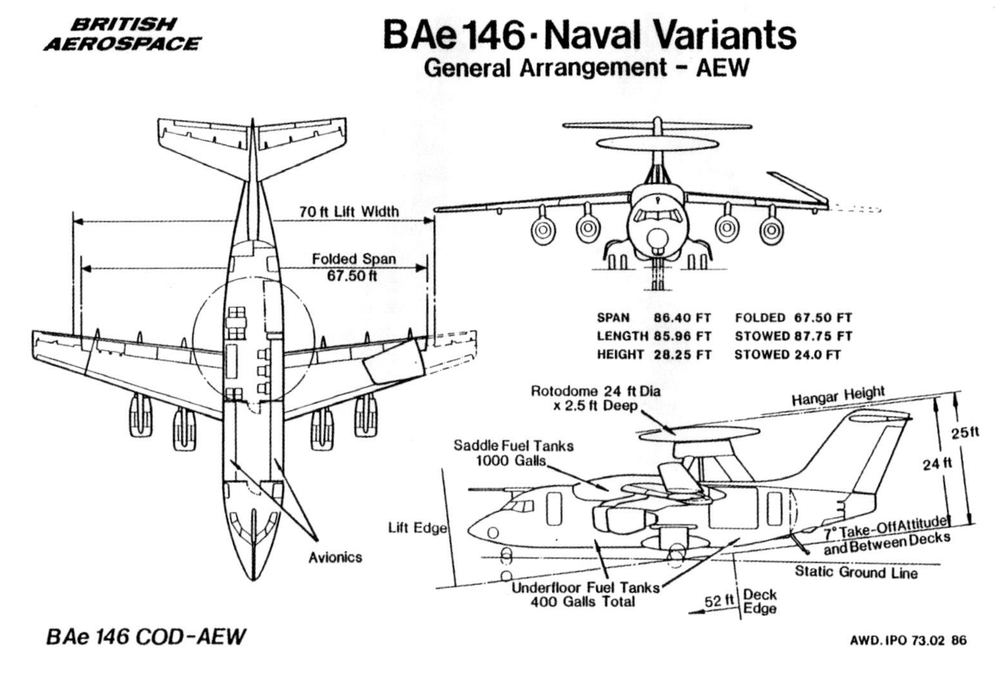

Derek Taylor

CHAPTER 16

Bumps in the Road: Accidents & Incidents

It's always painful to document the loss of life and aircraft. Thankfully, the BAe 146 aircraft had an impeccable safety record while in service. As with any aircraft, there have been incidents and accidents, and it would be impossible to cover every one in which the BAe 146 was involved. A selection of accidents and incidents are discussed in this chapter, generally because of their severity, and what follows is in no way a complete list of accidents or incidents.

December 5, 1987: PSA BAe 146-200 E2039 N177US

In 1987, a series of "firsts" would occur, and none of them were positive for PSA or the BAe 146. In fact, earlier in the year, the airframe involved (N356PS reregistered N177US) experienced an uncontained engine failure, the first for the BAe 146. The second major incident involved a ground collision between it and an Eastern L-1011 between Terminal 1 and Terminal 2 at Los Angeles International Airport on December 5 at 11:15pm. The damage to the aircraft was substantial, and even though staff quickly spray painted the PSA name from the tail and fuselage, it wasn't hard to figure out who owned the damaged aircraft with a big smile under the nose. The conditions at the airport were foggy, and this was the third incident that occurred at LAX that day, including a near miss between two other aircraft on approach, and a near collision on the runway, neither of which involved a 146 or PSA.

There were two theories behind what had occurred. A former BAe rep who was based at LAX with PSA said that the two mechanics onboard were taxiing the aircraft to the maintenance hangar, and had to keep 'riding the brakes' as the aircraft was rather spritely on the ground. To avoid this, they shut down the #2 and #3 engines which had the negative effect of shutting down the hydraulic pumps needed for steering and braking. They activated a DC pump but it didn't come up fast enough. The result was the front nose wheel was locked in a turning radius (they were turning the aircraft when they shut the pumps down) which put the 146 on a collision course with an Eastern L-1011. The crew then shut down the remaining engines in an attempt to halt the aircraft's inertia, but it was too late and the 146 plowed right into the Eastern L-1011. The nose went under the wing of the L-1011, and the horizontal stabilizer of the L-1011 sliced right through the vertical stabilizer of the PSA 146, nearly severing the tail. Had the forward fuselage collided with the tail skid, it was theorized that the cockpit would have been sliced open like a tin can.

The second theory as reported by Eastern spokesman Colin Gates, said PSA officials told him that the plane was awaiting service for a brake problem, and when the mechanics took it for a test taxi and attempted to do a 180-degree turn, they lost control. Nevertheless it would seem logical that if brake issues were suspected, the aircraft would be towed to an unoccupied area prior to tests being undertaken. Neither aircraft were occupied by passengers. The 146 had to have the tail removed before it could be moved. A large forklift held the tail up while a mechanic cut through the control cables, severing the tail completely and allowing it to be removed in one piece. The aircraft was then towed to a hangar owned and operated by Delta Airlines, while British Aerospace dispatched a team from Hatfield to repair the aircraft and restore it for service. After British Aerospace crated up and shipped a new vertical stabilizer to the West Coast of the USA, it took nearly four months for the small crew to fully repair, paint, and restore the aircraft to service.

Had the 146 not stopped, the lower horizontal stabilizer fairing would have sliced open the cockpit like a can opener.

The port wing of the Eastern L-1011 sliced deep into the PSA BAe 146-100.

All photos on these pages courtesy Derek Taylor

Because of the accident and subsequent investigation, the space between Terminal 1 and Terminal 2 was closed forcing some aircraft to deplane outside the main apron (note the Northwest Orient DC-10 in background).

Once the control cable was severed, the tail fin came right off.

December 7, 1987: PSA BAe 146-200 E2027 N350PS "The Smile of Stockton", Flight 1771

Two days after the collision involving PSA BAe 146-200 N177US, the third 'first' for the BAe 146 occurred: total loss of aircraft. Unlike subsequent accidents outlined in this chapter, this one was not the result of pilot error or a failure of the aircraft: it was murder, plain and simple.

Enter David Augustus Burke, aka David Alexander Burke, born May 1952 in the United Kingdom to Jamaican immigrants. Living in Rochester, New York, David had worked his way up the ranks at USAir to become a Ticketing Supervisor. Well-liked by co-workers and having numerous friends and family, no one was aware of the trouble David was facing at the time. He had frequent mood swings which was a detriment in his life, and coming under investigation did not help his demeanor. He was listed as one of 18 persons picked up for having a suspected role in trafficking marijuana and cocaine from Jamaica through Miami Airport. He was also under investigation for involvement in the theft of Mercedes-Benz vehicles. A friend of David did say "I know for a fact that, at one time, he was supporting other peoples (drug) habits." He continued on, "He was no petty dealer, he was dealing in quantity. He was using his airport connections to move the stuff". Burke was reported to say, "Dope is for dopes, and I make money off them." In 1985, David was again a suspect of drug trafficking. Local law enforcement had a theory that baggage would arrive from Miami, and David would retrieve it and put it in the office tagged as 'lost luggage', then later a taxi would pick it up and deliver it to its final destination. David was never formally charged in any of the investigations, but he decided the risk outweighed the benefits, and change was in order. In December of 1986, he moved to Long Beach, California.

David Alexander Burke.

David took a reduction both in salary and title, losing his management status as a supervisor and having to go back to ticket sales, working at recently acquired PSA in Terminal 1 at Los Angeles International Airport (LAX). Being demoted further pushed him away from the Station Manager job he coveted, and David's life began to spiral out of control. He was seen driving expensive cars, having many visitors late at night who were never the same, and only stopped by for a couple minutes. His temper nor mood swings subsided, and it was suspected he was dealing in drugs again. David, who had been married more than once previously (fathering seven children with five different women), entered into a relationship with another USAir agent, Jacqueline Camacho. Over the span of a year with all the abrupt changes in his life, David found himself in a very dark place. Just days before his life with USAir changed, he suspected Jacqueline of having an affair, attempting to choke her twice and deliberately damaging her car. She obtained a restraining order barring him from coming within 100 yards of her or her home.

The beginning of the end came to a head on November 8, when he was filmed in an office via hidden cameras stealing cash from onboard alcohol sales. He was arrested as a result. A total of $69 was taken and David was suspended on November 15, and four days later on

November 19 he was formally terminated. In a bout of depression, he traveled to San Francisco to borrow a gun from fellow USAir employee Joseph Drabik, reportedly with plans to use it to threaten someone who owed him money. Over Thanksgiving, he traveled to a family reunion in Stone Mountain, Georgia, during which friends and family acknowledged he seemed fine and upbeat. Upon his return to California, on

APPLICATION TO MUTUAL OF OMAHA INSURANCE COMPANY

Policy Number T40AV A386499	Name of Insured DAVID BURKE
Principal Sum $195 000 — Premium $ 6.50	Address of Insured 21 W. SPRING ST. #C
Flight Number #1769 PSA	City LONG BEACH State CA. ZIP 90806
Departure City LAX	Name of Beneficiary
Returning from SAN FRANCISCO	Address of Beneficiary
Returning to LAX	City ROCHESTER State N.Y. ZIP 14619
Policy to be Effective Date 12/7/87 Hour 2:30 ☐ a.m. ☒ p.m.	Trip to be made on: ☐ One-way Ticket ☒ Round-trip Ticket
Countersigned by Licensed Resident Agent	Insured is owner unless checked here: ☐ Beneficiary is owner.
	Personal Signature of Insured David Burke

Form 1474M App

November 30 he received a call from Ray Thomson, his former supervisor and station manager, confirming that a grievance procedure regarding his termination had been scheduled for December 3. On December 2, David began preparing for the end to everything in his life, having updated his personal will (last updated seven years prior). David never showed up for his grievance hearing. But instead forced his girlfriend and six-year-old daughter to drive around town at gunpoint for six hours into the early hours of December 4. Jacqueline filed a criminal report with Hawthorne police a day later, but refused to identify David as the assailant.

SCHEDULE

T18BA Policy Number 05802 B	Name of Insured DAVID A. BURKE
Capital Sum $ 100 000	Address of Insured: Number and Street 21 W. SPRING ST #C; City L.B. State CA. ZIP 90806
Principal Sum $ 500 000	Name of Beneficiary
Term of Coverage 1 Days — Premium $ 8.50	Address of Beneficiary: Number Street; City ROCHESTER State N.Y. ZIP 14619
Countersigned by	Effective at: Hour 2:30 ☐ a.m. ☒ p.m. Date 12/7/87
Licensed Resident Agent	Place LAX
	David Burke

This Policy Is Nonrenewable and Provides Benefits for Loss of Life, Limb or Sight and Loss Due to Medical, Surgical, Nurse and Hospital Expense Resulting from Accidental Bodily Injuries. Such Benefits Shall Be in the Amounts, for the Periods and to the Extent Herein Limited and Provided.

MUTUAL OF OMAHA INSURANCE COMPANY
Mutual of Omaha Plaza
Omaha, Nebraska 68175
(Herein called the Company)

Hereby insures, subject to the provisions, exceptions and limitations of this policy, the person named as Insured in the Schedule against certain specified losses resulting from injuries. The term "injuries," as used in this policy, shall mean accidental bodily injuries received while this policy is in force and resulting, independently of all other causes, in loss covered by this policy.

PART A. CLASSIFICATION OF INJURIES. For the purpose of determining the amount of benefits payable under Part B and Part C, injuries are classified as follows.
(1) SCHEDULED AIRLINES. Injuries received while riding as a passenger in, or boarding or alighting from, an aircraft operated: (a) by a scheduled airline of United States registry holding a Certificate of Public Convenience and Necessity issued by the Civil Aeronautics Board (or its successor) of the United States of America, (b) by an intrastate scheduled airline of United States registry maintaining regular published schedules and licensed for the transportation of passengers, by a duly constituted authority having jurisdiction over civil aviation in the state in which said airline operates, or (c) by a scheduled airline of foreign registry maintaining regular published schedules and licensed for transportation of passengers by the duly constituted governmental authority having jurisdiction over

Form T18BA – Series 3655S

Ray Thomson was described as a gruff and blunt man with Texas mannerisms and a strong disdain for small talk, all of which alienated him with co-workers and employees. David called Ray and asked to schedule another hearing, which was set for 2:00pm on December 7. On the day at 1:10pm David arrived at LAX airport and parked his car in Lot D, section A-2. He took the shuttle bus to the terminal, arriving at the office at 1:50pm. After having two cups of coffee and making small talk with Ray's secretary, at 2:03pm Ray invited David in along with other associates. During the meeting, Burke admitted to having a $300 a week cocaine habit, as well as drinking two pints of alcohol a day over the past five or six years. David also claimed he wasn't being treated fairly at USAir, believing racism had something to do with his being passed over for promotion more than once. David claimed all of these events put him in a state of depression, and he asked not only for his job back, but that USAir provide him with counseling treatment for his addiction. Ray did not buy any of it, denied all of David's requests and upheld the termination. David left, and Ray prepared to head home on PSA flight 1771, which he did every day as he commuted from his home in Tiburon, north of San Francisco.

The two insurance policies David Burke purchased moments before he boarded the flight.

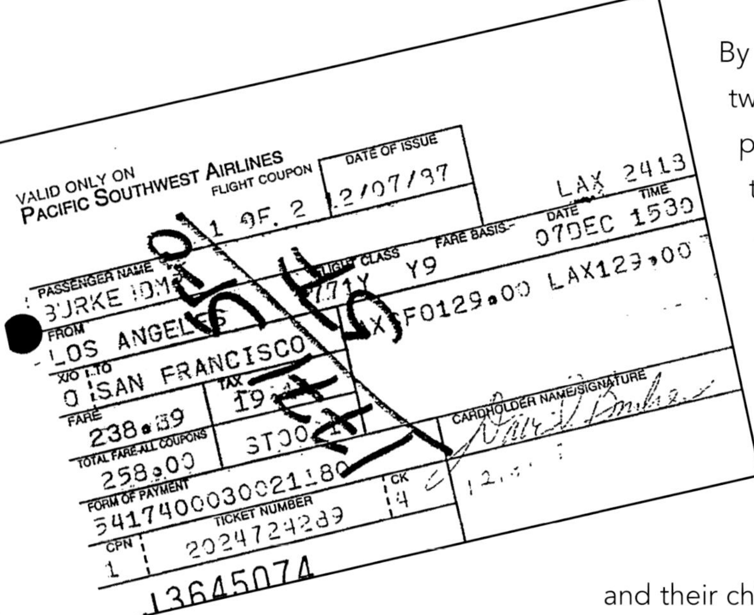
VALID ONLY ON PACIFIC SOUTHWEST AIRLINES
FLIGHT COUPON
DATE OF ISSUE
PASSENGER NAME BURKE DM
FROM LOS ANGELES
X/O TO SAN FRANCISCO
FLIGHT CLASS 1771Y Y9
FARE BASIS
DATE 07DEC TIME 1530
LAX 2413
FARE 238.39
TAX
TOTAL FARE-ALL COUPONS 258.00
FORM OF PAYMENT 5417400030021180
TICKET NUMBER 2024724289
CPN 1
CK 4
CARDHOLDER NAME/SIGNATURE
13645074

The ticket and credit card receipt for the ticket David Burke purchased.

By 2:30pm, David had purchased two life insurance policies over the phone from Mutual of Omaha, the first offering coverage of $195,000 for a single roundtrip flight, and the second covering $100,000 for any accident in a 24-hour time frame. A third policy providing $75,000 coverage was purchased at a Tele-Trip counter. All three named one of his former wives and their children in Rochester, New York as beneficiaries. David was the only passenger to purchase insurance on flight 1771. FBI found records that he had been looking at travel insurance policies prior to December 7, including one for $500,000 through New Jersey based Chubbs and Sons.

At 3:05pm, David called his girlfriend and left a farewell message on the answering machine: "Jackie, this is David. I'm catching Flight 1771. I love you. I wish I could tell you more. I just wanted you to know I love you." Immediately afterward, at 3:10pm, he was seen on security cameras in the baggage area of Terminal 1 transacting a narcotics deal. But David was not done and he made two further phone calls, the first believed to be to his children in Rochester during which he left the message, "I wish I could see you, but I'm not going to see you again. I'm sorry I messed things up. I love you. I love you. I love you." A call to his 12-year-old daughter living in the Los Angeles area included a message that he wouldn't be returning from San Francisco that night.

At 3:20pm, Burke went to the PSA ticket desk and purchased a ticket for flight 1771. Burke then proceeded to use a second ID identifying him as a PSA employee, enabling him to bypass security and enter the terminal through a series of access-controlled doors after signing in – employees were not required to pass through security screening to enter the passenger terminal. By then it was 3:30pm, and his girlfriend Jacqueline watched Burke board flight 1771 with the last of the passengers. The load was light, with a total of 43 people onboard including crew. Unbeknownst to Ray, who was occupying aisle seat 3C, Burke was only two rows back sitting by the window in seat 5F. The BAe 146-200 pushed back at 3:36pm, and was wheels up at 3:45pm.

Once the aircraft was airborne and at cruising altitude, Burke got up, proceeded towards the lavatory in the front of the plane, and dropped a note written on an air sickness bag on Ray's lap as he passed by. The note read: "Hi Ray. I think it's sort of ironical that we end up like this. I asked for some leniency for my family. Remember? Well, I got none, and you'll get none."

The entire event that unfolded in the air, from the first gunshot until the aircraft hit the ground at more than MACH 1, happened in approximately 90 seconds. David Burke, aged 35, took the lives of forty-two innocent persons, and rode the aircraft all the way to impact. Along with the crew listed above, David took the lives of Deborah Neil, Debra Vuylsteke, the working PSA Flight Attendants on board 1771. Additionally, Julie Gottesman, who was 20 and a trainee Flight Attendant on her first orientation flight was tragically lost. Julie was to be based in San Francisco a week later. Julie and First Officer James Nunn had just joined PSA in 1987.

The BAe 146 slammed into the hillside of a cattle ranch in the Santa Lucia Mountains, near Paso Robles at a 70-degree angle. PSA tech Clive Nicholson estimated the plane impacted at 860MPH at a 110-degree nose down pitch, noting that the plane hit fully intact and didn't fall apart at such high speed, a testament to the aircraft quality. There was little of the airframe or contents left, beyond very small fragments. Because of the nature of the event, the crash fell under the jurisdiction of the Federal Bureau of Investigation (FBI). Emergency crews were dispatched along with recovery teams to begin collecting all the debris and human remains. In the meantime, news of the crash was being splashed all over the news, with no information other than it crashed. In what can only be described as despicable human nature, prank phone calls poured into PSA about how and why the crash occurred. But one phone call struck the FBI as alarming. It was received from a male with a Middle-Eastern accent at around 5:00pm the day of the crash, saying he was calling because there was a gunman on board, planted by an Iranian terrorist group associated with the Ayatollah Khomeini. Normally such a call would have been dismissed, it was not yet public knowledge that there was a suspected gunman on board. In fact, this would not come out until much later in the night, with newspapers reporting the next

The following times are from the CVR (cockpit voice recorder) and do not correlate to exact local times. CAM refers to the cockpit microphone, and RADIO is when air traffic control is in the loop. CAM+R is when both the cockpit microphone and RADIO records the sounds. [?] is unintelligible on the recording. Each time is in seconds, not minutes.

:39 RADIO "Center, PSA 1771. Any reports on the ride ahead? We've had a little continuous light chop" (PSA First Officer James Nunn)

:46 CAM Bump-like sound (likely the bathroom door opening and hitting the cockpit door). Burke was exiting the bathroom, believed to have his gun drawn.

:46 RADIO "Ah, Rocky, well, tell you what, ah, high altitude says that, ah, I guess about the last, ah half hour they've been coming down 22 to get out of it. So, ah, matter of fact, I do have one now at 22. I'm not (?) about the minute, but, ah, he hasn't, ah, I guess requested (?), so it's not too bad." (FAA Oakland Center)

{Approximately 4:13pm the chain of events begins}

:48 CAM 1st high level gunshot-like sound

:49 CAM "Oh my God." (PSA FO)

:50 CAM Second high level gunshot-like sound

CAM "That's a gun." (PSA FO)

:51 CAM "Yeah, I know." (PSA Captain Gregg N. Lindamood)

:55 CAM "Tell 'em we got a problem." (PSA CPT)

:57 CAM 'Squawk 77. Squawk 77." (7700 is an Emergency)

:59 CAM+R "PSA, yeah, we've got a problem. We've had a gun fired on air board, on board the aircraft." (PSA FO)

:07 RADIO "Beg your pardon, sir?" (FAA)

:09 RADIO "We're squawking 7700 and we've had a gun fired on board the aircraft." (PSA FO)

{AT THIS MOMENT, AIRSPEED DROPS FROM 335Kt to 313Kt abruptly}

:15 RADIO "Okay. Do you want to go to Monterey? Could you make it, sir?" (FAA)

:16 CAM "Tryin' (?) to get lower." (PSA CPT)

:17 CAM Door opening (flight deck)

:19 CAM "(?) Captain!" (Female PSA Flight Attendant)

:20 RADIO Aircraft alarm (?) followed by "I've got a problem" (reportedly Burke's voice)

:21 CAM+R 3rd high level gunshot-like sound.

:21 CAM "Mm" (believed to be PSA FO shot, moaning)

:22 CAM+R 4th high level gunshot-like sound

:22 CAM "Ah" (believed to be PSA CPT shot, moaning)

:23 CAM+R Metallic clinking sound (believed to be Burke pushing throttles forward)

:24 RADIO "1771 descend at your discretion to one zero thousand." (FAA)

:28 CAM+R 5th high level gunshot-like sound (believed to have shot PSA FA)

:43 CAM+R Door opening/closing (?) sounds

:44 CAM Sound of increased air speed

:45 RADIO Aircraft alarms (?)

:45 CAM Low level sounds

:15 CAM+R 6th high level gunshot-like sound (believed to be PSA Chief Pilot Douglas Arthur who it is believed tried to save the aircraft before he was shot)

:15 CAM Thud-like sound, wind/static-like sounds, sounds of distant voices screaming and crying (believed to be Douglas falling, and distant voices the passengers realizing the impending doom)

:15 CAM+R Break up of recording

:33 CAM+R End of recording. Impact at 4:14:50pm

day of the suspected murder-suicide. The FBI was already aware of the exchange between PSA 1771 and Air Traffic Control. There was no connection with the caller in the end. Meanwhile, nighttime followed by rain the next day slowed the recovery events and there was a pause at the crash site to avoiding contamination of the scene. The flight data and cockpit voice recorders were found and hand-carried to Washington D.C. later that day.

While the accident was being investigated, and the collision with an Eastern L-1011 a few days earlier, the FAA had inspectors all over PSA offices and were specifically taking notice of any issues with the BAe 146. As a workaround, PSA technicians answered the phone with the opening line "Hey (insert name), it's a lovely day here" which was a coded phrase for "there's an FAA inspector standing here right next to me – I can't talk about anything sensitive." By Tuesday December 8, all of the print and television media was reporting that PSA 1771 had been brought down by a gun and that the perpetrator was former employee David Burke who had recently been fired.

On December 9 a portion of the gun, the .44 magnum, with a right thumb still attached was located in 2-3 inches of mud, and determined to be Burke's after matching the finger print. It was found next to a piece of Burke's Mastercard and in proximity to a wallet belonging to First Officer Nunn. The FBI wrapped up its investigation of the crash site quickly, and PSA had hired local college students to comb through remaining bits of debris at the crash site. On December 10, the note written to Ray Thomson was found at the crash site, reportedly intact. The following day one of the college students had found a portion of the badge Burke used to bypass security, complete with his picture on it. USAir claimed it took David's badge during his termination, but that it was possible for employees to have more than one badge. Spokesperson and Manager of Public Relations for USAir Nancy Vaughan stated that Burke turned in his identification after he was fired, and it was destroyed at corporate headquarters. Of all the tiny fragments recovered from the crash site, no one expected the murderer's finger to be found still on the gun trigger, nor the note to actually survive the impact.

(READER WARNING: GRAPHIC TEXT)

Due of the speed the aircraft impacted the ground, there was little in the way of human remains left. In the end, a total of two hundred pounds of human remains were turned over to the San Luis Obispo coroner's office, very little of which was identifiable. Requests went out to the family of each victim, asking them to submit personal items that the victim would have touched or used (generally toiletries), in hopes of finding DNA matches with remains. The FBI issued instructions that victims' families were not to be told about the acute fragments found at the crash site. Additionally, medical and dental record requests went out to physicians that had treated the victims in a previous capacity, in hopes a match could be made. Counter requests were made, with victims' spouses requesting items such as wedding bands be returned if found, and describing them in detail.

The tragedy took the lives of the President of Chevron USA (oil company) and three of the company's public affairs executives. Additionally, there were three officials of Pacific Bell, one of the largest phone networks in the United States, on board. The events that unfolded resulted in many large corporations revising or creating policies that prohibited multiple executives and senior management traveling on the same flight. The deliberate murder of the flight crew and subsequent crash resulted in the Federal Aviation Administration (FAA) requiring all airline employees to go through the same security screening as passengers when going to work, with no exceptions.

There is a memorial for flight 1771, the "Garden of Hope" in a section of the Los Osos Valley Memorial Park, and a number of the passengers and crew are buried in the cemetery.

* * * * * * * FLIGHT INFORMATION * * * * * * * *
ARRIVALS DEPARTURES
CITY SKED ON IN BLOX SKED PUSH OUT OFF A/C CAP DLY-CODE
LAX ORIG * * * 1530 1530 1532 1536 350 083
SFO 1643 1643E TERM * * * * * ***/****
CLASS OF SERVICE TOTAL
SEG AUT BKD Y Q B M V MS NP SA SEG LEG AGT
LAXSFO 98 33 17 11 1 4 1 0 0 5 39 39 00
RECV...08DEC 0101 CEO

QU SANAFPS
.SANAFPS 080101 9918
CHECK-IN LIST MON DEC 07 FLT 1771 FROM LAX SC083

CKIN NMBR	Q I	NAME	P T	A T	W I	DEST CODE	CKIN AGNT	SEAT NMBR	INB CONNECTS	OUT CONNECTS	CONN DSTN
G009		WEBB/EARL	Y	S		SFO	0000	06F			
G011		STUEDEMANN	Y	G		SFO	1268	03F			
G012		SCAFIRE/CA	Q	G		SFO	1224	04C			
G013		SAUR/ERIKA	Y	T		SFO	1346	07A			
G014		SAUR/ERIKA	Y	T		SFO	1346	07C			
G015		GIULIANO/D	V	T		SFO	1346	16D			
G016		KEKAI/THER	M	T		SFO	2418	05A		1969	RNO
G017		ROSEEN/JMR	Y	T		SFO	1207	04D			
G018		GOTTESFMAN	N	G		SFO	1224	01D			
G019		ADDINGTON	Q	G		SFO	1224	06C			
G020		CONE/STEVE	Y	T		SFO	1207	06D			
G021		SYLLAS	Q	G		SFO	1224	02A	MURPHY		
G022		SYLLAS	Q	G		SFO	1224	02C			
G023		SWANSON/A	Y	G		SFO	1224	15D			
G024		KEMPE/JG	Q	G		SFO	1224	08C			
G025		Rettinghaus, J.	Y	G		SFO	1235	02E			
G026		Siegfried, Linda	Y	G		SFO	1235	02F			
G027		FOX	Q	G		SFO	1224	09A			
G028		FOX	Q	G		SFO	1224	09C	Dealan		
G029		WEBB, NMR	M	G		SFO	1235	10A			
G030		WEBB, MAS	M	G		SFO	1235	10C			
G031		ROSENBERG	Y	G		SFO	1224	02D			
G032		BURKE/DMR	Y	T		SFO	2418	05F			
G033		Kaun, Karen	Y	G		SFO	1235	07F			
G034		SHIBA	Y	G		SFO	1224	08D			
G035		Conte, JOHN	N	G		SFO	1235	09F			
G036		RHEE/CMR	Y	T		SFO	1207	11A			
G037		CORDOVA, Anthony	Q	G		SFO	1235	05D			
G038		Phelan, Kevin	Q	G		SFO	1235	10D			
G039		NELSON, WAYNE	B	G		SFO	1235	12C X			
G040		WINTERS, L.	M	G		SFO	1235	11F			
G041		HOAG, D.	Y	G		SFO	1235	12D			
G042		Carroll, JIM	Y	G		SFO	1235	04E			
G043		PERRY, CLIFF	Y	G		SFO	1235	04F			
G044			N	G		SFO	1235	04A X			
G045		MIKA, KATHY	Q	G		SFO	1235	07D			
G046		ENGSTROM, SHARON	N	G		SFO	1235	08A			
G047		RAVIN, TOM	Q	G		SFO	1235	03A			
G048		Thomson, Ray	N	G		SFO	1235	03C			

RECV...08DEC 0101 CEO

QU TSTXMPS
.TSTXMPS 080101 9900
PS1771/07DEC 14F
* * * * * * * * FLIGHT INFORMATION * * * * * * * *
ARRIVALS DEPARTURES
CITY SKED ON IN BLOX SKED PUSH OUT OFF A/C CAP DLY-CODE
LAX ORIG * * * 1530 1530 1532 1536 350 083
SFO 1643 1643E TERM

**DEC 08 '87 13:18 PSA COMP SVS

Authors Note: Much of the information for flight 1771 was obtained from the FBI under the Freedom of Information Act. What drove this request was the author's disbelief that a suicide note would survive the impact, intact, while everything else was in fractional pieces (e.g. the gun, badge, credit card, etc.). I'm not a conspiracy theorist, but requesting this piece of evidence was key. The FBI supplied me with more than 300 pages of reports, evidence examples, media reports, the cockpit voice recorder (CVR) transcript, and more. But the note to Ray Thomson was not provided, and the letter from the FBI said a lot of evidence was destroyed after 1998. Why some copies of evidence were provided (and not destroyed), yet one of the key pieces to the investigation leads to speculation on the veracity of it. The CVR transcript was also somewhat different from what is still being reported today by media. What is reported publicly differs enough to be more of an ending you'd hear from a Hollywood movie villan versus actual reality. That said, it doesn't change the fact that David Burke committed this horrible atrocity. PSA and former BAe rep Clive Nicholson pointed out what mostly survived the impact was paper, and photographs of the crash site support that fact.

The passenger manifest for flight 1771 including their seat assignments.

February 20, 1991: LAN Chile, BAe 146-200 E2061 CC-CET

Barely a year after delivery to the airline, CC-CET (E2061) flown by LAN Chile slid off the runway at Puerto Williams (Guardiamarina Zanartu Airport). It was operating a charter flight that had originated in Punta Arenas and was originally expected to land on Runway 26 with wind at 180 degrees and 4 knots. Air traffic control radioed an update that the wind had increased to 6 knots at 160 degrees, and the captain requested a landing on runway 08 instead. There was light rain, but the aircraft was clear to land on Runway 08 (4,724 feet) at 3:15pm and touched down at the 1,400-foot mark, well past the touchdown zone, at 112 knots (target touchdown speed should have been 103 knots). The wet surface, negative slope, late touchdown with higher speed and very little braking action resulted in the aircraft sliding off the runway and into Beagle Channel, where it was completely submerged.

The number two engine broke off when it hit a railway tie used to secure barges at the entrance to the channel, and the airframe pivoted approximately 15 degrees to the port side. The fuselage was pierced in multiple places from collisions with objects and debris, the bottom of the fuselage scraped a bed of stones puncturing the skin, allowing water to enter the lower part of the cabin and sinking it quickly. Of the 73 persons on board (66 passengers, 7 crew), 53 managed to get out and stood on top of submerged aircraft with only the wings and the very top of the fuselage above the frigid waters. Twenty tourists (from the United States) were unable to evacuate in time and drowned. The Chilean Navy onboard the Society Explorer, as well as airport staff and local citizens, rushed to offer aid. Using Zodiac boats, the passengers were whisked aboard the Society Explorer.

The LanChile BAe 146 is being towed out of the water.

July 23, 1993: China Northwest BAe 146-300 E3215 Flight 2119

The first BAe 146 loss due to a mechanical failure came when China Northwest flight 2119 was taking off from Yinchuan Airport in the North-central part of China (close to Mongolian border).

After an initial aborted take-off at 2:20pm, the pilots attempted a second take-off 20 minutes later at 2:40pm. The China Northwest BAe 146-300 B-2716 (E3215), delivered only seven months earlier, was on its second take-off roll when the flaps retracted after the right-side flap actuator failed. The aircraft was past velocity 1 (V1) and was traveling too fast to safely abort a take-off. The captain realized he had to get the aircraft in the air, but could not rotate the aircraft at VR speeds because of the lack of flaps and kept the aircraft accelerating in hopes of generating enough lift despite the rapidly diminishing runway.

Instead, when he did attempt to rotate the nose into the air, the tail dragged on the runway. Eventually the 146 did get airborne, but reached only 32 feet above the ground before losing lift again and colliding with raised banks. The aircraft broke up on impact and ended up in a lake just beyond the airport boundary. The tanks were full of fuel but there was no explosion. The crash resulted in the deaths of 58 passengers and one crew member.

September 25, 1998: Paukn Air BAe 146-100 E1007

Paukn Air flight 4101 EC-GEO departed Malaga, Spain at 7:25am with 34 passengers and 4 crew members. The flight was headed to Melilla, a resort destination and autonomous Spanish city on the Moroccan north coast of Africa. Around 22 nautical miles from the airport, the flight descended through 3,000 feet which was below the minimum sector altitude (MSA) of 4,000 feet. The aircraft continued to descend in instrument meteorological conditions (IMC) and in mountainous terrain. At 7:49am local time, the ground proximity warning (GPWS) alert sounded "terrain, terrain" as the hills were very close. Moments later at just over 800 feet, the GPWS sounded again with the audible alert a pilot never wants to hear: "(whoop-whoop) pull up! (whoop-whoop) pull up!". Seconds later, the aircraft hit terrain at an elevation of 886 feet, killing everyone onboard.

It was clear that the pilots had not followed the arrival procedure and descended below the minimum safe altitude, resulting in controlled flight into terrain (CFIT).

8
10
11

November 24, 2001: Crossair Avro RJ100 E3291 HB-IXM Zurich Switzerland Flight 3597

Crossair flight 3597 departed Berlin-Tegel at 8:01pm on November 24, 2001. During the descent towards Zürich just after 8:58pm, the aircraft was cleared for a VOR/DME approach to runway 28. The pilots of a flight that landed just ahead of the Crossair flight notified the tower that the weather was close to the minimum for the runway. At 9:06:10pm, Crossair reached the minimum descent altitude (MDA) of 2,390 feet, and the captain said he had visual contact with the ground and continued the approach. Seconds later at 9:06:36 the Avro RJ100 collided with treetops and crashed into the ground, bursting into flames. Of the 33 persons on board, 21 passengers and three crew members died at the scene of the accident.

The crash investigation determined that the captain deliberately descended below the MDA of the standard VOR/DME approach without having the required visual contact with the approach lights or runway, and the co-pilot made no attempt to prevent the continuation of the flight below MDA, a failure of cockpit resource management (CRM). The approach of the runway at Zurich has no alarm system available if a minimum safe altitude warning (MSAW) is triggered. Additionally, Crossair was criticized for not making correct assessments of the captain's competency or flight performance, and where weaknesses were found corrective action was not taken. The fatigue the captain was experiencing also played a role in the poor decision making, and the range of hills the aircraft came into contact with was not marked on the approach chart the flight crew referenced.

January 8, 2003: Turkish Airlines Avro RJ100 E3241 Diyarbakir Turkey TC-THG "Konya" Flight 634

Turkish Airlines flight 634 took off uneventfully from Istanbul airport at 6:43pm with 75 passengers and five crew members. Just over an hour later and in final descent into Diyarbakir Airport, air traffic control cleared the aircraft for landing. Approximately eight nautical miles from the airport at an altitude of 5,000 feet, air traffic control radioed flight 634 to continue approach and report back when visual contact had been established. Thick fog was in the air and visibility had dropped considerably, impeding the flight deck from making visual contact with the runway. To make matters worse, Diyarbakir airport was not equipped with an instrument landing system (ILS).

In clear violation of Turkish Airlines flight procedures, the captain continued the approach to one mile and descended below 500 feet, violating the MDA of 2,800 feet. At 200 feet and just one mile from the runway, the GPWS triggered alarms, and after eight further seconds the crew aborted the landing and initiated what they thought was a go-around. However, the aircraft struck the ground at 8:19PM, 3,000 feet away from the runway and 100 feet away from the approach lights, at more than 131 knots.

Sliding along the ground the aircraft began to break up immediately and then hit a slope where the fuselage separated into three pieces and caught fire. Everyone on board with the exception of five passengers perished in the accident. The post investigation determined no abnormalities with the aircraft, nor with the flight crew. The rescue teams onsite noted the dense fog, with visibility dropping to as low as 3 feet at times, and the fire could not even be seen until the crew was onsite. Inclement weather and the flight crew not acting upon the GPWS immediately was found to be the cause of the crash.

Aftermath of Crossair flight 3597.

October 10, 2006: Atlantic Airways BAe 146-200 E2075 Flight 670

Atlantic Airways flight 670 was a private charter carrying workers from Stavanger to Molde via Stord Airport. The load was light with just 16 people on board, four of which were the flight crew and flight attendants. Upon landing at Stord Airport at 7:30am, the lift spoilers did not deploy and the flight crew called for full braking. Because of the speed of the aircraft and lack of functioning spoilers, the wings were still generating lift which reduced brake effectiveness and degraded the aircraft's stopping. The duty officer at the airport noticed the 146 was producing wake vortices, something that had not been seen before. The aircraft failed to slow down, even with the use of emergency brakes, which locked up causing the aircraft to skid down the damp runway. A likely contributor was that the approach speed per the flight manual of 112kt was not adhered to, and instead Atlantic Airways flight 670 touched down at 120kt. The aircraft began to hydroplane, but the captain was able to keep the aircraft on the runway until it reached the end. Still traveling at more than 15 knots, the aircraft plummeted down a steep rocky slope into trees, large rocks and light poles.

The fuselage buckled in many areas, trapping the pilots in the cockpit and jamming the front door that would enable egress. On the starboard side of the aircraft both front and rear doors were blocked by the rocky terrain and could not be opened. Fuel began pouring into the cabin from the wing tanks due to puncturing, and a small fire ignited spill which spread throughout the cabin, killing three passengers and the flight attendant in the forward part of the fuselage. The remaining passengers and flight attendant were able to escape using the rear port side door and climbed out onto the rocks. Because the firefighting staff had been watching the event unfold, they were able to reach the scene quickly after the 146 departed the runway.

Accident Investigation Board Norway (AIBN)

During the investigation it was revealed that should the spoilers fail to deploy, the landing distance of the 146 increases by over 40%. Additionally, fragments of 'boiled' rubber were found on the runway, a condition known as 'rubber reversal' which occurs after a cushion of steam forms due to friction created by a sliding tire. The runway's lack of grooving increased the chances of hydroplaning and contributed to this event. The accident report suggested that while the excursion could have possibly been avoided if the flight crew understood the failure of the lift spoilers and the aircrafts stopping distance, it criticized the airport for contributing substantially to the cause of the accident. The lack of a grooved runway as cited, as well as the safety area not being in accordance with Norwegian BSL E 3-2 and not extending far enough past the stop end. Additionally, the slope was also cited as being steeper than prescribed and the terrain obstructed fire-fighting and rescue work.

Accident Investigation Board Norway (AIBN)

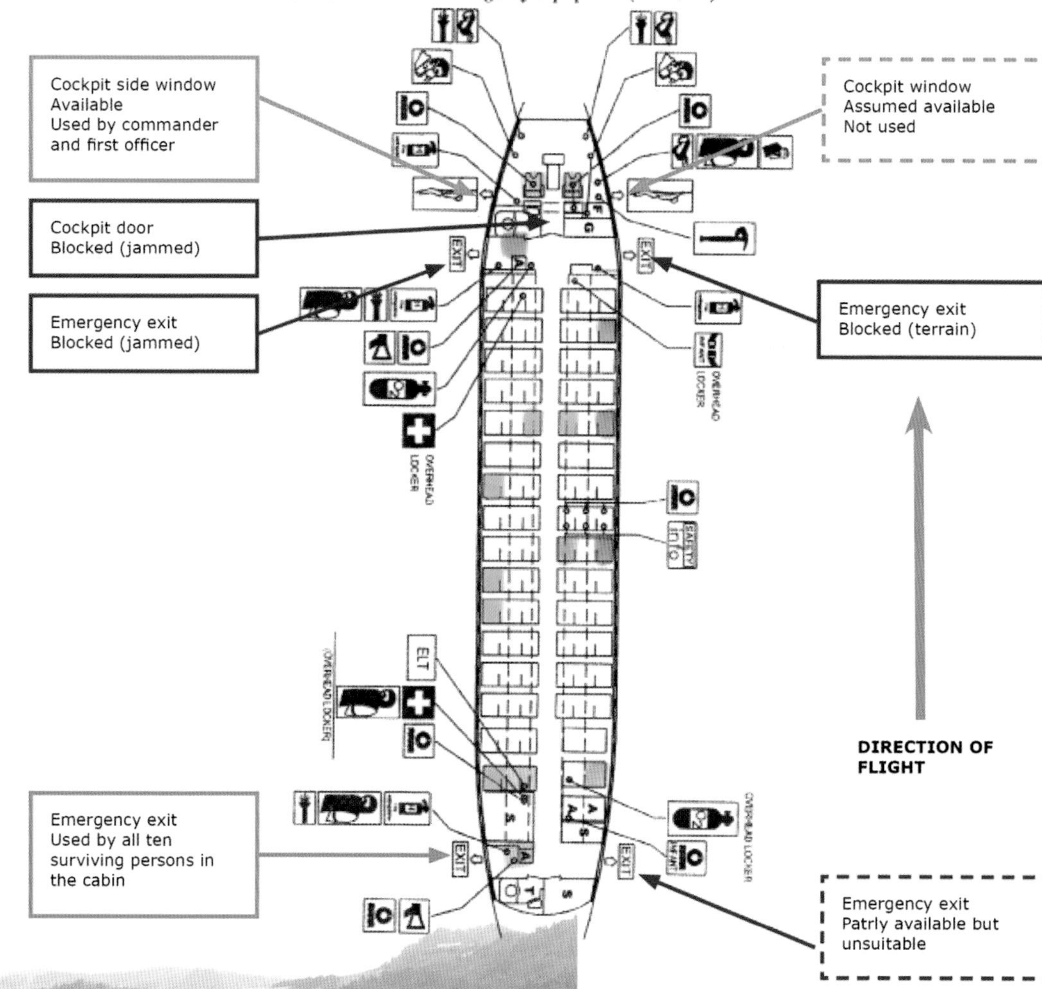

The seating layout from the crash report showing where pople were seated, and obstacles everyone faced. Green squares represent an uninjured passenger, while an orange square represents an injured passenger. Pink squares represent fatalities.

As you can see, there was a very steep drop once the aircraft left the runway.

Accident Investigation Board Norway (AIBN)

This photo shows the brakes locked up and skidding down the runway.

Accident Investigation Board Norway (AIBN)

Accident Investigation Board Norway (AIBN)

What was left of the BAe 146 after the fire was put out.

April 9, 2009: PT. Aviastar Mandiri Airlines BAe 146-300 E3189 PK-BRD

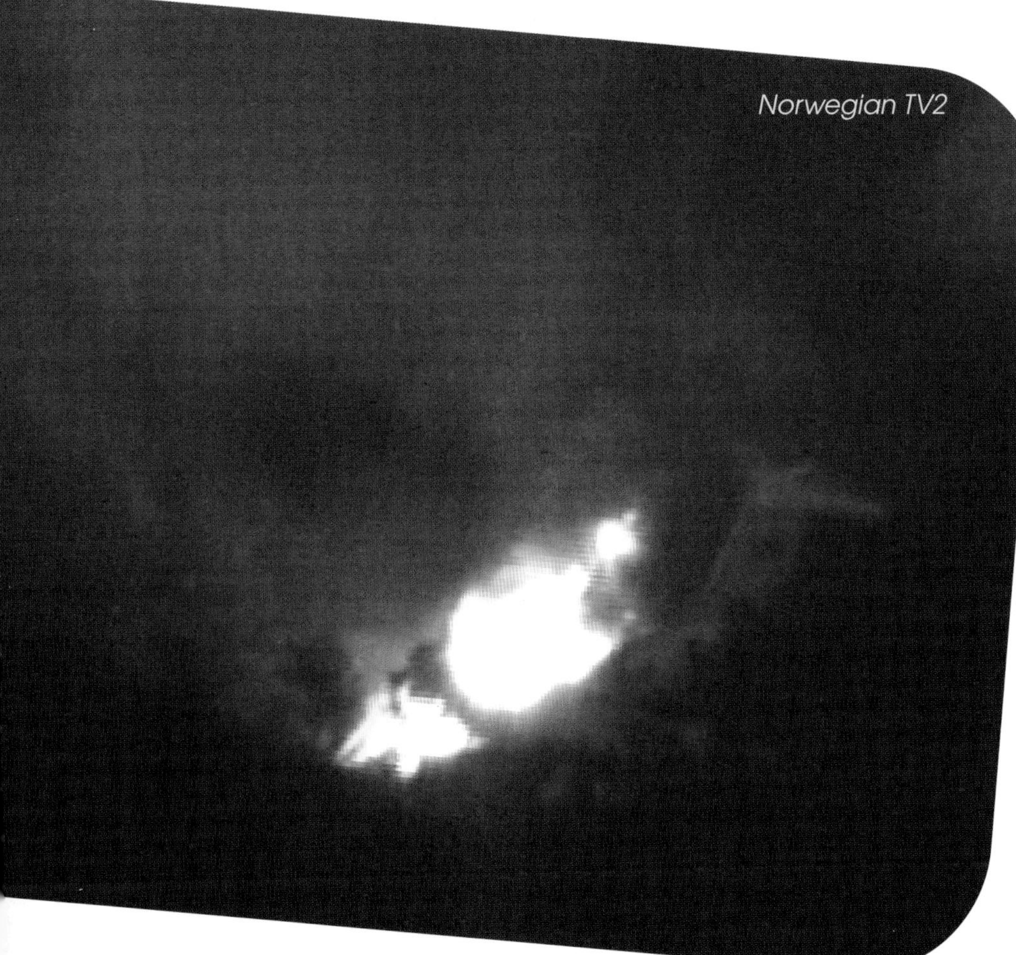

The aircraft was captured on fire by a local TV station.

BAe 146-300 E3189, flown by Aviastar Mandiri Airlines, was a passenger aircraft that was also able to carry cargo following a modification for combined passenger/cargo operations that was not manufacturer approved. The aircraft was configured in a combi-like configuration, with 42 passenger seats and space in the rear for freight such as rice and flour. The flights operated with not just two pilots, but also a loadmaster, engineer and two flight attendants. Originating in Papua New Guinea, the trip was operating as a cargo flight under instrument rules for a sector that departed from Sentani.

Upon arrival into Wamena, it completed a visual approach and landing as there was no ILS at the airport. The airfield is in mountainous terrain in the Baliem Valley at 5,085 feet. A missed approach was executed after the crew determined that the 146 would overshoot the runway. Along the way, passing through 5,600 feet, a series of alerts were issued including 'whoop-whoop, pull up' to 'low terrain' and 'don't sink' warnings. The aircraft overflew the airport and lined back up to make a second attempt. The same warnings occurred with the co-pilot and the engineer shouting instructions out "there's a hill on the left" and "sir, sir, sir, open, sir, left, left". After a series of banks and the final turn which was beyond 40 degrees, the aircraft entered a 10-degree nose down pitch. The co-pilot kept repeating to the captain "don't sink" and then the EGPWS sounded "terrain, too low, bank angle, bank angle, terrain, terrain." The copilot began calling out "sir, sir, sir" when the aircraft was in a 49 degree left bank turn and 6-degree nose down pitch. As the warning systems were sounding, the aircraft reached a 16-degree left bank with 12-degrees of nose pitch up. The aircraft impacted Pikei Hill on Tengah Mountain with the landing gear half-way through its deployment, killing all six crew aboard.

Factors contributing to the crash included the unapproved modification to the aircraft. Additionally, the center of gravity (CofG) calculations used for its operation were based on a 110-passenger configuration and not the actual configuration and therefore, an incorrect weight and balance trim sheet was used. The 146 was equipped with an EGPWS – enhanced ground proximity warning sensor – but the device was operating only as a GPWS, and it is not clear if the pilots or the airline turned off the additional functionality. Quite simply, the pilot in command failed to maintain situational awareness of the terrain in the area, and the EGPWS being turned off was a contributing factor.

View from starboard side of aircraft.

Footloose Tomcat

The tide has come in and engulfed the front of the aircraft.

Footloose Tomcat

October 19, 2013: SkyJet Airlines RP-C5525 BAe 146-200 E2031

SkyJet Airlines (Magnum Air) is a local commuter airline based in Manila, Philippines. It had a fleet of four BAe 146 aircraft (two series 100s, two series 200s), but has suffered a number of incidents over the years. A recurring incident at Skyjet Airlines seems to be runway excursions, with two involving the BAe 146 in 2018. This particular incident was more significant as it involved an exclusive charter to the Balesin Island Club, which maintains an airfield adjacent to a high-end resort.

The aircraft set out for the island carrying 68 passengers and seven crew members but encountered delays en route due to weather. Passengers reported circling the island three to five times which involved some very steep turns, climbing and descending very quickly. The travellers onboard also noticed an extended flight time, as well as the abrupt aircraft movements. Panic slowly manifested itself, with one passenger requesting oxygen. Some others began to sense an accident approaching, and complained to the flight attendant when they found there was no life safety vests under their seats. After three aborted landings, the aircraft descended and finally touched down. Capt. Maximo Rosales miscalculated the length of the runway and the stopping distance required, resulting in a runway excursion.

Passengers described a brief moment when the aircraft bounced and then stopped immediately, thrusting them all forward as it came to a halt. The 146 came to rest on a shallow reef and beach that was exposed while the tide was out, although the nose gear collapsed as the aircraft drifted sideways after leaving the airstrip. None of the passengers or crew were injured, but with the floor buckled, the pilots were trapped in the cockpit. Flight attendants directed passengers to not panic, and after seeing the height of the tail and the aircraft slide, began directing them to exit from the front of the cabin. Some passengers smelled fuel, and instead departed the aircraft from the rear using the inflatable slides. Later in the day, the tide rolled back in submerging the lower front of the 146. Twenty-four hours afterwards the aircraft was removed from the beach and was pulled up into a more secluded area, with a sand wall built around it to keep people from taking photographs. The 146 was written off after being submerged in saltwater, and in a sense of irony the airline did not refund any of the passenger's airfare because they reached their destination.

Footloose Tomcat

Aircraft was subsequently towed in.

Remains of the Lamia Avro RJ.

November 28, 2016: LaMia Avro RJ85 E2933

LaMia flight 2933 was a charter flown with an Avro RJ85, that carried the Brazillian Chapecoense football (soccer) squad and support staff from Santa Cruz de la Sierra, Bolivia to Medellín where the team was to play in the 2016 Copa Sudamericana Finals. With 73 passengers and four crew members, the flight departed Viru Viru International Airport at 5:50pm and included a planned fuel stop at Cobija-Captain Anibal Arab Airport near the Bolivian border due to the total distance being just less than 1,600 miles. As a result of a late departure and the closing of Cobija-Captain airport before the expected arrival time, the refueling stop could not be accommodated.

The crew, realizing that the non-stop distance to Medellín was the same as the aircraft's maximum range, began to petition Bolivian officials with the AASANA (Airports and Air Navigation Service Administration) to approve a flight plan with no fuel stop. Each time a new plan was submitted it was rejected. Realizing that the flight would plan not be approved without a fuel stop, a revision was submitted with a fuel stop in Bogotá. The distance from Santa Cruz to Bogotá was 1,486 miles, and after refueling, the last 'hop' to Medellín was a mere 116 nautical miles. The flight plan with the fuel stop in Bogotá was approved by the AASANA official.

The doomed aircraft finally departed, but the flight crew had no intention of heading to Bogotá for a refueling stop. They were confident the aircraft would arrive at Medellin with the fuel loaded, based on the aircraft's maximum range published in the flight manuals. At 9:42pm, the pilot reported an electrical failure and fuel exhaustion in the airspace between Sonson and Abejorral, approximately 45km from Jose Maria Cordova airport, in Medellín.

The aircraft had begun a descent just thirty minutes prior, but another aircraft in the area was diverting to Medellín because of a suspected fuel leak. LaMia 2933 was put into a holding pattern along with two other flights, completing trips around the hold which added 54 nautical miles to the flight path. At 9:49 the flight crew asked for priority landing because of 'problems with fuel', and air traffic control responded that clearance should be expected in about seven minutes. At 9:52, the crew of LaMia flight 2933 declared a fuel emergency, requesting immediate clearance to land. One minute later, engines 3 and 4 flamed out due to fuel starvation. Two minutes later at 9:55pm, engines #1 and 2 flamed out, turning the RJ85 into a glider. The RJ85 impacted the side of a mountain, Cerro Gordo, with a base of 8,500 feet at 9:59PM, with the wreckage split between the southern and northern sides of the crest.

Heavy fog prevented access with an air rescue helicopter, although ground rescue crews arrived two hours after the crash had occurred. Four hours after impact, the first survivor arrived at the hospital for treatment. Seven people were found alive at the crash site quickly. The final survivor was found at 5:40am, nearly eight hours after the impact. One of the survivors was a flight attendant, who reported that the captain's final words were "There is no fuel." The cockpit voice recorder stopped recording nearly 90 minutes before the crash for unknown reasons. Aside from the obvious issue that pushing the aircraft to its limits without factoring in any unplanned deviation could starve the aircraft of fuel, one contributory event was the dropping of the landing gear earlier than necessary, increasing drag on the airframe and as a result fuel consumption. In the end, the airline bore the responsibility from poor flight planning, pushing the aircraft to its operational limits, not declaring an emergency sooner, and failure to follow the fuel management rules set by Bolivian DGAC. It was a tragic accident that should never have happened.

Two weeks later, U.S. Spanish language media company used data from Flightradar24 to determine that LaMia had violated fuel and loading regulations on almost half of the past 23 flights it operated. This included two from Medellín to Santa Cruz, during which the aircraft would have had to use some of the its mandatory fuel reserves (45 additional minutes flying time). In short, these flights were flying without true reserves, which was instead included in the normal flight fuel requirements. The Bolivian government suspended a number of DGAC and AASANA officials including the director of the DGAC National Aeronautics Registry, who was the son of one of LaMia's owners. It also filed criminal charges against government and airline officials that were complicit or did not intervene to stop the tragedy from unfolding.

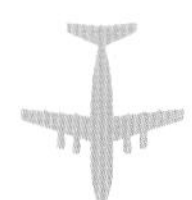

Colombia Civil Aviation Authority

CHAPTER 17

The Original Customers of the BAe 146 / Avro RJ

Fifty-four (54) airlines ordered and took delivery of new build aircraft direct from British Aerospace. This chapter documents the carriers and the aircraft they received direct from the manufacturer. It does not include operators that received aircraft second hand, nor aircraft originally operated by BAe then subsequently redelivered in various capacities.

A grand total of three hundred ninety-four (394) aircraft were built and delivered, with the last produced airframe E2394. Due to the economic effects of September 11, 2001, airframe 386, 387, 393, and 394 were stored for nearly a year and a half before being taken by Blue 1 towards the end of 2003. BAE Systems also had seven additional RJX airframes at various stages of build that were never completed and were assigned production numbers of E2395, E2396, E3397, E2398, E3399, E2400, and E3401. Airframes E3402, E3403, and E2404 were provisionally allocated, but no components were ever produced for these aircraft. Including the 394 production aircraft, the seven RJX airframes in various stages of completion, and the five structural test airframes, more than 400 aircraft were produced or in the state of production..

The order in which the airlines appear in this chapter is based on delivery of their first aircraft. All information was derived from a production list maintained by Ken Haynes, and was accurate at the time of printing (2021).

Dan Air

Based in the United Kingdom, Dan Air became the first airline operate the BAe 146. Initially taking delivery of two -100 aircraft, it would be six years before it acquired four series -300 aircraft. A hybrid charter/scheduled service airline, Dan Air would eventually be sold to British Airways in 1992.

Dan-Air's first BAe 146-300 ordered new from BAe.

LINE	SERIES	DELIVERY	REGISTRATION
6	100	5/23/83	G-BKMN
7	100	6/18/83	G-BKHT
126	300	6/2/89	G-BPNT
155	300	8/15/90	G-BTNU
183	300	6/30/92	G-BUHB
193	300	7/1/92	G-BUHC

Air Wisconsin

United States based Air Wisconsin was the first airline to place an order (that was delivered), and was the launch customer for the series -200 and the series -300. Air Wisconsin signed an purchase agreement prior to Dan Air, operating the BAe 146 from June 1983 until April 2006, marking 23 years of service. The 146 represented Air Wisconsin's first foray into jet service and was used to open up new markets initially in its own name, and later as a feeder carrier for United Airlines, under a scope clause exemption agreement with the US major carrier. By 2000, United Airlines pilots' union agreed that Air Wisconsin could operate the BAe 146-200 or Avro RJ85 (or similar aircraft) on United's behalf providing it had no more than 85 seats and did not exceed 90,000 MGTOW (maximum gross take-off weight). As a result, Air Wisconsin retired its BAe 146-300 aircraft earlier than expected, retaining just the series 200s.

Air Wisconsin subsequently became an US Airways Express operator. The scope clause with its new partner prevented it from operating any aircraft that the mainline airline ever operated. Because PSA, acquired by US Airways, had flown the BAe 146 previously, no US Airways Express operator could operate the type and this brought operations of the ageing BAe 146 with Air Wisconsin to a close after nearly 23 years of service. Air Wisconsin continued to operate the BAe 146-200 until Easter Sunday, April 16, 2006, when the final flight departed Chicago O'Hare Airport (ORD) to the airlines home base in Appleton.

LINE	SERIES	DELIVERY	REGISTRATION
12	200	5/25/83	N601AW
14	200	9/29/83	N602AW
18	200	12/17/83	N603AW
20	200	2/23/84	N604AW
16	200	7/21/84	N605AW
33	200	2/27/85	N606AW
52	200	1/15/86	N607AW
49	200	4/24/86	N608AW

LINE	SERIES	DELIVERY	REGISTRATION
70	200	6/19/87	N609AW
82	200	9/16/87	N610AW
120	300	12/16/88	N611AW
122	300	12/23/88	N612AW
132	300	5/24/89	N614AW
141	300	9/9/89	N615AW
145	300	11/4/89	N616AW

The flight and ground crew after the last commercial flight of Air Wisconsins BAe 146 aircraft.

Royal Air Force

The Royal Air Force (RAF) purchased two series -100 aircraft for trials as potential replacements for The Queen's Flight Hawker Siddeley Andovers. These two aircraft were put on a series of 'missions' and evaluated for suitability for the RAF requirements. After the trial operations, they were traded in to BAe for two new build aircraft that would constitute the 32nd.

LINE	SERIES	DELIVERY	REGISTRATION	NAME
4	100	9/16/83	ZD695	
5	100	6/14/83	ZD696	

TABA

One of the first regional airlines in the five territories of Brazil, Transportes Aéreos da Bacia Amazônica was an air taxi operator. TABA handled the region in the north and central west regions of the country and was one of the earliest customers of the BAe 146, with its livery being the same as the colours carried by the BAe demonstrator.

10	100	12/8/83	PT-LEP	
11	100	12/17/83	PT-LEQ	

The Mali Government

The Mali Government's BAe 146 was the first to be purchased with extended range Pannier tanks. They operated a single BAe 146-100, primarily as a Presidential aircraft. It served for two years, but only flew some 600 hours before BAe repossessed it on behalf of the funding bank, when the Mali Government defaulted on its finance payments and could no longer fly the aircraft.

9	100	10/15/83	TZ-ADT	"Nioro du Sahel"

AirPac

Air Pac was the first North American airline to operate the BAe 146 into an airfield with unpaved runways (Dutch Harbor, Alaska) via scheduled flights; the first to bring jet services to Dutch Harbor; and one of the earliest 146s to offer hot meal service. Transporting passengers and fresh seafood in the main cabin, the 146 was operated into a number of airports in Alaska and Washington state.

LINE	SERIES	DELIVERY	REGISTRATION	NAME
13	100	3/3/84	N146AP	"The Alaska Connection"

PSA

Based in San Diego, California, PSA put the BAe 146 on the map with a record (at the time for BAe) 20 aircraft order with 24 options. PSA took delivery of all 20 ordered aircraft, and exercised four options. PSA named it's aircraft the "Smile of (city/name)" after the destinations they served.

In PSA's later years, experiments with the livery took place. The cheatline no longer curves under the nose, and the titles are a different font (and solid gray color).

LINE	SERIES	DELIVERY	REGISTRATION	NAME
22	200	6/13/84	N346PS	"The Smile of PSA"
23	200	06/13/84	N347PS	"Tri-Cities"
24	200	7/14/84	N348PS	"Medford"
25	200	11/9/84	N349PS	"Reno"
27	200	12/7/84	N350PS	"Stockton"
28	200	12/19/84	N351PS	"Eureka"
31	200	3/27/85	N353PS	"Orange County"
30	200	3/29/85	N352PS	"Eugene"
34	200	5/24/85	N354PS	"Concord"
36	200	6/8/85	N355PS	"Fresno"
39	200	8/8/85	N356PS	"Disneyland"
40	200	8/20/85	N357PS	"Springfield"
41	200	9/10/85	N358PS	"Monterey"
42	200	10/8/85	N359PS	"Fresno"
43	200	10/8/85	N360PS	"Oakland"
44	200	10/31/85	N361PS	"Eugene"
45	200	11/18/85	N362PS	"Los Cabos"
46	200	12/3/85	N363PS	"Orange County"
47	200	12/10/85	N364PS	"San Diego"
48	200	12/19/85	N365PS	"Yakima"
72	200	6/10/87	N366PS	
73	200	6/10/87	N367PS	
74	200	6/19/87	N368PS	
75	200	7/2/87	N369PS	

Aspen Airways

Named after the tree (not the town), Aspen Airways operated the first commercial jet service into the city of Aspen, Colorado using the BAe 146, which was certified for the steep climb and descent required for safe operations. Flying between Aspen and destinations around Colorado as well as Texas, the carrier eventually became a United Express operator and expanded its reach into California and Illinois. Unlike the colorful Convair 580s that Aspen operated in shades of green, blue, orange, and brown, the three BAe 146 were painted in the orange livery only.

LINE	SERIES	DELIVERY	REGISTRATION
15	100	12/14/84	N461AP
17	100	6/7/85	N462AP
63	100	12/18/86	N463AP

Ansett W.A. / Ansett Australia

Ansett ended up taking seven BAe 146-200 aircraft for its Western Australian operation before being integrated into Ansett Australia, which deployed a total of 21 BAe 146-200 and -300 aircraft. The first of 11 -300 aircraft were ordered for its New Zealand affiliate in 1989 and equipped with 15 business and 75 economy class seats in a five abreast layout. Ansett Air Freight also operated two cargo -200 variants.

Despite being acquired by Air New Zealand in 2000, Ansett Australia collapsed due to ongoing financial losses in September 2001. Air New Zealand transferred the New Zealand based 146s to regional affiliate Mount Cook Airlines where after a year of operations, they were replaced with Boeing 737-300 aircraft.

LINE	SERIES	DELIVERY	REGISTRATION
37	200	4/22/85	VH-JJP
38	200	6/22/85	VH-JJQ
93	200	10/27/88	VH-JJS
98	200	11/20/88	VH-JJT
110	200	3/1/89	VH-JJW
113	200QT	5/4/89	VH-JJY
114	200QT	6/3/89	VH-JJZ
116	200	7/15/89	ZK-NZA
119	200	10/9/89	ZK-NZC

AirCal

Local Southern California airline AirCal, based at noise sensitive John Wayne Airport (SNA), was facing competition on its home turf from PSA. To counter the additional slots PSA garnered from the airport and to alleviate the payload restrictions its new Boeing 737-300s had incurred operating at SNA, AirCal purchased six BAe 146-200 aircraft. Before it took delivery of all the aircraft, the carrier was purchased by American Airlines. American Airlines continued to operate the BAe 146-200s until November 1, 1990 when the last was withdrawn from service on flights to San Francisco from SNA. It pulled out of the Orange County market citing the high operating costs of the BAe 146 and low load factors on the ultra-competitive route.

LINE	SERIES	DELIVERY	REGISTRATION
51	200	3/5/86	N141AC
53	200	3/22/86	N142AC
54	200	5/17/86	N144AC
55	200	6/8/86	N145AC
57	200	8/1/86	N146AC
58	200	10/4/86	N148AC

RAF – Queens Flight

Following successful flight trials, the RAF purchased three BAe 146-100 aircraft to replace the aging Andover prop aircraft allocated to transport the Royal Family, as well as VIPs and dignitaries.

21	100	4/23/86	ZE700
29	100	7/9/86	ZE701
124	100	1/14/91	ZE702

Pelita Air Service

Pelita is a non-scheduled carrier that operates on behalf of Indonesian oil company Pertamina, providing executive transport. In practice its 146 aircraft were used for Presidential and VVIP flying around the country

LINE	SERIES	DELIVERY	REGISTRATION	NAME
50	200	6/28/86	PK-PJP	
239	200	1/11/94	PK-PJJ	"Wamena"

Handover ceremony of Presidential's first BAe 146-200.

Presidential

A regional carrier operating out of Washington D.C., Presidential started out carrying passengers between small cities that were served by turboprops, eventually becoming a Continental Jet Express affiliate and later a United Airlines affiliate.

LINE	SERIES	DELIVERY	REGISTRATION	NAME
59	200	8/8/86	N401XV	"Franklin Pierce"
60	200	8/28/86	N402XV	"James Buchanan"
61	200	9/13/86	N403XV	"Abraham Lincoln"
64	200	11/20/86	N404XV	"Andrew Jackson"
66	200	12/21/86	N405XV	"Ulysses S. Grant"
62	200	3/19/87	N406XV	"Teddy Roosevelt"
69	200	6/26/87	N407XV	"Franklin D. Roosevelt"
77	200	8/4/87	N408XV	"Harry S. Truman"

CAAC / China Eastern Airlines / Air China

Civil Aviation Administration of China (CAAC) took delivery of ten aircraft before the airline was split up. The airline division of the government's aviation authority was split up into six separate airlines in 1988 (Air China; China Southwest Airlines; China Eastern Airlines; China Northwest Airlines; China Southern Airlines; China Northern Airlines.

LINE	SERIES	DELIVERY	REGISTRATION
19	100	9/10/86	B-2701
26	100	10/14/86	B-2702
32	100	11/21/86	B-2703
35	100	12/13/86	B-2704
68	100	1/24/87	B-2705
71	100	2/18/87	B-2706
76	100	4/17/87	B-2707
81	100	6/19/87	B-2708
83	100	7/15/87	B-2709
85	100	8/6/87	B-2710

TNT Airways

TNT Airways fell far short of the 72 BAe 146QT aircraft it said it was going to order.

Air Foyle, based in Luton, became the first operator of the BAe 146QT aircraft (on behalf of TNT Airways). Air Foyle ultimately ended up flying 19 of the type on behalf of TNT Airways although the contract ended after 12 years and the aircraft were transferred back to the parent company. Air Foyle also had a 25% stake in BAe 146 operator CityJet in 1999, but sold its stake to Air France. Malev Hungarian Airlines, the flag-carrier of Hungary, flew mostly Soviet-era aircraft on passenger operations. By 1988, Soviet aircraft were being replaced with Western built aircraft, and Malev was rumored to be a possible large 146 customer but the only development was operating a single 146QT on behalf of TNT Airways. Pan Air was a cargo airline based at Madrid-Barajas International Airport in Spain, running freighter service on behalf of TNT Express. Euralair Horizons, based out of France, was a charter operator that deployed BAe 146-200QT aircraft on behalf of TNT Express from the summer of 1989. Malmo Aviation, based in Sweden and operating domestic routes from Stockholm-Bromma Airport, originally began flying the BAe 146-200QT on behalf of TNT from 1988. Mistral Air operated the Italian registered aircraft. One of the negative effects of TNT flying horses was the corrosion of seat rails. The boxes carrying the horses were supposed to be water tight, and TNT filed many warranty claims as a result.

LINE	SERIES	DELIVERY	REGISTRATION
56	200QT	5/5/87	G-TNTA
67	200QT	9/20/87	G-TNTB
78	200QT	12/3/87	I-TNTC
86	200QT	2/11/88	SE-DEI
89	200QT	3/25/88	G-TNTH
100	200QT	9/30/88	G-TNTJ
102	200QT	10/24/88	EC-198
105	200QT	11/18/88	HA-TAB
109	200QT	12/13/88	SE-DHM
112	200QT	1/27/89	F-GTNU
117	200QT	5/10/89	F-GTNT
150	300QT	12/21/89	SE-DIM
153	300QT	6/8/90	G-TNTE
154	300QT	9/13/90	G-TNTF
182	300QT	1/1/92	G-TNTG
186	300QT	2/3/92	G-TNTK

LINE	SERIES	DELIVERY	REGISTRATION	NAME
168	300QT	2/10/92	G-TNTL	
166	300QT	2/28/92	G-TNTM	

Zimbabwe Government (Air Zimbabwe)

Air Zimbabwe was the national carrier for the South African country, with its base in the capital Harare. Running a very small fleet of aircraft, the BAe 146-200 was purchased to serve domestic routes (including unpaved airfields) as well as an ad-hoc VIP transport, with the front of the cabin that could be quickly modified with eight VIP business class seats.

LINE	SERIES	DELIVERY	REGISTRATION	NAME
65	200	11/29/87	Z-WPD	"Jongwe"

Air UK

AirUK was one of the few airlines that operated all three passenger versions of the BAe 146 (-100,-200,-300) with deliveries for its series -200 beginning in 1987, and the last in March 1995. Two years later in 1997, KLM took full ownership of Air UK, rebranding it as KLM UK. The 146 fleet was repainted into KLM colors, but after two years, the aircraft were moved to a new and wholly owned subsidiary, the low-cost carrier Buzz. Buzz did not last and was sold to Ryanair. Over the course of eleven years (1987-1998), 22 BAe 146s wore Air UK colors.

LINE	SERIES	DELIVERY	REGISTRATION
79	200	11/27/87	G-CNMF
88	200	3/30/88	G-CHSR
94	200	4/7/88	G-CSJH
123	300	2/28/89	G-UKHP
125	300	3/10/89	G-UKSC
142	300	11/20/89	G-UKAC
157	300	3/27/90	G-UKID
162	300	1/4/91	G-UKAG
158	300	3/9/91	G-UKRC

The final livery worn by AirUK

WestAir (United Express)

Founded in 1978 as a regional airline based in Fresno, California operating turboprop aircraft, WestAir became a United Airlines affiliate, serving a number of smaller destinations as a feeder carrier. It purchased six new build aircraft in 1987-1988.

LINE	SERIES	DELIVERY	REGISTRATION
80	200	12/16/87	N290UE
84	200	12/19/87	N291UE
87	200	12/22/87	N292UE
97	200	4/15/88	N293UE
107	200	8/5/88	N294UE
108	200	8/27/88	N295UE

UAE

The government of the United Arab Emirates purchased a single BAe 146-100 to serve as a VIP transport. It was often used to fly VIP's and their prize Hawk birds into remote parts of countries such as Morocco and Pakistan so that they could indulge their passion for Hawking.

LINE	SERIES	DELIVERY	REGISTRATION
91	100	12/21/87	A6-SHK

Malmo Aviation

Based in Sweden and operating domestic routes from Stockholm-Bromma Airport, the carrier began operating a 146QT on behalf of XP Parcels, followed by passenger services from 1992.

151	300QT	2/13/90	SE-DIT

Air BC

Air BC was a regional airline based in British Columbia, Canada, serving the west of the country including Alberta, as well as providing some flights into the Pacific Northwest (United States). Mainly a de Havilland Canada DHC-6/7/8 operator, the BAe 146 was the first and only jet aircraft Air BC operated shortly after Air Canada became the majority shareholder in the airline. It eventually was combined with other Canadian regional carriers to form the airline Air Canada Jazz. Air BC operated five BAe 146-200 aircraft.

LINE	SERIES	DELIVERY	REGISTRATION
90	200	5/13/88	C-FBAB
92	200	6/8/88	C-FBAE
96	200	6/12/88	C-FBAF
111	200	1/24/89	C-FBAO
121	200	2/24/89	C-FBAV

AirBC's route map with the BAe 146 (routes marked 'jet' in orange).

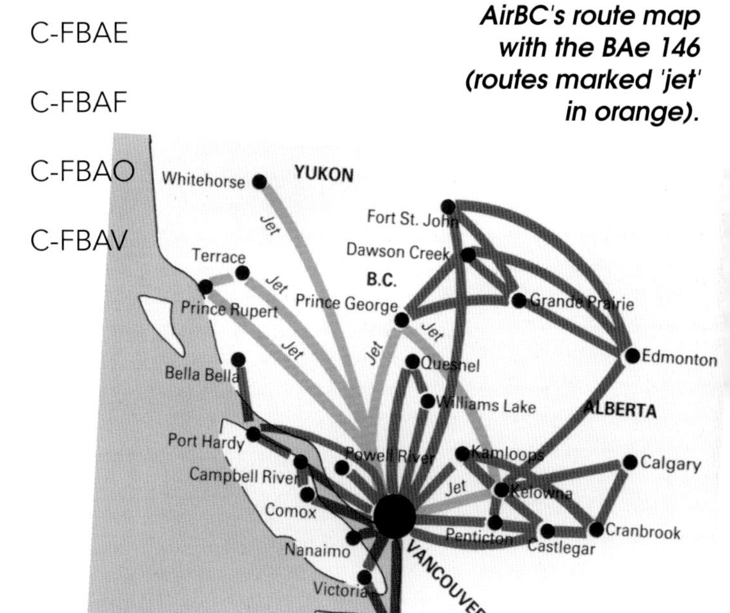

Loganair

Based in Scotland at Glasgow Airport, Loganair was a domestic regional airline providing services in the United Kingdom. It eventually became an affiliate of British Airways, and later Flybe.

LINE	SERIES	DELIVERY	REGISTRATION
99	200	7/19/88	G-OLCA
103	200	5/89	G-OLCB

Druk Air

Based at Paro Airport high up in the mountainous terrain of the Kingdom of Bhutan, Druk Air is the flag carrier of the tiny nation. It was started in 1981 using Dornier Do 228 prop aircraft, but later it purchased a single BAe 146-100 which was delivered in November 1988. The purchase also marked the first time the country obtained a commercial loan in the amount of $21.8 million USD from the UK Export Credit Guarantee Board. Load factors were well above 50%. With growth occurring, a second BAe 146-100 was purchased to serve New Delhi, Bangkok, and Kathmandu. The 146 was perfect for the airline's base at Paro airport, given the mountainous terrain, steep approach and a departure in a valley 7,300 feet above sea level. The mountains the 146 weaved through to get to the airport were as high as 16,000 feet, and as a result of the zig-zag approach visual flight rules (VFR) prevailed at all times. In 2001, corrosion was discovered in the wing fuel tanks of A5-RGD. It was sent back to Woodford where the wings were replaced and re-entered service in 2001. Druk Air was going to be the launch customer for the RJX85 before the program was cancelled, as it would have given the airline the ability to serve Hong Kong non-stop.

95	100	11/16/88	A5-RGD
395	RJX85	-----------	G-ORJX (temp registration)
396	RJX85	-----------	Components arrived, not assembled

Signing ceremony for Druk Air's first BAe 146-100.

Artists rendering of RJX85 launch customer Druk Air.

Air Nova

Air Nova, operating from a base in Nova Scotia, Canada, purchased the BAe 146 as its first jet aircraft to serve routes in Eastern Canada and the Northeastern United States. It started using two leased second-hand series -100s until the -200s that had been ordered were ready for delivery. Air Nova was eventually merged with Air BC, Air Ontario and Canadian Regional Airlines to form Air Canada Jazz.

LINE	SERIES	DELIVERY	REGISTRATION
106	200	1/11/89	C-GRNZ
115	200	1/24/89	C-GRNY
130	200	7/13/89	C-GRNX
133	200	9/29/89	C-GRNV
140	200	12/7/89	C-GRNT
139	200	12/21/89	C-GRNU

Air Botswana

The state-owned flag carrier of Botswana purchased two BAe 146-100 aircraft for domestic regional services. The airline had struggled since inception to achieve profitability, and subsequently grounded one aircraft leaving only one in service for over a year due to lack of money for parts to repair it and return it to service.

LINE	SERIES	DELIVERY	REGISTRATION
101	100	1/14/89	A2-ABD
160	100	8/15/90	A2-ABF

Thai Airways

Thai Airways engaged in significant expansion in the late 1980s, and with that came the decree to begin replacing turboprop aircraft with jets on many domestic routes. Thai Airways ultimately took delivery of ten aircraft from British Aerospace in a combination of purchases and leases, including all three passenger variants of the BAe 146 (series -100, -200, and -300).

The aircraft operated for less than a decade, and were disposed of when Thai Airways was looking to streamline its operations. In 1998, restructuring of the fleet began and the BAe 146-300s were retired after eight years of service.

LINE	SERIES	DELIVERY	REGISTRATION	NAME
128	200	4/28/89	HS-TBK	"Chiang Kham"
131	300	6/23/89	HS-TBL	"Sukhirin"
129	300	11/22/89	HS-TBM	"Watthana Nakhon"
104	100	12/2/89	HS-TBO	"Lahan Sai"
149	300	1/12/90	HS-TBN	"Mukdahan"
185	300	6/4/91	HS-TBK	"Chiang Kham"
191	300	6/15/91	HS-TBJ	"Chon Buri"
181	300	12/11/91	HS-TBL	"Sukhirin"
189	300	12/23/91	HS-TBO	"Lahan Sai"
206	300	2/8/92	HS-TBM	"Watthana Nakhon"

Ansett New Zealand

With ties to Ansett Australia, Ansett New Zealand commenced operations with Boeing aircraft in 1987, but by had 1988 replaced its Boeing 737-100s with BAe 146s. Ansett New Zealand flew domestic routes and eventually also operated a single QC aircraft on freight services at night.

LINE	SERIES	DELIVERY	REGISTRATION	NAME
127	200	7/29/89	ZK-NZB	"Rotoroa"
134	300	12/5/89	ZK-NZF	"City of Wellington"
135	300	12/9/89	ZK-NZG	"City of Christchurch"
143	300	12/19/89	ZK-NZI	"City of Dunedin"
137	300	12/22/89	ZK-NZH	"City of Auckland"
147	300	2/19/90	ZK-NZJ	"City of Nelson"

Discovery Airways

Discovery Airways second BAe 146-300, E3165 (N886DV assigned).

Neil Brant

Discovery Air probably represents the carrier that had the shortest duration in business (five months from first flight to shut down) with the largest fleet of new build aircraft (three). The carrier, which was based in Hawaii, did not last long owing to irregularities with its operating certificate and alleged foreign ownership in violation of U.S. law. Three more aircraft were built and painted in Discovery Airways colors (3163/N885DV, 3165/N886DV, 3169/N887DV – all series 300 aircraft), but were not delivered before the airline was shut down.

LINE	SERIES	DELIVERY	REGISTRATION
136	200	12/16/89	N882DV
138	200	12/23/89	N883DV
156	200	1/18/90	N884DV

P.T. National Air Charter

Indonesian charter airline National Air Charter purchased one new build BAe 146-200 and also acquired two second-hand aircraft. Although completed and painted the aircraft never went into service and after a long period of storage was sold for executive use.

LINE	SERIES	DELIVERY	REGISTRATION	NAME
144	200	6/22/90	PK-DTA	

National Jet Systems

National Jet Systems (now part of Cobham) was one of four Australian domestic carriers operating under the Qantaslink brand that operated BAe 146 aircraft. Only National Jet Systems purchased new build aircraft, acquiring examples of all three passenger variants of the 146 direct from British Aerospace.

152	100	2/19/92	VH-NJR	Australian Airlink
176	200QC	7/19/91	VH-NJQ	
198	300QT	10/6/92	VH-NJF	Australian Air Express
213	300	10/25/94	VH-NJL	Qantas Airlink
217	300	10/28/94	VH-NJN	Qantas Airlink

National Jet operating as a Qantas Airlink affiliate.

Sagittair

Sagittair, based out of Italy, ordered two new build series 300 aircraft for charter operations. The airline's plans included serving Rome, Milan, and Naples as well as Cairo. Unfortunately, the business did not last beyond a year and shut down in 1991.

146	300	6/21/90	I-ATSC
159	300	7/10/90	I-ATSD

Princess Air

Formed by local travel agency Burstin Travel, the airline was started in 1990 with a single 200QC aircraft to run charter flights during the day, and cargo operations at night. Two aircraft were ordered but unfortunately only a single 146 was delivered before the company shut down in 1991.

LINE	SERIES	DELIVERY	REGISTRATION	NAME
148	200QC	6/8/90	G-PRIN	"Princess Alison"

Delta Air Transport (Sabena)

Delta Air Transport (DAT) was based in Brussels, Belgium, and ran scheduled and chartered flights over short haul routes both independently and as an affiliate of Sabena. Shortly after September 2001, Swissair (which it owned 49% of Sabena) filed for bankruptcy after running out of money, and subsequently was unable pay its liabilities to Sabena. This had a knock-on effect with Sabena going into liquidation as it too did not have the capital to continue flying. SN Brussels Airlines was born from DAT and changed its name accordingly in 2002. In 2005, it merged with Virgin Express and rebranded as Brussels Airlines. DAT was one of the largest customers for the BAe 146/Avro RJ aircraft.

164	200	7/27/90	OO-DJE
167	200	8/14/90	OO-DJF
172	200	12/24/90	OO-DJH
180	200	12/24/90	OO-DJG
192	200	4/26/91	OO-MJE
196	200	6/28/91	OO-DJJ
271	RJ85	12/6/95	OO-DJK
273	RJ85	12/6/95	OO-DJL
275	RJ85	12/18/95	OO-DJN

LINE	SERIES	DELIVERY	REGISTRATION
279	RJ85	12/22/95	OO-DJO
287	RJ85	5/7/96	OO-DJP
289	RJ85	6/27/96	OO-DJQ
290	RJ85	7/31/96	OO-DJR
292	RJ85	8/22/96	OO-DJS
294	RJ85	9/26/96	OO-DJT
295	RJ85	10/23/96	OO-DJV
296	RJ85	11/4/96	OO-DJW
297	RJ85	11/22/96	OO-DJX
302	RJ85	3/1/97	OO-DJY
305	RJ85	4/13/97	OO-DJZ
308	RJ100	6/27/97	OO-DWA
315	RJ100	10/16/97	OO-DWB
322	RJ100	2/27/98	OO-DWC
324	RJ100	4/1/98	OO-DWD
327	RJ100	5/27/98	OO-DWE
332	RJ100	8/12/98	OO-DWF
336	RJ100	11/6/98	OO-DWG
340	RJ100	12/16/98	OO-DWH
342	RJ100	1/22/99	OO-DWI
355	RJ100	7/30/99	OO-DWJ
360	RJ100	10/14/99	OO-DWK
361	RJ100	11/3/99	OO-DWL

Eastwest Airlines

Eastwest had been a long-time target for BAe, but the company had elected to continue its relationship with Fokker. When the airline was acquired by Ansett, BAe saw its opportunity and finally made a breakthrough in 1990 when an order was placed for eight BAe 146-300 aircraft (and 4 options). The first five aircraft went into service by late 1990 following a ceremony at Sydney Airport. Eastwest redesigned its livery for the new jets. A new split sphere logo had yellow on top representing the sun and green on the bottom half representing the fertile coastline, with a sliver of red splitting the two halves representing Australia's Red Centre. The first five aircraft went into service by late 1990 and had replaced the Fokker F.28s by mid-1991. The 146s, nicknamed "Leisure Jets", were configured in a 96-seat layout and served Hobart, Brisbane, Cairns, Canberra and Melbourne.

The low noise profile allowed Eastwest to live up to being a 'good neighbour'. It phased out all non-reusable materials from flights and used paper products made from 100% recycled material. Eastwest eventually became part of the TNT organization.

LINE	SERIES	DELIVERY	REGISTRATION
171	300	8/25/90	VH-EWI
173	300	8/25/90	VH-EWJ
175	300	9/29/90	VH-EWK
177	300	10/28/90	VH-EWL
179	300	12/1/90	VH-EWM
190	300	4/26/91	VH-EWN
195	300	6/12/91	VH-EWR
197	300	6/30/91	VH-EWS

Handover ceremony for the first aircraft for Eastwest.

Makung Airlines (now Uni Air)

The blue painted outline marks the Type 2 Emergency exit and the path down the main gear doors to escape in the event of an emergency.

Local regional carrier based in Taiwan, Makung Airlines took delivery of five new-build series 300 aircraft with high density seating resulting in the fuselage having a mid-cabin type 3 plug-emergency exit door. It was the only airline that took aircraft with this feature, one that BAe marketed in the late stages of the series 300's life as well as for the RJ115 (which no operator took up). Three 146-300 aircraft were fitted with the emergency exits, but only two were taken up. The 146 series was the first domestic jetliner introduced into Taiwan once the government implemented an "Open Skies" policy in 1988. These aircraft were hybrids and did not have ALF502 engines, but instead were powered by non-FADEC versions of the ALF507. They were reportedly the most reliable of all the BAe 146 aircraft.

LINE	SERIES	DELIVERY	REGISTRATION	NOTES
161	300	8/3/90	B-1775	First with emergency exit
174	300	1/3/91	B-1776	Emergency Exit fitted
205	300	11/5/92	B-1777	
209	300	5/4/94	B-1778	
202	300	1/11/95	B-1781	Last new 146 delivered.

Meridiana (now Air Italy)

Italian carrier Alisarda changed its name in September 1, 1991 to mark an order for new aircraft and route expansions across Europe. Based in Milan, Meridiana took delivery of 11 aircraft to operate both scheduled and charter services. It initially introduced the series 200 and then – as part of expansion via a related but separate company, Meridiana Spain – the series 300. The series 200 aircraft were configured for 79 passengers with 40 in "Electa" business class with winged seating, while there were 39 in the economy class "Marco Polo" cabin.

The Spanish division was unsuccessful and closed down but the Italian operation continued unaffected.

LINE	SERIES	DELIVERY	REGISTRATION
184	200	5/25/91	I-FLRV
178	200	7/25/91	I-FLRW
165	300	8/8/91	EC-807
170	200	8/15/91	I-FLRX
169	300	8/30/91	EC-839
163	300	11/6/91	EC-876
187	300	12/20/91	EC-899
204	200	4/7/94	I-FLRU
210	200	5/21/94	I-FLRE
227	200	6/23/94	I-FLRO
220	200	7/1/94	I-FLRI

Crossair (now Swiss)

Crossair was founded in Basel, Switzerland during 1978, right at the time the development of the BAe 146 was given the go ahead. Operating very short flights with small turboprops (including some Fairchild Metro II aircraft) and a single Cessna 551 jet, the carrier grew over time. In 1990, it began to take its first true commercial jetliners with an order for three used BAe 146-200 aircraft (ex-US Air) after it was shown that they could operate out of Lugano, which no other jet could use. Despite the steep approach and very short runway and surrounded by mountainous terrain, the 146 excelled at operating a full load without penalty. The airline continued to grow and added service to destinations to Europe, and the 146 was used to open routes to London City Airport, for which the aircraft was ideally suited. Crossair became the first airline to fly the BAe 146 into London City Airport (LCY) in 1992.

Swissair acquired a majority stake in Crossair in 1992. The 146 was dubbed the "Jumbolino" to promote the concept of it being a miniature jumbo jet, and they even featured a flying elephant logo applied under the cockpit windshield on the port side. The 146 carried the original Crossair livery, while newer Avro RJs carried a revised and refreshed paint scheme starting with airframe 226.

Given its success with the 146, the airline placed orders for 20 updated Avro RJ series aircraft. In 2002, following the collapse of Swissair, Crossair changed its name to Swiss International Airlines. The airline changed its livery again to feature an all-white fuselage with large titles on the front, and a simple red tail and white cross representing the flag of Switzerland. The Avro RJ85 and RJ100s continued to operate for the airline to destinations that didn't warrant larger aircraft, as well as to those demanding short field capability.

Swiss International Airlines operated its final Avro RJ100 flight (LX7545) with the last of the type in the fleet (HB-IYZ) on August 15, 2017, marking the end of nearly 26 years of BAe 146/Avro RJ services. It was the fourth largest customer for British Aerospace regional aircraft. Its aircraft have since been replaced with new Bombardier CS100s, which has since been rebranded as the Airbus 220.

LINE	SERIES	DELIVERY	REGISTRATION
118	300	9/18/91	HB-IXZ
231	RJ85	5/13/93	HB-IXG
233	RJ85	6/26/93	HB-IXH
235	RJ85	7/10/93	HB-IXK
226	RJ85	4/23/93	HB-IXF
262	RJ100	10/19/95	HB-IXX
272	RJ100	11/1/95	HB-IXW
274	RJ100	12/15/95	HB-IXV
276	RJ100	12/22/95	HB-IXU
259	RJ100	1/9/96	HB-IXT
280	RJ100	2/8/96	HB-IXS
281	RJ100	2/29/96	HB-IXR
282	RJ100	3/27/96	HB-IXQ
283	RJ100	4/30/96	HB-IXP
284	RJ100	5/27/96	HB-IXO
286	RJ100	7/22/96	HB-IXN
291	RJ100	8/23/96	HB-IXM
338	RJ100	11/24/98	HB-IYZ
339	RJ100	12/21/98	HB-IYY
357	RJ100	9/6/99	HB-IYX
359	RJ100	10/8/99	HB-IYW

Air Jet

Operating two 200QC aircraft, this France based charter operator ended up flying five BAe 146 aircraft with two being new-build aircraft.

LINE	SERIES	DELIVERY	REGISTRATION
188	200QC	10/7/91	F-GLNI
211	200QC	6/21/94	F-GOMA

Conti-Flug

Conti-Flug named its first aircraft "Alexander von Humboldt.

Conti-Flug was founded in 1964 and ran scheduled flights from Tempelhof, Germany and London City Airport as well as shuttle between Hamburg and Toulouse for Airbus employees. The carrier ceased operations in 1994, but had a brief resurgence again before shutting down again in 2001.

200	200	10/22/92	D-ACFA
201	200	12/18/91	D-AJET

British Air Ferries

A local airline based at London Southend Airport began scheduled passenger flights to Germany in 1977, although the flights transferred to British Island Airways in 1979. The airline was originally expected to purchase the BAe 146-100, even creating a brochure for it, and BAe painted one aircraft in the carrier's colors during route trials. However, the airline had gone into receivership more than once, and was sold in 1983. It finally operated a new-build BAe 146-100 for six months of charter operations (for the Malaysian government) that had previously been used as the RJ demonstrator, and before becoming British World Airlines it flew a series 300 aircraft on behalf of British Airways. The airline ceased operations after the events of September 11, 2001. While both aircraft were actually titled to their respective airlines and never owned by BAF, BAF is listed here because it operated the aircraft from first delivery.

LINE	SERIES	DELIVERY	REGISTRATION
199	100	9/5/92	G-RJET
203	300	6/10/92	G-BTTP

China Northwest Airlines

Operating the BAe 146 on domestic flights out of Xian, China, China Northwest Airlines eventually merged with China Eastern, and the aircraft were subsequent repainted in China Eastern livery.

207	300	12/1/92	B-2711
212	300	12/14/92	B-2712
214	300	12/23/92	B-2715
215	300	12/29/92	B-2716
216	300	3/26/93	B-2717
218	300	4/22/93	B-2719
219	300	8/25/93	B-2720
222	300	1/12/94	B-2718

Mid Pacific Cargo

Mid Pacific Air was a Hawaiian-based regional airline offering services between the islands. It also operated a single 146QT on behalf of TNT Express carrying cargo between New York and Bermuda. The carrier ceased operations in 1995.

LINE	SERIES	DELIVERY	REGISTRATION	NAME
194	300QT	12/8/92	N599MP	

Photographer Unknown / Photo provided by Stefano Pagiola

Turkish Airlines

The flag carrier of Turkey, previously known as Turk Hava Yollari (THY), became Turkish Airlines and placed an order for new Avro RJ100s (and later RJ70s) to run domestic services.

LINE	SERIES	DELIVERY	REGISTRATION	NAME
232	RJ100	7/22/93	TC-THA	"Denizli"
234	RJ100	8/12/93	TC-THB	"Erzurum"
236	RJ100	9/7/93	TC-THC	"Kayseri"
237	RJ100	10/21/93	TC-THD	"Van"
238	RJ100	11/4/93	TC-THE	"Gaziantep"
240	RJ100	3/9/94	TC-THF	"Sanliurfa"
241	RJ100	3/23/94	TC-THG	"Konya"
243	RJ100	4/14/94	TC-THH	"Kutahya"
264	RJ100	6/22/95	TC-THM	"Siirt"
265	RJ100	6/28/95	TC-THO	"Tokat"
230	RJ70	3/29/96	TC-THJ	"Usak"
252	RJ70	5/3/96	TC-THN	"Mus"
249	RJ70	6/24/96	TC-THL	"Kahramanmaras"

Paul Seymour

Business Express Airlines (BEX)

Boston, Massachusetts airline BEX was a Delta Express affiliate running flights out of Boston and New York to Detroit, Milwaukee, Norfolk and other regional destinations. Flying smaller feeder routes, it began to compete with other regional carriers for traffic and placed a large order for 20 RJ70s. British Aerospace placed former Discover Airways aircraft with BEX until its own RJ70s were delivered. Unfortunately, the carrier failed to become profitable and only took delivery of three of the RJ70 aircraft. It cancelled its further orders and eventually replaced the BAe and Avro aircraft with smaller aircraft prior to being acquired by American Airlines AMR Eagle Corp.

LINE	SERIES	DELIVERY	REGISTRATION
223	RJ70	9/11/93	N832BE
224	RJ70	10/1/93	N833BE
225	RJ70	11/12/93	N834BE

Barry Lloyd

Air Malta

Based on the Mediterranean island nation, Air Malta was a local European regional airline. It ordered four Avro RJ70 aircraft after BAe successfully demonstrated the type's capabilities with extended range Pannier tanks that enabled service to London. It was a hard-fought battle with Fokker, and was a cash deal. The aircraft also served routes to Catania, Palermo, Tunis, and Monastir, and featured a number of changes over other production aircraft including a galley that shared commonality with the Airbus A320 fleet and a rear restroom ahead of the rear passenger door. As a state-owned company there were inevitable political dimensions to the deal with Malta seeking to join the European Union and wanting British assurances of support. BAe encouraged a visit by the British Foreign Secretary which undoubtedly helped the things along, and there were other sweeteners attached such as a $300,000 USD credit to allow the Maltese Defense Force to buy a used Britten-Norman Islander for their Air Arm, plus the trade in of the carrier's four Boeing 737-200 aircraft. Air Malta wanted to establish a hub to serve southern Europe and Northern Africa, but these plans never really developed.

LINE	SERIES	DELIVERY	REGISTRATION	NAME
254	RJ70	9/21/94	9H-ACM	"Philippe Vielliers de L'isle Adam"
258	RJ70	10/21/94	9H-CAN	"Antonio Manoel de Vilhena"
260	RJ70	12/20/94	9H-ACO	"Claude de la Sengle"
267	RJ70	3/8/95	9H-ACP	"Jean de la Valette"

SAM Colombia

SAM was a Columbian regional airline that leased nine RJ100 aircraft because of the type's short field capability and hot and high performance. The aircraft served domestic routes as well as some international destinations such as Aruba, Costa Rica, Panama, Ecuador, and Curacao. The Colombian economy plummeted in 1999 affecting the airline, and the aircraft were returned to BAe. The resulting aircraft needed wing spar repairs due to corrosion from the humid environments they operated in.

242	RJ100	8/23/94	N505MM
244	RJ100	9/14/94	N506MM
247	RJ100	9/14/94	N508MM
245	RJ100	9/21/94	N507MM
248	RJ100	10/1/94	N509MM
250	RJ100	10/1/94	N510MM
255	RJ100	12/1/94	N511MM
221	RJ100	12/20/94	N504MM
263	RJ100	9/23/95	N512MM

Lufthansa (Cityline)

Following discussions with RJ85 operator Crossair regional carrier Lufthansa Cityline, based in Frankfurt, Germany, selected a new Avro fleet after a hard-fought battle between BAe and Fokker. The UK manufacturer took three of Lufthansa's Fokker 50s on trade-in basis. Cityline served destinations across Europe, feeding traffic into Lufthansa's international network. The airline became the fifth largest operator of passenger BAe/Avro RJ aircraft, ultimately taking 18 after initial scope-clause concerns with the Lufthansa pilot's union were addressed. Eventually Lufthansa combined Lufthansa Cityline with Eurowings and Air Dolomiti to form Lufthansa Regional, and operated a total of 37 BAe 146/Avro RJ aircraft.

Paul Seymour

LINE	SERIES	DELIVERY	REGISTRATION
246	RJ85	10/18/94	D-AVRO
256	RJ85	11/18/94	D-AVRA
253	RJ85	12/9/94	D-AVRB
251	RJ85	3/3/95	D-AVRC
257	RJ85	3/23/95	D-AVRD
261	RJ85	3/31/95	D-AVRE
269	RJ85	6/30/95	D-AVRF
266	RJ85	9/8/95	D-AVRG
268	RJ85	10/12/95	D-AVRH
270	RJ85	12/7/95	D-AVRI
277	RJ85	2/22/96	D-AVRJ
278	RJ85	3/21/96	D-AVRK
285	RJ85	3/28/96	D-AVRL
288	RJ85	5/24/96	D-AVRM
293	RJ85	9/12/96	D-AVRN
303	RJ85	2/21/97	D-AVRP
304	RJ85	3/21/97	D-AVRQ
317	RJ85	12/9/97	D-AVRR

Azzurra Air

Based in Milan, Italy and set up in 1996 by investor Air Malta which owned 49% of the equity, Azzurra Air operated charters to Spain, Greece, the Netherlands, and Portugal. Although Air Malta owned only 49%, it called the shots including the fleet acquisition. Azzurra Air RJ85s were sold in an Air Malta management shakeup and were subsequently replaced by three Air Malta Avro RJ70s.

LINE	SERIES	DELIVERY	REGISTRATION	NAME
299	RJ85	11/26/96	EI-CNI	"Lombardia"
300	RJ85	12/6/96	EI-CNJ	"Piemonte"
306	RJ85	5/8/97	EI-CNK	"Lazio"

Uzbekistan Airways

The flag carrier of the nation of Uzbekistan was established in 1992 and purchased three new-build RJ85s for local short haul routes, replacing Soviet built Yak-40 aircraft. UK-80001 configured with a dedicated VIP interior for the Uzebekistan President. Former salesman Peter Connolly recalled traveling to the country to discuss a sale of Avro RJ aircraft. He flew Uzbekistan Airways, and the inflight magazine had an article stating the airline was going to buy RJ85s before a deal had even been discussed with BAe.

309	RJ85	6/7/97	UK-80002
312	RJ85	12/19/97	UK-80001
319	RJ85	12/24/97	UK-80003

Jerome Dawson

Mesaba (Northwest Airlines)

Mesaba derived its name from a work in the Ojibwe language which means "Soaring Eagle". Started in 1944, it was initially primarily an employee shuttle for a paper mill company. Expanding over the years, it grew into a large east coast regional operator. As with the aircraft flown by Air Wisconsin / United Airlines, the Mesaba RJ85s were operated for Northwest Airlines under scope clause agreements, but restricted to no more than 70 passenger seats. Additionally, the Mesaba aircraft had the co-pilot steering tiller removed because of the scope clause. Mesaba operated the RJ85 with a small first-class section, which reduced capacity to 69 seats, the number advertised for the RJ70. The Avro marked Mesaba's entry into jet operations, and it was the single largest operator of passenger RJs (and BAe 146s) with 36 on its books.

LINE	SERIES	DELIVERY	REGISTRATION
208	RJ85	4/24/97	N501XJ
The very last aircraft built at the Hatfield factory.			
307	RJ85	5/24/97	N502XJ
310	RJ85	6/30/97	N503XJ
311	RJ85	7/25/97	N504XJ
313	RJ85	8/28/97	N505XJ
314	RJ85	9/19/97	N506XJ
316	RJ85	10/24/97	N507XJ
318	RJ85	12/4/97	N508XJ
321	RJ85	1/30/98	N509XJ
323	RJ85	3/11/98	N510XJ
325	RJ85	4/11/98	N511XJ
326	RJ85	4/27/98	N512XJ
329	RJ85	6/23/98	N513XJ
330	RJ85	7/1/98	N514XJ

LINE	SERIES	DELIVERY	REGISTRATION
333	RJ85	8/20/98	N515XJ
334	RJ85	9/24/98	N516XJ
335	RJ85	10/3/98	N517XJ
337	RJ85	10/29/98	N518XJ
344	RJ85	2/18/99	N519XJ
345	RJ85	2/25/99	N520XJ
346	RJ85	3/12/99	N521XJ
347	RJ85	3/24/99	N522XJ
348	RJ85	4/1/99	N523XJ
349	RJ85	4/29/99	N524XJ
350	RJ85	5/11/99	N525XJ
351	RJ85	5/27/99	N526XJ
352	RJ85	6/10/99	N527XJ
353	RJ85	6/29/99	N528XJ
363	RJ85	12/9/99	N529XJ
364	RJ85	1/13/00	N530XJ
365	RJ85	1/29/00	N531XJ
366	RJ85	3/2/00	N532XJ
367	RJ85	3/9/00	N533XJ
370	RJ85	4/27/00	N534XJ
371	RJ85	5/9/00	N535XJ
372	RJ85	5/26/00	N536XJ

Jerome Dawson

Interior of Cityflyer Express Avro RJ100 with 6-abreast seating.

Cityflyer Express

Based out of London's Gatwick Airport, Cityflyer Express was a British Airways affiliate, operating aircraft painted generally in British Airways colors with staff wearing BA uniforms. It became the first U.K. based airline to operate the RJ100, of which 16 were purchased new. The aircraft were painted in BA livery as a result of the affiliation and code share agreements. Cityflyer might have been an RJX customer if the RJX had been available earlier. British Airways eventually acquired Cityflyer Express in 2000.

LINE	SERIES	DELIVERY	REGISTRATION
298	RJ100	4/5/97	G-BXAR
301	RJ100	5/1/97	G-BXAS
320	RJ100	1/10/98	G-BZAT
328	RJ100	6/12/98	G-BZAU
331	RJ100	7/25/98	G-BZAV
354	RJ100	7/16/99	G-BZAW
356	RJ100	8/19/99	G-BZAX
368	RJ100	3/30/00	G-BZAY
369	RJ100	4/14/00	G-BZAZ
373	RJ100	6/16/00	G-CFAA
377	RJ100	11/30/00	G-CFAB
379	RJ100	12/15/00	G-CFAC
380	RJ100	1/26/01	G-CFAD
381	RJ100	2/26/01	G-CFAE
382	RJ100	3/23/01	G-CFAF
384	RJ100	6/8/01	G-CFAH

Aegean

Founded in 1987 in Greece as a charter operator, the carrier eventually started scheduled operations in May 1999 with the Avro RJ100 being the first commercial aircraft it purchased. Ultimately expanding to a fleet of six RJs, the airline continued to grow and partnered with other European airlines. Aegean was one of the last three new airlines to purchase new build Avro RJ aircraft.

LINE	SERIES	DELIVERY	REGISTRATION	NAME
341	RJ100	4/30/99	SX-DVA	"Deukalion"
343	RJ100	4/30/99	SX-DVB	"Ion"
358	RJ100	9/24/99	SX-DVC	"Helene"
362	RJ100	12/14/99	SX-DVD	
374	RJ100	6/29/00	SX-DVE	
375	RJ100	7/12/00	SX-DVF	

Air Botnia

Originally an independent regional airline, Air Botnia was purchased by SAS Scandinavian Airlines in 1998. It purchased five new build Avro RJ85 aircraft to replace ageing Fokker F28s. The new aircraft wore SAS colors but "operating with Air Botnia" titles. In 2001, the carrier's name was changed to Blue1.

383	RJ85	5/10/01	OH-SAH	"Suur Saimaa"
385	RJ85	6/20/01	OH-SAI	"Pihlajavesi"
388	RJ85	8/15/01	OH-SAJ	"Pyhäselkä"
389	RJ85	9/27/01	OH-SAK	"Näsijärvi"
392	RJ85	11/29/01	OH-SAL	"Orivesi"

Bahrain Government

The government of Bahrain (Defense Force) placed an order for a single RJ85 in May 2001. It took delivery of a VIP version of the Avro RJ85, the only one to be delivered to Bahrain new, in November 2001. It was the first civilian aircraft purchased and deployed by the Bahrain Defense Force. The BDF subsequently took delivery of a second aircraft, a second hand RJ85 originally supplied to Azzura Air.

LINE	SERIES	DELIVERY	REGISTRATION	NAME
390	RJ85	11/10/01	A9C-BDF	

E2394, the last Avro RJ to be delivered alongside E3001 now flying with FAAM, just prior to formal handover of both aircraft.

Blue 1

Changing its name from Air Botnia, Blue1 added further aircraft, this time RJ85s, taking delivery of the very last built Avro RJ aircraft, line numbers 393 and 394. These aircraft were unsold after the events of September 11, 2001, and remained stored as white tail aircraft until they were sold to Blue1. The aircraft were used by the airline to feed traffic between Finland and Sweden, mainly into SAS hubs.

386	RJ100	10/22/03	OH-SAM	"Pyhäjärvi"
387	RJ100	10/30/03	OH-SAN	"Päijänne"
393	RJ100	11/5/03	OH-SAO	"Oulujärvi"
394	RJ85	11/24/03	OH-SAP	"Pielinen"

The second to last Avro RJ "Oulujärvi" to be delivered to Blue 1 after being stored for over a year.

CHAPTER 18

Ryan Coulter / MachImages.com

The 146 Today

More than 35 years after the first flight of the BAe 146, and in an era where four engine aircraft have quickly fallen out of favor, the 146 perseveres. While airlines that have operated the aircraft in large numbers have since retired their fleets, the 146 still soldiers on. It is still possible to find 146s and Avro RJs operating flights in and around Europe with charter operators like Jota Aviation that provide ACMI (aircraft, crew, maintenance, insurance) services. There are also 146s still running cargo and passenger flights to remote locations in western Australia, flown by Cobham and Pionair. And long after LanChile planned to operate flights into Antarctica, that is now finally happening with Aerovías DAP.

Minden Air engaged in test runs with retardent.

BAE Systems Regional Aircraft

After the shutdown of the Avro RJ/RJX production line, all support for BAE Systems regional aircraft was transferred from Woodford and Toulouse to Prestwick, Scotland. From there, BAE Systems Regional Aircraft continues to provide support for all legacy types including the 748, ATP, Jetstream 31/32/41 and the BAe 146/Avro RJ series. BAE Systems still holds and manages the aircraft's type certificates as well as the ongoing responsibility for maintaining continued airworthiness for the types, as well as providing support for everything from spare parts through training and technical assistance to operators around the world. This will continue until the very last aircraft stops flying. Spare parts are kept at a vast, modern warehouse facility at Weybridge, Surrey, England, a location that harks back to manufacture of types such as the Vickers VC10 and Viscount, illustrating the rich heritage of BAE Systems. Responsibility for the sale and stock management of spare parts is handled by Prestwick with shipments being made directly from Weybridge. BAE Systems is no longer manufacturing new parts or ordering new stock from vendors but is refurbishing parts from aircraft on exchange, and offers items recovered from airframes that are being parted out. BAE Systems will even acquire and break up aircraft that are no longer viable to recover additional spare parts. The manufacturer works with Avalon Aero based in Bedfordshire, England on C-Checks as well as decommissioning aircraft for spares recovery. When an aircraft is parted out, the interiors are often disposed of since they tend to be worn or are customer specific. The engines and landing gear have the most value, along with pumps, valves and anything that can be repaired (known as rotables – rotated out, rebuilt and reinstalled). At the time this book was written, BAE Systems was working with GKN on producing one more batch of windscreens before it stops manufacturing altogether. Nevertheless, BAE Systems still maintains five-year production agreements with suppliers to ensure parts are available.

E2376 Avro RJX85 G-ORJX now resides at BAE Systems Regional Aircraft, located in Prestwick, Scotland.

One area that poses some complications is Iran, where there are now a number of 146 operators. Mahan Air now has the largest fleet of Avro RJ aircraft, having acquired them through agents and other countries that do not abide by the sanctions imposed on Iran. The carrier has also been accused of sponsoring terrorism, including

A local company that recycles aircraft parts into furniture built an oversized chair for BAE Systems reception area from a former engine nacelle.

transporting soldiers and weapons into Syria to support the Civil War. It comes as no surprise that Mahan Air needs parts or technical support from time to time. However, BAE Systems is not permitted under the sanctions to provide any support or information, including documents or processes that may be needed. BAE Systems even noted that there have been a few times when a representative from the airline has called posing to represent another owner, just to get basic operational or maintenance information. In the end, every single part supplied to both domestic and foreign operators must be documented. BAE Systems even has to take precautions to ensure it does not sell parts or provide information to brokers who will shuttle the items to Iran.

BAE Systems reports that the 146 series is still achieving a 98-99% dispatch reliability rate even though production ceased nearly 20 years ago. Regular issues with the engines and flaps have long since been sorted out, and groundings are no longer tied to those primary issues, but instead to small items and glitches such as valves that fail. BAE Systems offers operators several options for technical and spares support contracts covering a range of specific parts and a term that suits each operator. As the aircraft are still viable assets for many carriers and due to their sturdy construction, BAE Systems decided that there was a business case for initiating a life extension program for the type. This extends the lifetime limit from 40,000 cycles to 60,000 cycles, and as long as the aircraft passes detailed inspection including X-Ray evaluation, it can continue flying. BAE Systems also provides a range of modifications to ensure that the aircraft meets the latest airworthiness regulations, or provide technological and mechanical enhancements. Examples of these are the mandated upgrades for the 7.1 TCAS (Traffic Collision Avoidance System) and ADS-B (Automatic Dependent Surveillance Broadcast) systems. It can be challenging to ensure an older aircraft meets current standards, but BAE Systems works with customers to provide these capabilities. BAE Systems gathers data on the spares consumption of the in-service fleet, and monitors it to predict future part requirements, allowing operators to keep the aircraft flying. The manufacturer also continues to hold annual conferences for flight operations and technical audiences, attracting a large number of operators from around the world. This allows BAE Systems and its customers to discuss various aspects associated with the ongoing operation of the aircraft as well as providing a valuable networking and social interaction.

BAE Systems was continually working on alternative missions for BAe/Avro RJ aircraft, like this concept for a refueling tanker (artists rendering).

Honeywell continues to have close involvement with BAE Systems to support the LF-502 and LF-507 powerplants. The LF-507 had been suffering from fuel control and hydro mechanical assembly issues, and the engine manufacturer did come up with a modification program that went a long way to substantially improving reliability. BAE Systems will work with operators on maintenance (subcontracting) if desired, but there are plenty of other independent MRO (Maintenance, Repair and Overhaul) organisations across the globe that handle the 146 series. AFI KLM MRO Global (Air France Industries/Koninklijke Luchtvaart Maatschappij) at Norwich, England now the largest 146/RJ MRO facility, followed by Avalon Aero, Tronos, and Chevron. BAE Systems will continue to support the aircraft, but margins are reducing as more 146s are removed from service and broken up, and the worldwide fleet inexorably shrinks.

Currently, BAE Systems is also the technical lead for the Airbus E-FANX test aircraft which is a hybrid electric demonstrator. The E-FANX is based on the Avro RJ, which was selected because of its improved performance over the 146 aircraft. BAE Systems is responsible for ensuring that the extensive modifications made to the demonstrator will meet airworthiness and safety requirements when they are completed.

TRONOS

Tronos is one of the largest suppliers of second-hand BAe 146/Avro RJ aircraft in the world. It was founded by Adrian Noskwith, the son of Rolf Noskwith, who was a member of the Bletchley Park team that deciphered the German Enigma machine during World War II and later set up Charnos, a luxury brand of lingerie in the United Kingdom. Adrian was given a choice by his father: join the family business, or go his own way. Sharing his father's entrepreneurial spirit, combined with his love of aviation, Adrian chose the latter and set up an aircraft charter and brokering business.

E2376 wings and stabilizer are stored right next to the fuselage.

Adrian became enamored with the BAe 146 after seeing a cutaway of the aircraft in Flight magazine, and loved the way it looked. In the early days of Asset Management Organization (AMO), the salesman tended to gravitate toward entrepreneurial types to place aircraft. Peter Dunlop of AMO introduced Adrian to Allan Trotter at Flightline, a BAe 146 operator. It was this introduction that put Adrian into the BAe 146 aircraft. AMO's initial goal was to, at a minimum, do half-sensible deals to get aircraft operational again quickly. When BAe 146 operator Debonair shut down, Adrian came along and took five of the 146s the defunct carrier had been operating. He thought "I know something about this airplane, and I know something about the customers." Adrian's start with leasing 146 aircraft began with Atlantic Airways, which was already operating one BAe 146. Not long after, Atlantic Airways took a second BAe 146, and Qantas became a customer for another 146 that Adrian had acquired. At this point, Adrian had begun to transition his business into a lessor of aircraft.

Tronos's operation began to grow, with aircraft being shuttled to the other side of the planet having placed aircraft with TAM in Bolivia. The business began to evolve, and Tronos began to carve out a niche pursuing deals that AMO would not touch. At the time, AMO salesman Steve Doughty knew Tronos would be able to help grow the market for the 146 for mutual benefit by dealing with airlines that would not survive AMO's tough risk scoring assessment. For example, Steve originally sold Aerovías DAP on the concept of the 146 for its fleet, but the fledgling carrier was not able to meet enough of the parameters the Deal Committee had set up as its minimum requirements back in the day.

Tronos then exploited the opportunity, because as an entrepreneur running a private company rather than a corporation, Adrian had more flexibility in his deal term requirements. Aerovías DAP has since expanded its fleet of aircraft to include Avro RJs and even the ex-Moncrief Oil VIP BAe 146-100, acquired via Tronos. Tronos has been able to carve out a very successful niche in the leasing business because the 146 is considered an unconventional aircraft, and most of its competitors have discounted the type altogether. Tronos's success has been tied directly to the 146 and the vision Adrian had for the aircraft and his company. Although Boeing 737s and Airbus A320s are more in demand (and plentiful), the margins on those types are very thin because of supply and competition, and they are a commodity aircraft. The BAe 146 on the other hand is a niche aircraft, there is a dwindling supply of viable airframes and only a couple of businesses that place used aircraft, leading to higher margins.

Adrian Noskwith, founder of Tronos (right), with Prince Edward Island Minister of Development and Technology, Mike Currie and Tronos Canada President Mike Everett in the center.

As Tronos grew, so did its need to store and refurbish aircraft. Sensing another opportunity, it set up shop in Northeastern Canada at the airfield in Slemon Park on Prince Edward Island (PEI). With low labor costs, empty hangars available, and because Adrian's Canadian wife's family had business interests locally, the island made a lot of sense. BAe 146 aircraft are maintained, repaired, refurbished, and even parted out at the PEI facility. Steve Doughty left AMO when BAE Systems sold what was left of it to Falko and later joined Tronos, having got to know Adrian as a result of selling him two ex-FlyBe BAe 146s that were converted to air tankers.

It was the air tanker business that rejuvenated interest in the type, and Tronos led the way by being the first organization to modify the BAe 146 with such capabilities after a false start at Minden Air. Tronos partnered with Richard Thomasson (formerly of BAe) in 2009, who helped design a payload system that allowed the BAe 146 to drop 30,000 lbs of retardant in under eight seconds. The team at Tronos PEI, led by CEO Mark Coffin, installed and successfully tested the design on airframe E2049, giving the BAe 146 a new lease of life and opening up the aircraft as a fire tanker delivery platform. E2049 now serves as aircraft T40 at Neptune Aviation.

But the future of the BAe 146 and Avro RJ is no longer looking rosy, even for Tronos. Carriers that are and have used the type for decades like CityJet and Cobham are rapidly removing them from their fleets. As of December 2020, Tronos had only one viable aircraft left on the books, with four additional aircraft (E2018, 2048, 2115, 2208) having been parted out. The BAe 146 series combined with Adrian's business sense led to Tronos's success, and nearly thirty BAe 146/ Avro RJ aircraft passed through the company. Tronos will soldier on supporting the BAe 146 as long as it makes business sense, and will expand to other aircraft types to maintain its aircraft leasing and sales legacy.

Air Tanker / Fire Bomber

One of the most exciting developments for the BAe 146 series was not a second life as a converted cargo aircraft, a fate that befell many post-passenger aircraft. In fact, BAE Systems did try to re-launch the cargo conversion for the 146 but without success. Instead, the 146 has been adopted as a very versatile and strategic air tanker (fire) bomber by multiple operators in North America. More than 15 aircraft have been converted to fire bombers to date, and are available for deployment around the continent as needed. The 146 was determined to be an agile performer, capable of dropping up to 3,000 gallons of fire retardant at slow speeds with pinpoint precision thanks to the high wing and the rear petal brake. Companies including Aero-Flite, Air Jet, Conair, and Neptune Aviation continue to add BAe 146 and Avro RJ aircraft to their fleets. The BAe 146/ Avro RJ has been certificated in Europe, Canada, Australia, and the United States as an air tanker. The Avro RJ has the advantage of 4,000 pounds of additional take-off weight, but the water/retardant payload is the same as the BAe 146 at 3,000 gallons U.S.

Aircraft lives are measured in 'cycles', which is the equivalent of one take off and one landing. BAe offered a life extension program to increase the number of cycles a 146 or RJ could fly to 60,000, albeit with a heavy caveat: one cycle as an air tanker would be the equivalent of between 4-7 flight cycles of 'normal' flight due to the stresses placed on the airframe during drops. With a 60,000 cycle extension program in place, this means in the best-case fire bombers have 15,000 real-world cycles available, or 8,600 cycles in the worst case on a new build aircraft. Because the air tanker conversions are from used aircraft, their lifecycles will vary greatly depending on the number of cycles the airframe had when it entered service as an air tanker. Since the 146 has entered service as an air tanker and armed with real-world results from fire-fighting operations, BAe can evaluate how the 146 airframe is continuing to support these operations and if, with specific maintenance protocols in place, further life extensions can be granted.

In 2004, it was clear the existing fleet of air tankers was getting older, dropping less, and some even involved in deadly accidents with aircraft failing during missions. BAe began to look at potential opportunities for the 146 series to replace older aircraft for air tanker operations. In conjunction with BAE Systems, Tronos (Canada) and Minden Air (which was flying Lockheed P2-V Neptune aircraft in 2004) worked with BAE to bring a 146-100 (formerly with Air China) to the U.S. and to demonstrate it as a potential platform. In Nevada, Minden Air's base, weight was added to the 146 to simulate 3,000 gallons of retardant, and it was flown along profiles that an air tanker would normally fly to demonstrate its capability. Minden Air clearly saw the benefits of the BAe 146, including its angle of attack and low speed handling characteristics despite it being a swept-wing jet aircraft, and began to seriously look at it as a next generation platform. It would also address the grounding of ageing air tanker aircraft by the U.S. Forest Service that began in 2002 following a series of serious accidents.

Tronos

The first prototype air tanker built initially by Tronos.

But instead of the series 100, Minden decided that the series 200 was needed for the additional payload it could carry. Minden partnered up with and worked briefly with Tronos on the conversion and engineering, but the two ended up parting ways after about a year, reportedly not on good terms. There was a bit of delay before Minden got started again but Minden acquired its first aircraft, former Air BC BAe 146-200 airframe E2111, from BAE Systems in

January 2009 and reregistered it N446MA. Unlike the other 146 air tankers, Minden's aircraft carried an additional 100 gallons (3,100 total) of water or retardant. Minden sought performance data assistance from BAE Systems during the conversion and the aircraft received certification from the FAA in 2013 after all the modification work, but that was only the start of the long process to obtain a contract to deploy the aircraft to a fire. From there, it needed to demonstrate that it could meet drop test requirements before it received certification from United States Forest Service (USFS) and the Interagency Air Tanker Board. This included both the spread (how wide the retardant drop was) and the flow rate (how slow or fast the retardant drops).

The typical drop speed was around 120 knots (138MPH) at an altitude of approximately 150 feet above the fire or drop zone. The time needed to reload retardant once the aircraft arrived back at the airport was estimated at less than 10 minutes. The aircraft had an unpressurized variable

The interior tank system that Minden Air used for their system which held 3100 Gallons.

Brian Wiklem

gravity flow system where the spread of the retardant could be controlled from the cockpit electronically. This included computer-controlled delivery along with a touch screen that enabled the crew to select various coverage levels and preset deliveries, as well as compute aircraft weight and performance. Minden also installed an interactive GPS system which coordinated with the flight management system to modulate the drop doors, eliminating the variable of determining actual coverage. This was the first of its kind. Minden's plan was to use the first 146 (and later the second aircraft it acquired, E2106) to replace its P2V aircraft which were generations older.

By 2014, Minden had completed the drop tests, and reportedly the performance was excellent. However, Minden shut down not long afterward, lacking the funds to continue. Len Parker, owner of Minden Air, during a brief phone interview in 2017 mentioned that getting contracts from CAL-Fire were incredibly difficult, going so far as to suggest there were more nefarious circumstances surrounding why Minden Air was not awarded additional contracts for its existing fleet of Lockheed P2V aircraft. In the end, Minden Air filed for bankruptcy in 2016, and both BAe 146-200 aircraft on hand (one fully converted) were seized by the banks. In early 2020, start up Australian freight operator PionAir acquired the aircraft in order to provide its own fleet with a fresh supply of spares and powerplants from the Minden aircraft.

In 2009 Neptune Aviation Services of Missoula, Montana began to work with Tronos to develop the BAe 146-200 into an air tanker. Based on ex-Air Wisconsin (United Express) aircraft line number 136, the airframe was modified with an internal 3,000 gallon water/ retardant system. Neptune's system also employs computer control to actively monitor the flow performance and adjust the exits to maintain selected coverage levels. This system takes into account aircraft ground speed, g-loading, pitch attitude and tank flow rate. Once fully integrated and tested, Neptune put the first aircraft into operation in 2011, and two years later began phasing out its

Paul Seymour

Lockheed P2V Neptune fleet. The performance of the BAe 146 enabled it to cover a larger area, it had faster response times, and afforded 50% better turnaround times than the P2V. Neptune now has a fleet of eight BAe 146-200s and one Avro RJ85 that it can deploy as needed around the U.S. and over the past few years the aircraft have spent quite a bit of time fighting massive fires in California and Oregon. The BAe 146 has had a dispatch reliability rate of approximately 98.3% versus 97.6% on the P2V. Other factors that have proved important to operators and contractors is the speed and range of the 146, whether it's dropping retardant,

Guy Van Herbruggen

Jonah Curtin

ferrying to a location, and even its ability to serve multiple fires in a single day. It can be anywhere in the U.S. from Montana within approximately 3 hours.

Conair Aerial Firefighting of Abbotsford, British Columbia, Canada and its U.S. subsidiary Aero-Flite are using a small fleet of Avro RJ85 aircraft for airborne firefighting. Instead of an internal tank and drop system installed inside the passenger cabin, the exterior of the aircraft have been structurally modified with external pannier tank wrapped around the fuselage under the belly, giving the Avro RJ a 'pregnant' look. Even with the extended girth of the tank, BAE found it had negligible aerodynamic effects on the fuselage. Having an external tanking system means that Conair can pressurize the cabin for pilot comfort and flying at higher altitudes over longer distances. The exterior tank is attached to the fuselage with minimal impact, and an aerodynamic fairing is installed on the front of the tank. The Conair designed tanker made its first flight on August 21, 2013 and went into service not long after.

By the end of 2018, BAe 146s and Avro RJs had run more than 6,000 tanker missions, and dropped in excess of 14 million gallons of retardant. The fleets of Conair/Aero-Flite, Neptune Aviation, and soon Air Spray will constitute the

Phos-Chek, the fire retardant produces a trail on the aircraft belly as it is released.

Guy Van Herbruggen

Mike Eliason/Santa Barbara County Fire Dept.

12
N476NA

Mike Eliason/Santa Barbara County Fire Dept.

largest BAe 146/Avro RJ air tanker platform in the world, all based in North America.

What made the aircraft appealing as a platform to the USFS was that BAe, as the manufacturer, was not only engaged in helping air tanker operators engineer the aircraft, but also provided technical support for each of the modifications. Each air tanker operator uses a completely different internal (or external) tank and drop system, and has its own a STC (Supplemental Type Certificate).

FAAM

Airframe E1001, the original prototype, was kept through the years by British Aerospace to act as a test bed, eventually being modified to become the series -300 prototype E3001. On June 6, 2000, it found a home as the new Atmospheric Research Aircraft for the University of Manchester Institute of Science and Technology. The organisation was interested in a BAe 146 because it had four engines for safety, as well as the STOL capabilities that would enable it to operate from remote locations. The aircraft was put through more than three years of modifications and retrofitting before it was ready to perform as the Facility for Airborne Atmospheric Measurements (FAAM). Instead of merely taking E3001 and beginning conversion, BAe 'zero-lifed' the airframe by refurbishing every aspect of it so that it was equivalent to a new-build aircraft. There were naturally changes to the production aircraft not incorporated into the prototype E3001 and as a result this airframe is a composite of all the other models British Aerospace built. Even the wing was refurbished, despite it not being identical to production aircraft. The LF502 engines were replaced with updated LF507-1H powerplants, similar to those installed on Avro RJ aircraft (and identical to China Eastern's BAe 146-300s which used FADEC-less LF507-1H engines).

Additional upgrades added to E3001 included pannier tanks in the wing roots, as well as an extended range tank in the rear cargo bay near the centre of the aircraft. This gave the aircraft a range up to 1,800 miles, and it was also cleared to cruise at 35,000 feet or as low as 50 feet above the ocean. A significant number of sensors were installed on the wings, along with additional devices such as LIDAR to measure volcanic cloud density and water content as well as external video cameras. Unlike the C-130 that performed the role previously, the 146 does not have a nose probe but instead uses sensors around the front of the nose to fulfill the same role. BAE Systems has to certify each instrument that is installed on the aircraft to EASA/FAR Type 21 standards, which ensure there will no issues inflight and that they can withstand the forces of a crash. The passenger cabin does not look like a passenger interior at all, and is outfitted with measurement equipment stations and seating for up to 19 scientific crew. Equipment is rack mounted, meaning that it can be easily swapped out for different measuring instruments or replaced in the event of a fault. The aircraft is also configured to deploy sondes, devices consisting of instrument packages that are dropped and then ascend under a balloon. They measure temperature, pressure, humidity, until the balloon bursts at high altitudes after which a parachute deploys to enable the sonde lands safely. The sondes are also capable of communicating information back to the aircraft.

After all of the equipment was installed but before the aircraft was made 'pretty' with an application of paint and delivered, it was taken on a series of flights to test all the systems and flight parameters. Pilots Alan Foster and Pete Lofts and flight engineers Colin Darvill and Tim Bartup took the heavily modified aircraft on its first trip aloft following the reconfiguration on October 1, 2003. Unfortunately, all the external sensors and modifications resulted in additional drag that exceeded the original and initial calculations (~10%). Retrofitting drag reducing design changes lowered the drag increase, and after painting the aircraft, it was delivered on May 10, 2004. The FAAM 146 was planned to fly an average of 500 hours per year, and is made available to other research organizations time and cost permitting, but it is currently averaging 300-400 hours. Because of the weight of all the onboard instruments (more than four tons), the fuel carried, and the number of persons on board, the FAAM 146 takes off at nearly MTOW at the start of each mission. And because of the sensors installed all around the aircraft, it is not permitted to land on unpaved fields due to risk of damage. However, the FAAM 146 does carry a fly away pack as well as a mechanic to deal with problems should the aircraft experience an issue in the field.

FAAM Head of Facility Alan Woolley says lightheartedly that they travel to "the weird parts of the world" during research projects.

In support of Red Nose Day, an charitable event put on by Comic Relief.

The first test flight with all the added scientific sensors and equipment mounted.

Locations like Fresno, California where they looked at mountain rotor activity (rotors are winds and humidity pushed up into the sky from thermodynamic activity against a mountain) during 2006 for Project TREX. Each mission generally has a creative designation weaved into it in order to obtain funding, and names have included Project Tic-Tac-Toe, and Project Vamos. One mission resulted in the FAAM 146 being called in to find a gas leak on the Elgin platform in the North Sea. FAAM pinpointed the gas leak on the oil derrick, including the gas flow leakage rate, something no one else was able to accomplish. Some missions, like one that visited India, take years of planning to finally accomplish. Despite two years of planning, once the FAAM aircraft and team arrived on site it took an additional three weeks before final sign off to begin the scientific flights was received. The military wanted to conduct a full inspection of every single piece of equipment from the proverbial "what is this? Why do you need it?" perspective. But the blistering mid-day heat in Delhi worked to the advantage of the FAAM crew and after about half of the items were inspected, the Indian military personnel (in full military regalia) were sweating profusely and quickly signed off on the papers to conclude the inspection. Other missions have had unique stipulations, like the requirement to carry a military observer during flights in Brazil. Sometimes flights have been delayed or even cancelled because someone was not consulted beforehand. Wherever the aircraft arrives, FAAM staff usually engage with local dignitaries who want a tour of the aircraft and to meet the staff.

The aircraft is owned by UKRI (United Kingdom Research & Information), which has appointed BAE Systems as the prime contractor to provide a range of services. Air Task, as the AOC holder, operates the aircraft on a civilian certificate under CAA regulations as a civil airliner. The BAe 146 is based at Cranfield University Airport alongside other stakeholders and is accommodated onsite within a hangar when not in use, with airport neighbor Avalon Aero as the EASA Part 45 maintenance provider. The aircraft has had some incidents during its life at FAAM. It is struck by lightning on average every two years, with the precursor to the strike being static on the radio 1-2 seconds prior. It was also struck by lightning while at Prestwick, with the nose being physically damaged and wicks on the wings being lost during the discharge. The conductor strip on the radome nose was vaporized and liquid metal hit the windscreen scratching it quite a bit. Some instruments were damaged and when the bill was tallied up, it came to just over $100,000 USD. But there was a benefit that arose out of the strike on the nose as two replacements were found, one which was white but did not fit well and another that was orange that did fit. The truck transporting it broke down, but when the nose finally arrived, it was fitted just before Christmas. Naturally, the comparisons to Rudolph the Red Nose Reindeer were made, and FAAM kept the nose on for "Red Nose Day" in March, a massive fundraising telethon put on by Comic Relief, during which the crew posed with red noses in front of the aircraft. The red nose and the "Golden Sick Bag" Award for Perseverance that is presented to whomever was impressively sick or conducted themselves in some way during adverse conditions shows that the crew enjoy their jobs and roles. The Golden Sick Bag was awarded to Rob King during Christmas 2018, following heavy oscillation of the aircraft over the Arizona desert that made Rob quite ill.

The FAAM 146 has been well received, and has achieved all its service and scientific goals. The only improvements the FAAM team wished the 146 could accommodate are the ability to fly at even higher altitudes, not to mention more range. The aircraft however is aging and there's always a push-pull when it comes to needs vs. capital expenditure. NERC (Natural Environment Research Council) wants to keep the aircraft on the bleeding edge of technology, as well as continue with upgrades but with the uncertain financial climate created by BREXIT still unresolved there will be an impact on those requests. The FAAM BAe 146 contract was renewed in 2019 for five years, with another five-year extension possible after 2024. It is expected that the FAAM BAe 146 will be flying at least another 10 years.

AIRLINES

Aerovías DAP - Chile

Aerovías DAP was founded in 1980 with DAP being the acronym of its founder, Domingo Andres Pivcevic, and is based in Punta Arenas, Chile from where it runs scheduled service as well as charters. It is the only airline that runs commercial jet charters to the continent of Antarctica and like its predecessor LAN Chile uses the British Aerospace 146 on the services. Aerovías DAP started out with de Havilland Canada Dash-7 turboprops, but in 2007 they began the search for a more advanced aircraft to expand its business. The key requirements were increased reliability, greater capability (capacity and ability for remote operations) and higher speeds which were being demanded to satisfy the growth in tourism.

Aerovías DAP's first BAe 146 aircraft (a series 200, registered CC-CZP) arrived in 2007 and operated chartered flights from Punta Arenas to King George Island (a portion of Antarctica claimed by Chile) during the summer months. The aircraft are branded Antarctic Airways (DAP Antarctica) when running flights to the continent. Operating to a remote airstrip on the Antarctic peninsula 630 nautical miles away from base demanded four engined reliability, unpaved field capability, and the range to turn back should adverse weather arise makes the BAe 146 ideally suited for this role. The block time for the route is approximately two hours.

Alejo Contreras / DAP

Walter Alvial / DAP

Aerovías DAP has reported reliability of its BAe 146s and Avro RJs (the first of which, CC-AJS, was added in 2013) as meeting or exceeding expectations, and has been equal across the fleet. Pilot response within the airline has been very favorable, including remarks that they like the maneuverability at low speeds, the wide array of runways it can operate from, and overall reliability. The airline now operates a mixed fleet of 146 and RJ aircraft, the most recent arrival of which is former Montex Oil BAe 146-100 N114M (CC-AXE) which joined the fleet in the latter half of 2018. This was the first series 100 Aerovías DAP acquired via Tronos, and it will be operated on exclusive charter and VIP programs while retaining its former VIP interior. It joined three Avro RJ100s, two Avro RJ85s, and two BAe 146-200 aircraft. August and December 2020 saw the addition of two BAe 146-200QT freighters, bringing the fleet total to ten.

In 2014, Aerovías DAP carried the heavy metal music act Metallica to perform a concert in Antarctica, making Metallica the only musical group to perform on all seven continents. In addition to tourism and support in Antarctica, Aerovías DAP has used the BAe series to run crew to remote locations for mining operations, sometimes serving high altitude and occasionally unpaved fields that the aircraft was designed for. DAP makes use of the aircraft to support emergency and rescue operations in response to natural disasters that have occurred in Chile, including earthquakes.

CityJet - Ireland

CityJet is one of the oldest and longest running operators of the BAe 146/Avro RJ series of aircraft. Founded in 1993 in Dublin, it began service between Dublin and London City Airport as a franchise operator for Virgin Atlantic Airways as Virgin CityJet. By 1996, it was operating flights under the Air France brand until 2000 when Air France fully acquired CityJet. Air France sold the airline to a consortium of Irish investors after a lengthy period of ownership and it resumed flying as an independent company. In recent years CityJet has focused more on ACMI services and has operated flights on behalf of a number of airlines including Aer Lingus.

CityJet was the dominant carrier at London-City Airport for a number of years, but in 2017, pulled out most of its scheduled flights. After a disastrous period operating Sukhoi SJ-100 Superjets for SN Brussels Airlines, and the 2020 Covid19 pandemic, it retired the last Avro RJ, replacing it with a fleet of Bombardier CRJ-900 aircraft.

Courtesy of London City Airport

Shaun Henry

JOTA Aviation – United Kingdom

Jota Aviation is a charter airline, based at London Southend Airport in the United Kingdom. It was founded in 2009 by Simon Dolan and Andy Green, originally to provide services to sister company Jota Sport. It expanded to meet ad-hoc demand as well as private charters and in 2014 acquired its first jet, a BAe 146-200 configured for passenger carrying. Jota charters its aircraft out to a variety of customers, but also provides ad-hoc on-demand lift to other airlines to cover for inoperable aircraft on an ACMI (aircraft, crew, maintenance, insurance) basis. In 2016, Jota added an Avro RJ85 and an Avro RJ100 to its fleet while in 2018, with the addition of three BAe 146-300QTs, it launched freight operations.

COBHAM - Australia

Cobham was originally formed as National Jet Systems in 1989 and flew scheduled services on behalf of Australian Airlines. National Jet Systems was expanding and acquired a small fleet of BAe 146-100 and -200 aircraft to fly under the Airlink brand for Qantas. As the airline grew it began operating under Qantas' regional brand QantasLink. National Jet Systems became part of the conglomerate Cobham which has businesses in the maritime, space, and military areas and began trading as Cobham Airline Services. Cobham's title font is a throwback to another airline long-gone in the United States: Braniff.

Cobham still serves as a QantasLink regional operator including carrying freight, but also charters its aircraft out to a variety of private businesses including some in the mining sector. This requires the aircraft to fly to remote destinations across Australia, often landing on unpaved runways. Today, Cobham operates a fleet of 12 BAe 146 / Avro RJ aircraft including the BAe 146-100, 200, 300, and 300QT, as well as the Avro RJ85 and RJ100.

Pionair

With a long history starting in the 1990s in New Zealand as a charter and tourist operator, Pionair expanded into the cargo market. But in 2008, with the worldwide financial crisis in full swing, Pionair fell on hard times. Steve Ferris, a successful aviation entrepreneur acquired the struggling airline in 2013. At the time, the airline was largely operating Convairs which were reaching the end of their service life. Steve began acquiring BAe 146 aircraft in 2015 to bring the airline current with operator needs. By the end of 2020, Pionair had a fleet of eight (8) BAe 146 aircraft including the original QT and QC prototypes, fulfilling a need of five (5) cargo aircraft and three (3) passenger equipped aircraft.

Steve's father Roy Ferris was heavily involved in supporting the Vickers Viscount, the 748, and the BAC One-Eleven fleets with BAC, relocating the family to Australia. In 1974, they again relocated briefly to Singapore so Roy could support the introduction of Concorde with Singapore Airlines. He left BAC/BAE by 1982 before the 146 literally took off, but in the later years ended up working for his son Steve for twenty years.

The aircraft are operating in their former operators liveries, however Steve has said when the time comes to repaint the aircraft, the Pionair livery will grace the aircraft starting with airframe E2050. The 146s join a small but growing fleet of Embraer E190s. Pionair's livery on the tails is reminiscent of Ansett Australia's which used the five multipointed stars to form the Southern Cross. By design, not coincidence.

Inal Khaev

Mahan Air - Iran

Mahan Air, based in Tehran, Iran is an international airline that operates flights to Asia, the Middle East, and Europe. It was late December 2006 when an Iranian carrier was finally able to procure and operate a BAe 146, with Taban Air acquiring a BAe 146-300. Shortly afterwards, in 2008 Mahan Air began introducing regional jets, sourcing aircraft registered in Kyrgyzstan in order to circumvent U.S. embargos on sales and transfer of technology to Iran. However, Kyrgyzstan is also on the list of embargoed countries and makes the acquisition, transfer, documentation and deployment of aircraft incredibly challenging.

The import of aircraft by Taban Air and Mahan Air did not represent the first time an Iranian airline had tried to acquire BAe 146s. Over many years BAe had made efforts to get into the Iranian market with the aircraft and in the Winter of 1991 the manufacturer was discussing sales to Iran. The crippling sanctions imposed by the United States ensured that no Iranian airline fleet would be updated by Boeing, McDonnell Douglas, Airbus, or any other manufacturers except those of Soviet origin. The ageing aircraft were taking a toll, but BAe engaged in conversations with the Iranian government that it hoped would lead to a deal for more than 50 BAe 146 aircraft valued at £680 million (6 aircraft initially, with the potential for up to 50 more). The first six aircraft were expected to be placed with the regional carrier Aseman Airlines.

With the world slowly recovering from a recession following the ending of the Gulf War, BAe was looking to sell aircraft – a lot of aircraft. Instead, the United States stopped the sale on the grounds that the sanctions prohibited the supply of aviation equipment of U.S. origin. The only exception to the sanctions were if the U.S. content was less than 20% and even then, the U.S. had to grant permission regardless. With wings, engines, and avionics produced in the U.S., the content pushed the 146 past the 20% limit and the U.S. State Department blocked the sale. This decision stung British Aerospace even more, given that it had made nearly 10,000 employees redundant over the past year.

BAe did not give up, and less than six months later was considering wet-leasing BAe 146 aircraft to Iranian operators on an ACMI basis to avoid the sanctions. The first six aircraft were scheduled to go a new regional carrier Bon Air which was founded in 1992. The group that set up the airline was called the Bonyad Mostazafan (foundation of the oppressed), and was formed in 1979 to manage the late Shah's assets within Iran. The U.S. was having none of it, and called out the United Kingdom on the basis that leasing did not make it acceptable to transfer aircraft to Iran. Any Western aircraft placement in Iran was going to be a non-starter for the U.S.

Jumping forward to 2008 and Mahan Air, Iran's first privately owned airline, began acquiring BAe 146-300 and Avro RJ100 aircraft. Additional Avro and BAe aircraft filtered into Iran between 2011 and 2015, when its fleet reached 17 airframes. The United States did not make Mahan Air's life easier, declaring in 2011 that the carrier was allegedly ferrying arms, fighters, and other materials to support the Islamic Revolutionary Guard Corps- Quds Force, and issuing

Captain Mohsen Rahnama, Mahan Air.

sanctions on the airline itself. During January 2019, Germany banned all Mahan Air operations, citing the airline's involvement in Syria as well as other security concerns.

The aircraft has been well received in general, but it does have its limitations in some hot and high locations within Iran. Captain Mohsen Rahnama went from being a First Officer on the Airbus A300 B2/B4, A300-600 and later the A340-600 to Line Training Captain and Ground Instructor on the Avro RJ series for Mahan Air. Captain Rahnama explained that going from Airbus to the Avro RJ series required an adjustment period. Terminology, for example, differed between the aircraft types. The doors are 'not open', instead they are 'NOT SHUT', while the cockpit door was not 'unlocked' but 'NOT LOCKED'.

Captain Rahnama cited two areas where the aircraft struggled to perform: hot and high, and icing conditions. At hot and high locations, he explained they might have to wait more than an hour to get favorable wind conditions or lower temperatures in order to take-off. This issue existed even after take-off, through climb and cruise because the LF507 engines didn't generate enough power, making the entire flight a challenge for the flight deck. Icing conditions caused a degradation of performance because of the altitudes flown during winter which were usually below 30,000 feet. As soon the as anti-icing and de-icing systems were activated the engine performance would suffer, causing the aircraft to stop climbing, level off or in some cases even to descend. Captain Rahnama says that if the engines had just a bit more power, the 146 would be a flawless regional jet. Another area where the aircraft struggles is with the cockpit and cabin temperature in extreme conditions, both of which can be difficult to keep warm in cold weather and cool in hot weather.

Despite the two issues pertaining to engine power, Captain Rahnama says the aircraft is pure joy to fly. Excellent handling characteristics are apparent from the first time a pilot gets to handle the aircraft. When it comes to high cross wind landings, it is possible to do "wonders" with the Avro RJ. The steep approach capability also serves Mahan Air well at some of its destinations, not to mention the obvious advantages of short field landings and take-offs. Rahnama even described occasions when they will take-off with flaps fully extended in landing configuration as they continue to provide lift instead of creating drag.

Airbus E-FAN

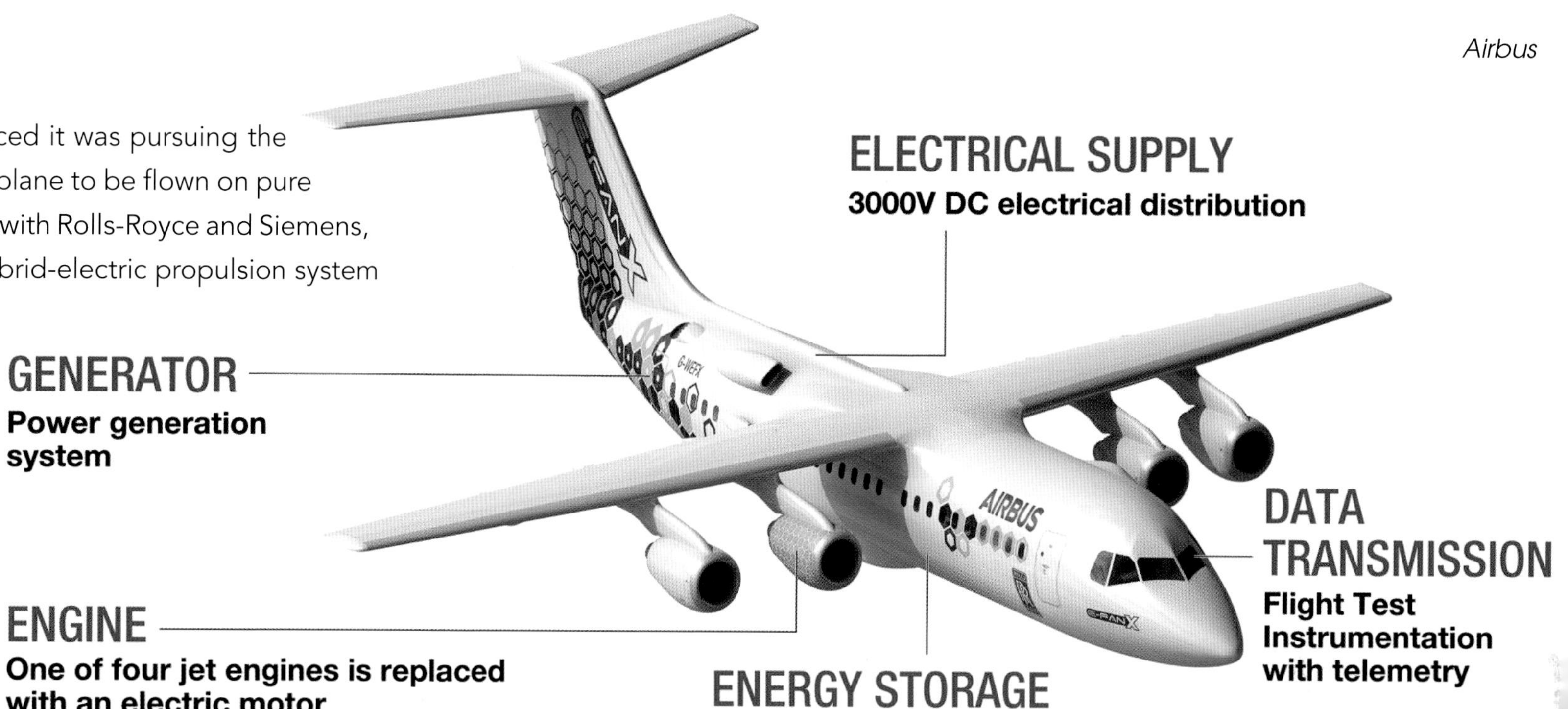

Airbus

In April 2016 Airbus publicly announced it was pursuing the E-FAN X, an environmentally friendly plane to be flown on pure electric power. Airbus, in partnership with Rolls-Royce and Siemens, will work towards demonstrating a hybrid-electric propulsion system for commercial aircraft that will hopefully lead to full electric power. Airbus will be responsible for the overall integration and architecture, Rolls-Royce the turbo-shaft engine and two-megawatt generator, and Siemens the two-megawatt electric motor and power system.

A BAe 146 would be used as a testbed to trial one of the four engines under electric power, although plans envisage eventually replacing a second engine with an electric motor once initial results are available. At the Farnborough Airshow in 2018, the UK government also announced financial support for the E-Fan X project. Construction of the various parts of the electric motor was underway and the first flight test was scheduled for 2021. However, the Covid19 pandemic and the economic toll it took worldwide resulted in the E-Fan program being cancelled.

Airbus

There are other operators of BAe 146s and Avro RJs throughout the world, ranging from air tanker operations to small independent airlines in Southeast Asia. Together they prove that there is plenty of life left in the remaining airframes and the death of four engine regional aircraft has not yet come, even though the list of commercial operators utilizing the types continues to decline. The depth and breadth of today's operations remain very impressive and they are a lasting testament to a design approaching forty years of age.

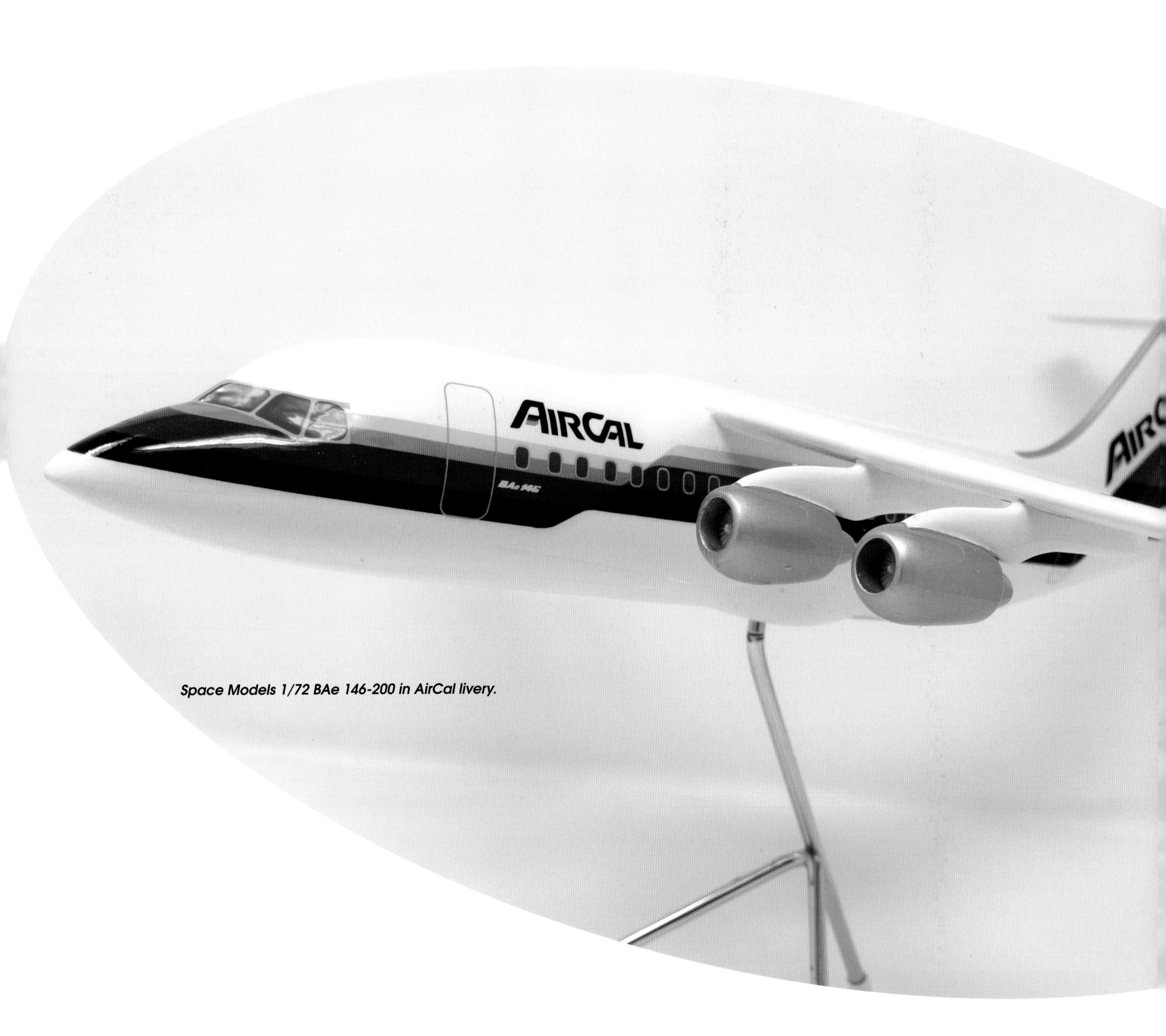

Space Models 1/72 BAe 146-200 in AirCal livery.

CHAPTER 19

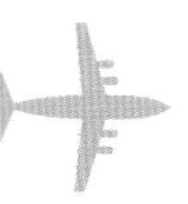

The BAe 146 for Enthusiasts: Collectibles and Special Projects.

Fans and enthusiasts of the BAe 146 have access to a number of collectible and branded items. Unfortunately, it would take an entire book to list every conceivable item that could be deemed a collectible or even desirable but those who hard enough can find pretty much anything from the Hawker Siddeley days when the aircraft was known as the HS146, the BAe 146 and Avro RJ/RJX items. Almost anything can be found from more commonly available safety cards, aircraft models of varying sizes, to British Aerospace (and airline) 146 specific brochures, flight and maintenance manuals, giveaways such as pins, posters, ties, hats, and other very limited tchotchkes like paper weights, keychains and more.

Most can be found through online auction sites such as eBay, but the options are endless. Etsy has 146 items available, and material is often available at local airline enthusiast shows. Online trading sites such as Craigslist and Facebook Marketplace can also produce items from time-to-time. Even Amazon will occasionally have sellers of 146 collectibles and memorabilia.

For aircraft models, the biggest producers were Space Models (1/72 was the most common size) who also produced sales models for BAe, Wooster PPC (1/100, 1/144), Jet-X (1/200, 1/400), Gemini Jets (1/200, 1/400), Herpa (1/200, 1/500) and Schabak (1/600). Atlantic Models and PacMin have produced large scale models. There are other model makers who produced the BAe 146, but those mentioned are the most commonly available. The author of this book was also the founder of Jet-X, and his love of the aircraft resulted in more than 110 models of the BAe 146 being produced (96 releases in 1/400 scale; 16 releases in 1/200 scale).

Safety Cards

BAe 146

Safety
Sécurité
Sicherheit
Seguridad

BAe-146

PRESIDENTIAL AIRWAYS

Please study this card and for the safety of other passengers, leave it on board the aircraft.

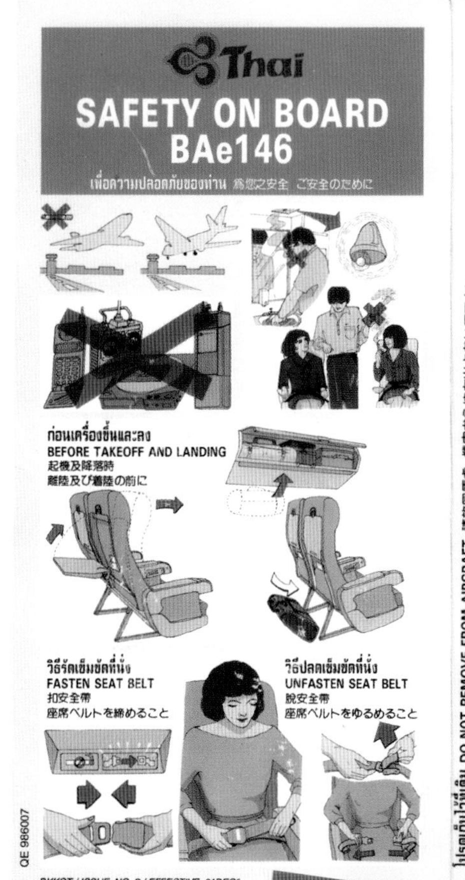

WHAT TO DO IN AN EMERGENCY

QUE SE DEBE HACER EN CASO DE EMERGENCIA

Safety
Sécurité
Sicherheit
Seguridad

BAe 146

Please study this card and for the safety of other passengers, leave it on board the aircraft

Safety
Seguridad
Sécurité
安全
Sicherheit

BAe-146-100/200

Federal law requires that all passengers must review this important information.
Follow these instructions, the directions of the crew, and posted signs.
Please do not remove this card from the aircraft.

Regulaciones Federales requieren que los pasajeros revisen esta información, la cual es importante para su seguridad.
Por favor siga estas instrucciones, las instrucciones de la tripulación, y las señales colocadas.
Por favor, no se lleve esta tarijeta del avión.

Pour leur propre sécurité, la loi fédérale demande aux passagers d'examiner cette importante information.
Veuillez suivre ces instructions, celles de l'équipage ainsi que celles qui sont affichées.
S.V.P. laisser cette carte dans l'avion.

連邦規定により義務づけられておりますので、乗客の皆様は 安全のため下記の注意書きを必ずお読みください。
この説明書、乗務員の指示、掲示に従ってください。
機外に持ち出さないようお願いいたします。

Laut gesetzlicher Vorschrift müssen sich Fluggäste mit diesen wichtigen Informationen zu Ihrer Sicherheit vertraut Machen.
Bitte folgen Sie diesen Anweisungen, der Beschilderung, und den Anweisungen des Bordpersonals.
Bitte, entfernen Sie diese Karte nicht vom Flugzeug.

AEGEAN
AVRO
RJ
BRITISH AEROSPACE
Regional Aircraft

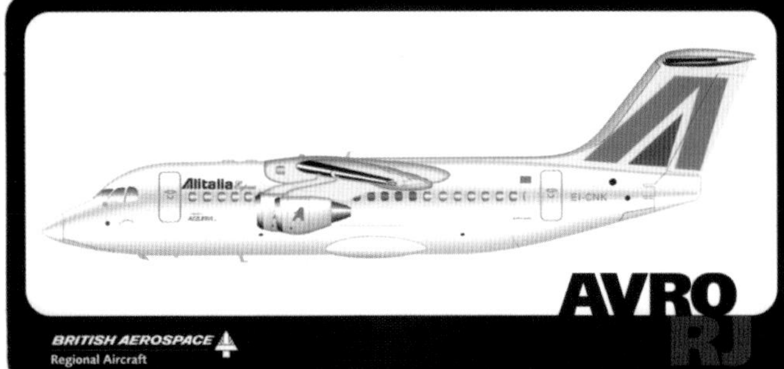
Alitalia
AVRO
RJ
BRITISH AEROSPACE
Regional Aircraft

Botnia
Air Botnia
AVRO
RJ
BAE SYSTEMS

debonair

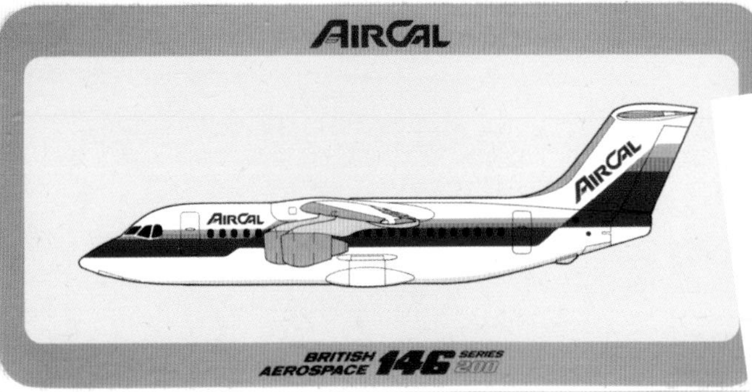
AIRCAL
AIRCAL
AIRCAL
BRITISH AEROSPACE 146 SERIES 200

Air Pac
AIRPAC
BRITISH AEROSPACE 146 SERIES 100

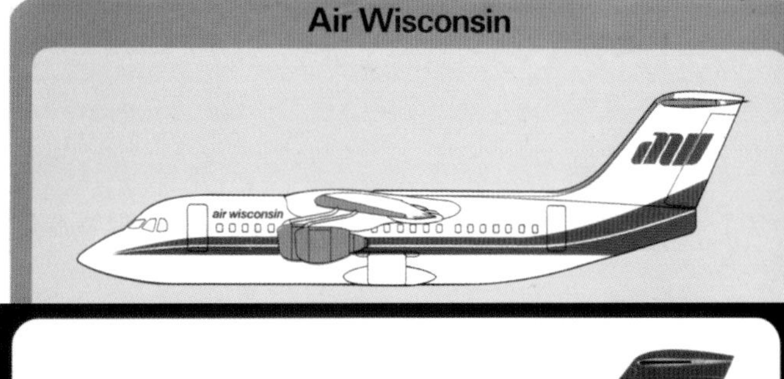
Air Wisconsin
air wisconsin

AVRO
RJX70
AVRO
RJX
BAE SYSTEMS

AVRO
RJX
AVRO
RJX85
BAE SYSTEMS

AVRO
RJX
AVRO
RJX100
BAE SYSTEMS

Princess Air Cargo
BRITISH AEROSPACE
146 200
Quiet Convertible
PSA
PSA
BRITISH AEROSPACE 146 SERIES 200

BRITISH CARIBBEAN
BRITISH AEROSPACE 146 SERIES 100
Dan-Air London
DAN-AIR LONDON

Revell Model Kit. Many are available including the 146STA, the Freighter, and -200/85 passenger.

Flying foam promotional slip-fit kits

FOR YOUR SAFETY

BAe 146

AmericanAirlines

SEAT BELT

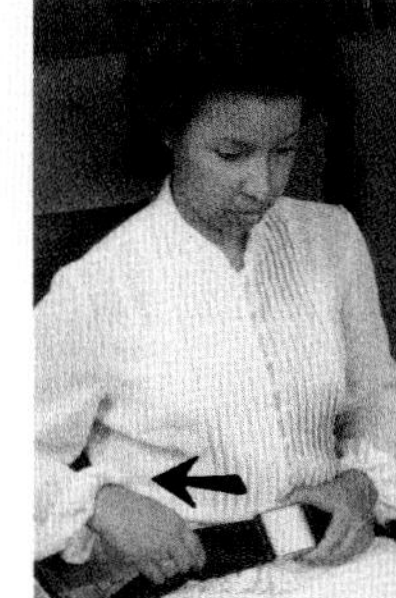
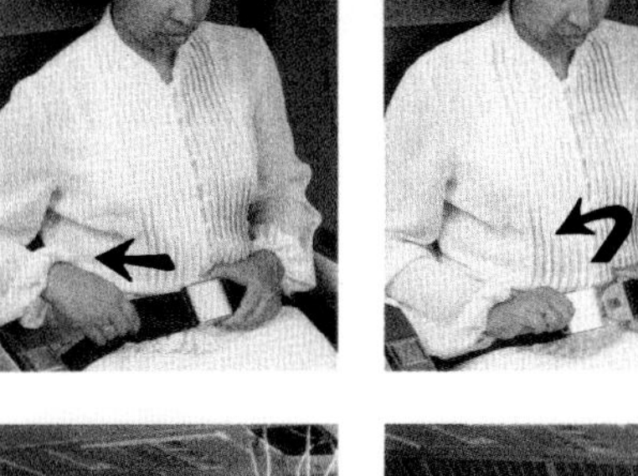
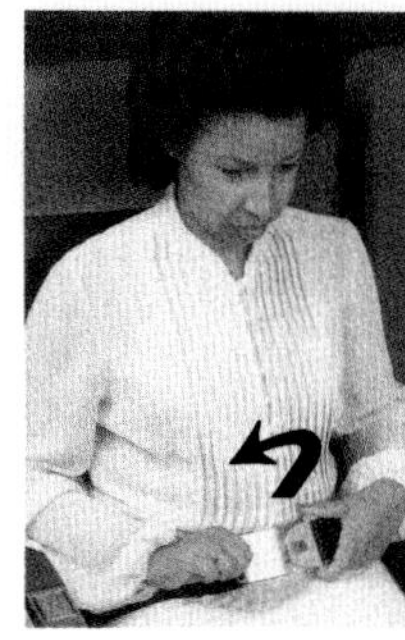

OXYGEN

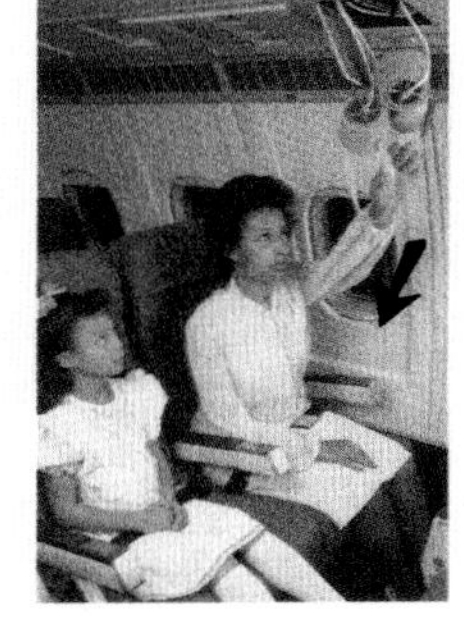
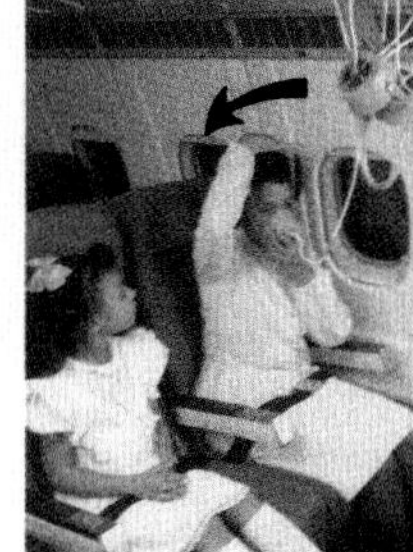
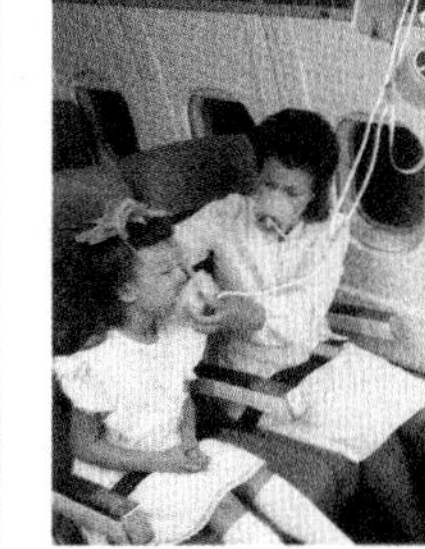
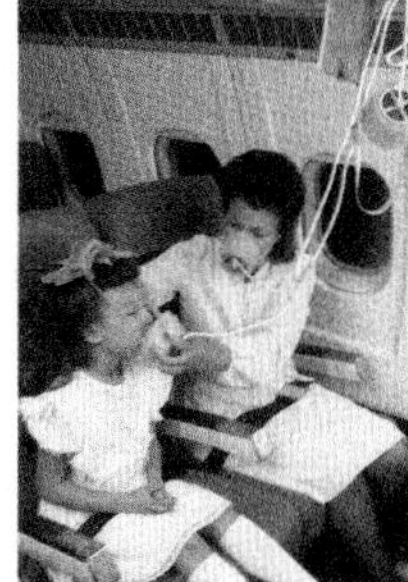

FLOTATION SEAT CUSHION

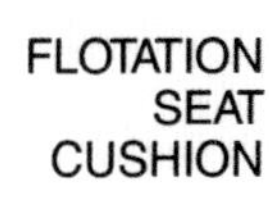

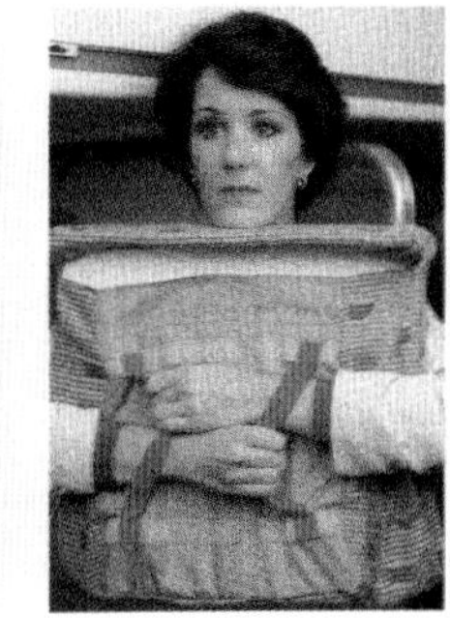

PLEASE DO NOT REMOVE THIS CARD FROM AIRPLANE

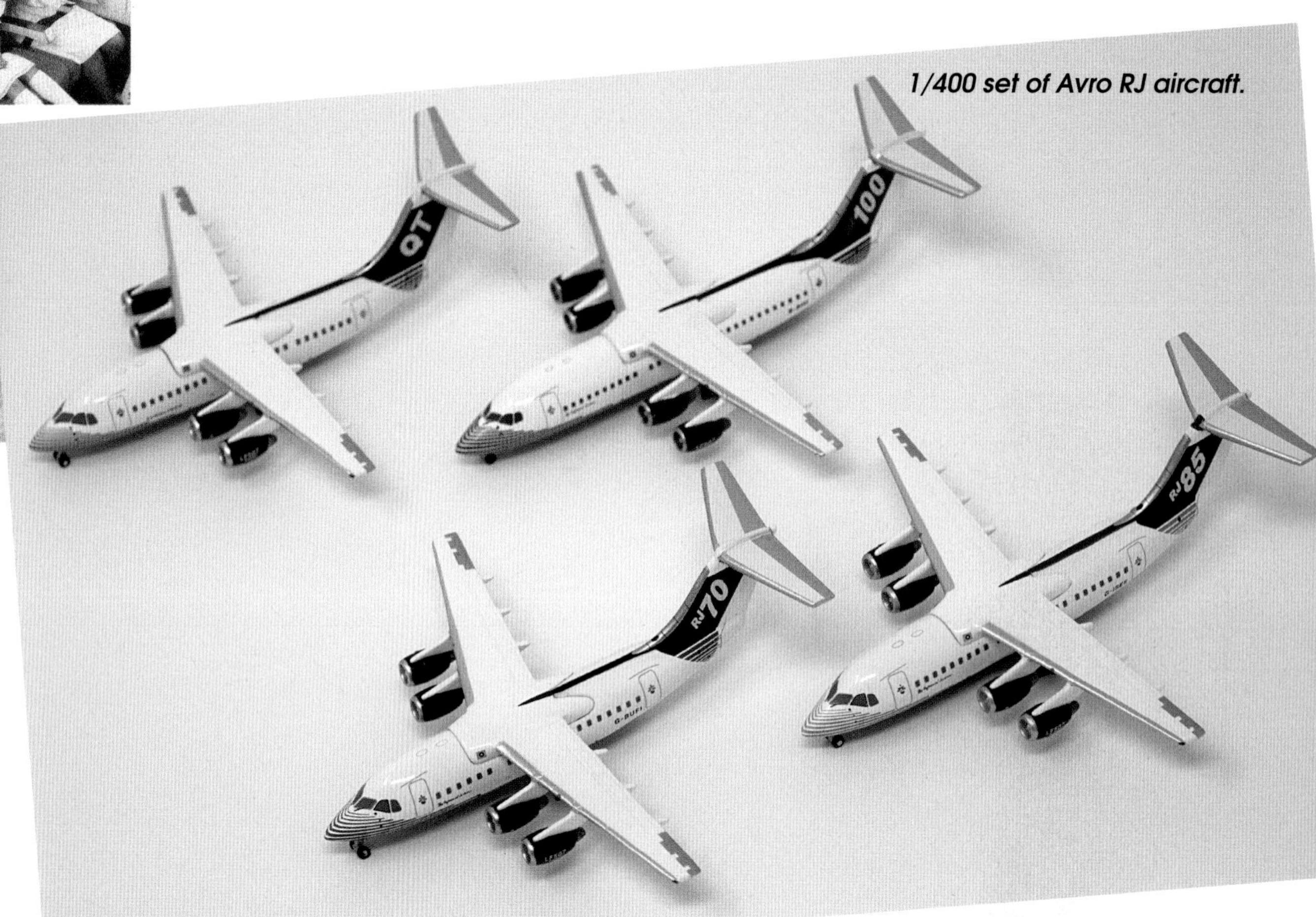

1/400 set of Avro RJ aircraft.

Jet-X released a number of 1/200 diecast metal BAe 146-100 and -200 models.

Jet-X 1/200 model of G-SSHH.

Gemini Jets has also released quite a few BAe 146-200 and Avro RJ85 1/400 diecast models.

PSA gave small bottles of wine featuring their BAe 146 on the bottle to passengers.

Special Projects

The cockpit was loaded onto a truck and transported to Vince's house.

Vince Essex

The 146 Garden Shed

Vince Essex, fellow BAe 146 enthusiast and co-founder of Cello Aviation (operator of the BAe 146 / Avro RJ) decided he wanted part of a BAe 146 to grace his home. In 2017, Vince purchased the cockpit of a former Swiss Avro RJ100 to turn into small summer house for his garden. Working with Stuart Abbott of Stu-Art Aviation Furniture, the forward fuselage was acquired, largely gutted of equipment that was no longer needed, and then it was sliced in half horizontally. Vince disposed of the baggage hold and the avionics bay which were not used in the finished product. This enabled 50% of the weight and height to be shed, making the cabin easy to enter without air stairs as well as easing transport on a flatbed truck. Vince ensured that his dream of owning a cockpit would come to fruition, and while his lovely wife Jennie was less thrilled about the idea initially (fear of a 'crashed airplane' look in the garden), she has come to enjoy the end result. This will undoubtedly become an inspiration for other aviation enthusiasts to come up with creative ways to salvage former aircraft for personal projects.

Vince Essex

Now installed in Vince's yard, ready to be enjoyed complete with power.

Juneau Projects 'Gleaners of the Infocalypse'

Juneau Projects, based in Birmingham, United Kingdom used the rear section of a BAe 146 fuselage including the tail for an art project. Featured at the Tatton Park Biennial festival in 2012, the theme for which was "Flights of Fancy", the aircraft was painted with an elk on the tail and titled "Gleaners of the Infocalypse". Formerly G-TBIC, it was parted out in March of 2012, and after being hauled up and painted, was displayed as a plane converted into a studio for wildlife artists. The piece imagines the aftermath of a global information and technological disaster which has brought society as we know it to an end, after which survivors re-use remnants of items from the 21st century.

As a result, inhabitants use part of a crashed aircraft as a canvas to paint images of the deer that surround them. The theme referenced the sci-fi novel "Snow Crash" written by Neal Stephenson, which was a contraction of 'information' and 'apocalypse'. The aircraft interior featured homemade easels and paintings of local wildlife alongside crash helmets decorated with painted camouflage and deers' heads which were used by the inhabitants to get closer to their subjects. As artist Phil explained "the idea is that these two wildlife artists are working in the structure and they've gone a bit feral, doing their thing in a new society."

Ben Sadler and Philip Duckworth began this project looking to source an aircraft for the art piece. They had initially approached Air Salvage International in Stroud, but at the time the aircraft available were either too small or cost more than the budget set for the project. They settled on an BAe 146 that was about to be parted out in Essex, and knowing the aircraft was located not built far from where the display would take place made it all the more desirable. This particular aircraft spent a lot of time traversing the airspace over Manchester and Birmingham, and tied in perfectly with the backstory.

All photos on this page courtesy of Juneau Projects.

The artwork was on display during the festival, but at the conclusion of the event, the BAe 146 presented challenges for all involved. Art collectors were not used to buying a piece as large as the rear of an aircraft, nor the longevity of the frame, or the maintenance involved to keep the piece from deteriorating (and losing value) – not to mention that the aircraft had to be moved. Richard Parr at Retro Aviation who helped Juneau Projects source the piece took delivery of it and stored it in Staffordshire. He eventually planned to sell it to the Royal Air Force for training, but it is unclear if this ever happened.

Scaffolding was installed so that the artists could finish the art on the tail.

Another local artist painted up the nose of a BAe 146 for the 2014 Glastonbury Festival

Ricardo Cavolo

CHAPTER 19

Snapshot of BAe 146 / Avro RJ Operators

It would be impossible to document in detail every single operator of the BAe 146/Avro RJ and every single aircraft in the confines of a single volume book (including each change in livery over the years). Before this book went to the publisher, it was already out of date with respect to the current operators of the BAe 146/ Avro RJ line, and the downside to a book is it can't be updated quickly (or without you buying another copy). The beauty of the internet is that it makes it possible to stay up to date with each and every airframe in operation. Through resources such as Facebook, Instagram, dedicated message boards and email lists, the BAe 146 is alive and well all over the world.

The BAe 146/Avro RJ was flown by a wide variety of operators around the world over its life, including airlines that never purchased new aircraft but instead procured them second-hand. North and South America, Africa, UK & Europe, Australia, parts of the Pacific, Southeast Asia, and even Antarctica have all seen the BAe 146 or Avro RJ operations. Here are just some of those operators.

RAF Red Arrows flew with Manx BAe 146 at the airlines Air Fair over Ronaldsway Airport, Isle of Man to celebrate 50 years of flying.

Handover ceremony for the first Manx BAe 146.

Paul Seymour

AirAtlantic
N405XV
AirAtlantic
G-BKMN
CARIBAIR
Paul Seymour
AIR JET
JET SERVICES
OB-1877-P
STAR PERÚ

British European flying Air France colors.

Brian Wiklem

KLM uk
KLM
G-UKAC

LAR
transregional

PARK
-RJA
G-6-341

ZK-NZM
QANTAS
NEW ZEALAND

TAS
I-ATSD
BRITISH AIRWAYS
TriStar

Airlink of South Africa chartered passengers onto the island of St. Helena, which just opened for commercial service in 2016, using an BAe 146-200 aircraft.

EPILOGUE

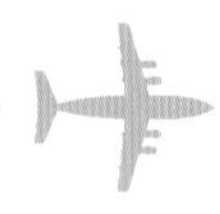

(with a contribution by Peter Connolly)

Success can be measured in many ways, commercially and technically being the most obvious. During its production life span the 146 was only rarely considered to be successful. The aircraft struggled for much of that time to achieve an economic rate of production, an acceptable in-service reliability and a good market reputation. Over hyped in its early days, it quickly went from hero to zero once in service and those early travails dogged it long after they were no longer valid.

BAe supported the programme throughout the 1980's against all the odds, and as it turned out, it was done so recklessly. When the reckoning came in 1991 and the company was almost sunk by its regional aircraft operations, a sea of change in attitude took place. Thereafter it was treated as a potential ticking time bomb whose dismantling and destruction were to be handled with extreme care and at a distance.

Yet ironically that seminal event had forged a new team and culture that was actually not just battle hardened, but innovative and competitive. The product too had reached a level of maturity that was increasingly being recognised and rewarded with a period of market success. However, burdened with the searing memories of just a short time ago, BAE's senior management was completely uninterested in the serious investment and attention required to exploit the opportunities that had been created. The real tragedy is that a more positive and strategic approach might have allowed them to sell the programme off and exit the business more profitably and without destroying the skilled and capable workforce. Contrast that with the way Bombardier has acted with its Dash 8 and regional jet programmes.

Seen in a wider context the 146 programme shared many of the characteristics of other British industries in the 1970's and onwards. Management was often amateurish without much market focus, whilst the labour force was in thrall to powerful unions that impeded and resisted change. Too many companies were living off past glories and failed to recognise how much the world had and was changing.

The Hatfield site, where the 146 was designed and assembled, embodied all these aspects. Old fashioned, socially stratified and arrogant it was ill-suited to make the 146 a success. Its fate was sealed with the disastrous decision to develop a four engined regional aircraft using an engine from a company with little experience or interest in the commercial airliner market.

And yet beneath this ossified structure there were people who were highly skilled and experienced. Designing, building and testing a commercial jetliner is a damn difficult business, and the team at Hatfield succeeded and that needs to be put into the reckoning. Anyone with any doubts might consider and contrast the prolonged efforts of Mitsubishi to develop a new regional jet family.

Looking back with hindsight and balancing all the factors the BAe 146 series, including the updated Avro RJ versions of the aircraft, was in many other respects a success. With nearly 400 built, it would be difficult to use the word 'failure' to summarize the project. The 146 series aircraft was a success. It served all over the world, and fulfilled a variety of functions: passenger, freighter, and VIP roles. There is no doubt the aircraft had issues, namely engine reliability and convincing airlines that four engines were economical in the early days. The market spoke, and some airlines like Air Wisconsin, Crossair/Swiss, Cityjet and DAT/Sabena operated the aircraft for more than 20 years, a testament that it was ideal for specific missions. An aircraft that is problematic, uneconomical, or downright unreliable doesn't stay in service for twenty years, especially in numbers.

In the commercial aircraft world, aircraft often find second lives beyond the passenger world as freighter aircraft. Today, airlines are not converting second life aircraft to freighters at the same rate as in the past. Instead they can buy or more often lease new build aircraft. Aircraft today appear to be disposable, with some build rates of Boeing and Airbus aircraft eclipsing 50 per month on certain lines. Airlines seem to buy/lease aircraft, run them for no more than ten years, retire or return them and repeat the process.

The BAe 146 on the other hand, has found a second life, but not as a freighter; instead, it found a new life as an air tanker, fighting fires in North America, and now Australia. This doesn't include other countries where the aircraft appear from time-to-time, but general year-round usage. The 146 performance figures made it an ideal air tanker that could fly at slow and precise speeds, but with the climb performance to get into and out of dangerous terrain safely. No one ever anticipated the 146 would be performing in this role today, and it's a fantastic one at that: saving homes and ultimately lives that can be attributed to Tronos's vision beyond passenger and freight service. The Avro RJ, specifically the LF-507 engines, finally brought the reliability that was missing during the first 10 years of the BAe 146s life. Problems with the LF-507 were fewer than the ALF-502 series, and the powerplant soldiers on today. Arguably, it was the LF-507 that kept airlines like Cobham, SN Brussels and Crossair operating the aircraft for over two decades. The 146 aircraft were built like the Swiss army knife of planes, able to serve nearly any mission or destination. It has also gone on to become an atmospheric testing research aircraft, and another is being prepared for a partial electric engine conversion in cooperation with Airbus for future airliner propulsion.

The 146 series has largely disappeared from major airline markets except Australia. It has continued on with small operators in South America, Africa, Southeast Asia, Australia, and occasionally North America shuttling passengers. North America is home to mostly 146 air tankers today.

But there's one thing that's absolutely clear, regardless of the troubles BAe went through to launch and maintain the aircraft: the BAe 146 was a success, albeit a qualified one. It established the beginning of the regional jet market, long before airlines ever considered operating jets on commuter routes. Now the reverse is true. Not only are regional jets running short routes, but they're also operating longer trips, sometimes two- and three-hour missions. The BAe 146 and Avro RJ are expected to be around another 10-20 years running passenger, freight and air tanker services, a testament to a well-built aircraft that continues to serve its owners well. 394 successful feeder-liners built and operated, and the last British built airliner.

I've spoken to as many people as possible involved with the BAe 146, both from British Aerospace as well as respective airlines (past and present). If something obvious appears to be lacking, or there was minimal coverage regarding certain carriers or events, it's because I was either unable to reach someone with first-hand experience, or information was limited. I've been surprised throughout this journey the number of amazing people who sat down to talk with me, in person, via email, messaging, or on Skype calls. They provided incredible stories and details, the totality of which a single book would be unable to hold.

I hope that you enjoyed the book and were able to gain a much more insightful understanding of the operational aspects of the aircraft from airlines' perspectives. If you have feedback please, by all means, feel free to reach out to me: brian@avgeektv.com

Who knows, there may be a sequel to this book: More from Four: More stories of the 146.

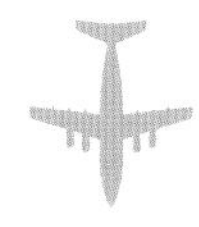

BIBLIOGRAPHY

There was a substantial amount of research that went into this book, largely derived from British Aerospace materials (both internal and external). Sources cited below.

British Aerospace / Hawker Siddeley:

- Sales and marketing brochures (HS146, BAe 146, Avro RJ/RJX);

- Press releases, Internal documents, memos, Service Bulletins, and letters

- Photographs and corporate videos

- Newsletters and corporate newspapers and magazines (e.g. Focus on Hatfield, 146 News, etc)

Books:

- Modern Civil Aircraft #11: BAe 146 M.J. Hardy, Ian Allan 1991

- BAe 146/RJ Britain's Last Airliners: Stephen Skinner 2005

- BAe 146 'Whisperjet': David Oliver, 2018

- Collision Course: Raymond Lygo, 2002

- TASS – The 146 Campaign 1991, Jim Greening 1985

- Wings for Sale, Barry Lloyd 2016

- The Aircraft of Air New Zealand, Paul Sheehan 2003

- No Frills: The Truth Behind the Low-Cost Revolution in the Skies, Simon Calder 2002

Interviews & Correspondence

- Derek Taylor, former Customer Support Field Rep BAe

- Roger Pascoe, former Customer Support Field Rep BAe

- Ian Atkinson, former Program Manager for BAe

- Clive Nicholson, former Customer Support Field Rep BAe

- Vince Essex, Cello Aviation

- David McLees, Arkansas Aerospace

- Chuck Ross, former BAe 146 pilot, PSA

- Jim Hamilton, former BAe 146-100 pilot, Royal West

- Maurice J. Gallagher, CEO Allegiant Air (former founder of Westair and Pacific Express)

- Kevin Govett – Montex Oil N114M maintenance (former AAR and BAe in Arkansas)

- Bob Lundt, Air Wisconsin

- John Sloan, former BAe 146-100 pilot, Royal West
- Nicholas Harambour, DAP
- Amber Biela-Weyenberg, Air Wisconsin
- Pat Doyle, former BAe 146 pilot for Air Wisconsin
- David Dorman, former BAe Marketing
- Jim Skinner, former pilot, Montex Oil Corporation
- Albert Boring, former AirCal
- David Gale, former RAF 146
- Graham Laurie, former Sqn. Ldr 32nd pilot
- Andrew Scott, Sr. Pr Manager, London City Airport
- Ben and Phil, Juneau Projects
- Howard Guy, CEO, DesignQ
- Art Beutler, former Honeywell Avionics
- Gary Ellmer, former COO BEX (former Royal West and Westair)
- David Banmiller, former CEO of AirCal
- Peter Connolly, former sales at BAe
- Steve Doughty, former sales at BAe
- Nick Godwin, former marketing BAE
- Nigel Benson, former sales at BAe
- Paul Stirling, former sales at BAe,
- Phil Bolt, former sales at BAe
- Derek Ferguson, former flight test engineer, BAe
- John Stevens, BAE Systems
- Mark McArdle, BAE Systems
- Terry Snow, BAE Systems
- Andrew James, BAE Systems
- Stephen Morrison, BAE Systems
- Art Beutler, former Honeywell Systems

Internet

- www.airliners.net
- Facebook (groups):
    - o BAe Hatfield
    - o BAe 146 / Avro RJ Appreciation Page
    - o Queen of the Skies BAe 146 Appreciation Page
    - o Woodford Aerodrome

    - o PSA Pacific Southwest Airlines
    - o Royal West Airlines
    - o Presidential Airways
- www.smiliner.com
- Airline Pilots Forum & Resource

Reports & Airworthiness Directives (AD):

- Aircraft Air Quality Malfunction Incidents: Design, Servicing, and Policy Measures to Decrease Frequency and Severity of Toxic Events – Richard Best, Susan Michaelis, School of Safety Science, NSW Australia 2005

- Aircrew Exposed to fumes on the BAe 146: an assessment of symptoms – Moira Somers, J Occup Health Safety 2005.

- Report SL 2012/04 Atlantic Airways BAe 146-200 Accident Investigation Board Norway 2012

- Accident COL-16-37-GIA Avro RJ85 CP-2933 Fuel Exhaustion Grupo de Investigacion de Accidentes GSAN-4-5-12-035 November 29, 2016

- British Aerospace 146-200A VH-JJP Rollback Investigation Report B/925/3042, Department of Transport Bureau of Air Safety Investigation (BASI) March 22, 1992

- John Wayne Airport – BAe 146 Operational History

- Air Accidents Investigation Branch report no. 10/90 Princess Air, July 31, 1990.

- CAA Paper 2004/04 Cabin Air Quality Report, April 2004

- FODCOM number 14/2001 'Malmo Incident', Susan Michaelis, April 2008

- The Right Product at the Wrong Time: The Downfall of European Regional Aircraft Manufacturers, Hans Heerkens, Erik J de Bruijin, Harm-Jan Steenhuis, 2008

- Removal of Altitude Limitation in Icing Conditions 11/99 CASA

- AD 91-08-11, Docket No. 90-NM-237-AD regarding polished metal aircraft, FAA, May 17, 1991

- NTSB report DEN00IA080 Air BC power rollback report, April 30, 2000

- Omission of Oil-plug Seals Leads to In-flight Engine Shutdowns, Flight Safety Foundation Aviation Mechanics Bulletin July-August 1999

- Air Safety and Cabin Air Quality in the BAe 146, report by Senate Rural and Regional Affairs and Transport References Committee, October 2000

- The Air Malta Avroliner Saga or: in search of birds of lead, Malta Independent, September 12 2010

- Ice Crystal Icing Engine Testing in the NASA Glenn Research Center's Propulsion Systems Laboratory: Altitude Investigation, Michael J. Oliver, 2014

- Aerotoxic Syndrome, University of New South Wales, Chris Winder, August 2010

- Prince gives up flying royal aircraft after Hebrides crash, Steve Boggan, July 20, 1992, The Independent

- Federal Bureau of Investigation (FBI) Freedom of Information Act.

Video:

- British Aerospace 146 The Quiet Profit Maker

- Broken Wings

- British Aerospace 146: The Making of an Airliner

Newspapers and Magazines

- LA Times

- OC Register

- Aviation Week and Space Technology

- Flight International

- Air International

- Airliner World

- Airways

- Airliners

- Air Wisconsin AWAC Insider

- 50 Years of the Kings Flight & Queens Flight

Aviation Agency Resources:

- Federal Aviation Administration (FAA)

- Civil Aviation Authority (CAA)

- National Transportation Safety Board (NTSB)

- Civil Aviation Safety Authority (Australia)

- European Aviation Safety Agency (EASA)

Miscellaneous Resources:

- BAe 146 Flight Attendant Transition Training Manual, PSA

- BAe 146 Normal & Emergency Procedure Checklist, PSA

- Air Atlantic BAe 146 Specifications for Airport Firefighting and Refueling Services

- AVCO-Lycoming Engine Brochures

- Avro Sales Team Product Benefits Guide

- British Midland Commuter Operations Manual Cabin Crew Manual

- Globe Hopping with the BAe 146, Sean Kelly, PSA magazine, June 1985

- San Diego Air and Space Museum (SDAS) – PSA Archives

May the Fours Be With You - Supporters of Fighting to Be Heard:

Thanks again – your support on Kickstarter was not forgotten and will always be appreciated.

RJ100 Club

Patrick Doyle, Michale Woodbury, George Andritasakis, Patrick Gannaway, Erin Valenciano, Adam Novish, Henry Harteveldt, Jim Cassou, Karl English, Kevin Trinkle, Andrew Oliva, Michael van der Plas, Michael Bewert, Michel Van Steenbergen, Francesco P., Maarten Van Den Driessche, David McLees, Jason Mackey, Joseph Zadeh, Tom Harris, Stephen Pagiola, John Glancey, Mark Seabrook, Gareth Batchelor, Mark Goodliffe, James Gleave, Simon Wise, Scott Jones, Ruben Baghdassarian, Obie Obenchain, Jerry Jessop, Derek Taylor, Scott Becker, Toby Vickers, David F Michener Jr., Julian Beames, Fabian Vanesch, Janice Featherstone, Phil Cooper, Paul Miller, Colin Hothersall, Michael Overton, Patrik Moller, David Fairbotham, Phillip Petitt, Alan McDonald, Cameron Hutchison, Bob Hemmings, Matthew Dovey, Paul Forest, Simon Wright, Oliver Harper, Yuri Rivera, Howard Obenchain, James Cassou, James McStay, Patrick Houghton, Ray Johnson, John Taylor-Green, Craig Hansen, Gregg Bender, David O'Brien, Peter Vincent, Harold Coghlan, Mark Strzesiewski, John Delapena, Glen Davis, Ryan Murray, Andrew Dorman, Aylius-Navid Arian, Christie Walker.

Chris Sedgwick of Falko Regional Aircraft Limited.

RJ85 Club

Nigel Randall, Mike Kelley, Graham Laurie, David Rowlands, Jonathan Boetig, Vince Essex, Brian Hinderman, Aaron Dwyer, Jim Skinner,

RJ70 Club

Graham Price, Mario Fabila, Eric Batson, Derek Ferguson, Stephen Buckingham, Chris Glancey, Ian Atkinson, Damian Murphy, Peter Merry.

Think about it....

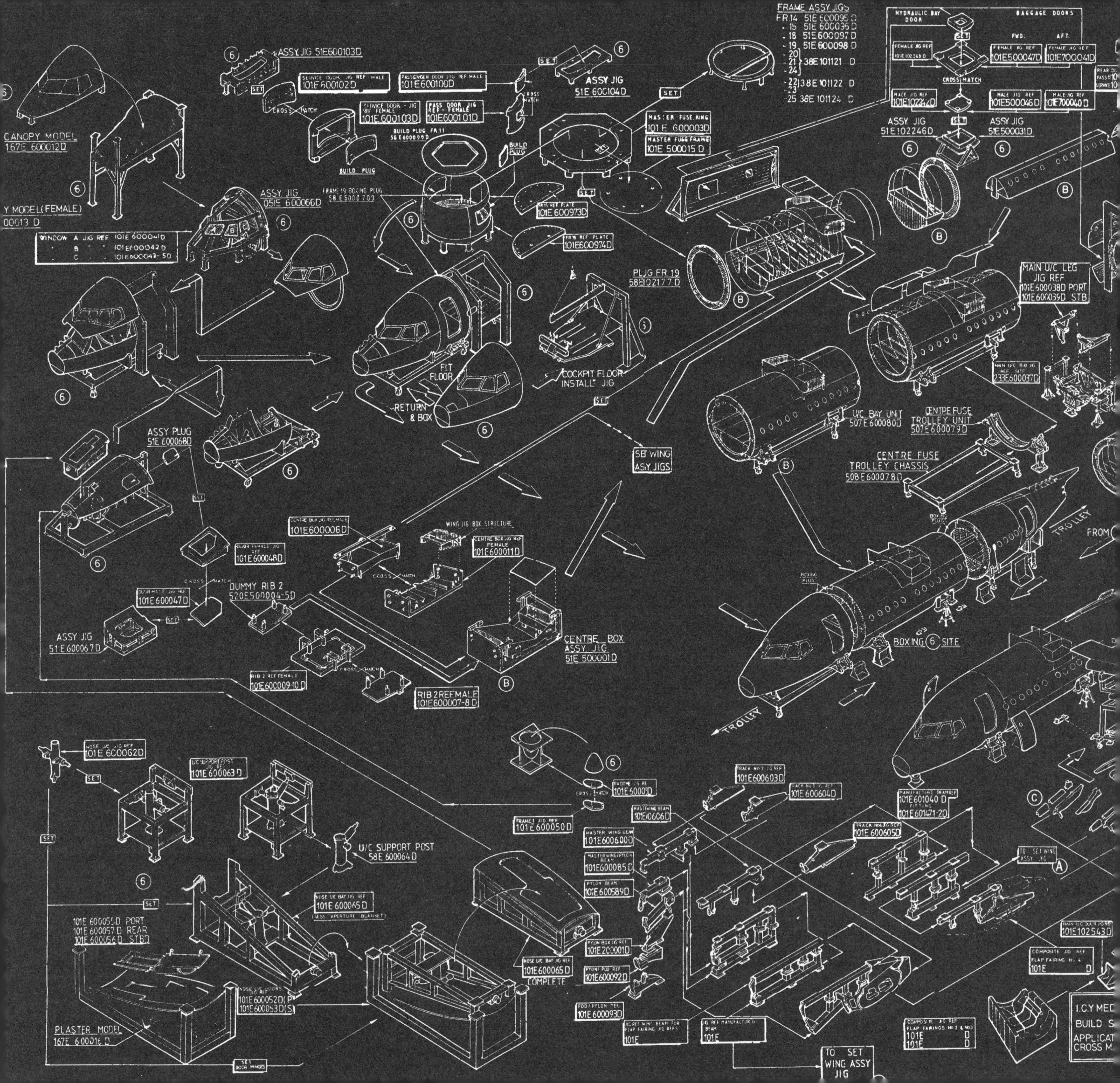

FRAME ASSY JIGS
FR.14 51E 600095 D
15 51E 600096 D
18 51E 600097 D
19 51E 600098 D
38E 101121 D
38E 101122 D
25 38E 101124 D
HYDRAULIC BAY DOOR
BAGGAGE DOORS
FWD.
AFT.
101E500047D
101E700041D
CROSS MATCH
101E500046D
101E700040D
ASSY JIG
51E102246D
ASSY JIG
51E500031D
ASSY JIG 51E600103D
SERVICE DOOR JIG REF MALE
101E600102D
PASSENGER DOOR JIG REF MALE
101E600100D
ASSY JIG
51E 600104D
SET
CROSS MATCH
101E600103D
101E600101D
BUILD PLUG
MASTER FUSE RING
101E 600003D
MASTER FUSE FRAME
101E 500015 D
CANOPY MODEL
167E 600012D
Y MODEL (FEMALE)
00013 D
ASSY JIG
051E 600066D
FRAME 19 BOXING PLUG
101E600973D
101E600974D
WINDOW A JIG REF 101E 600041D
B 101E600042D
C 101E600043-5D
PLUG FR 19
58E102177 D
MAIN U/C LEG
JIG REF
101E600038D PORT
101E600039D STB
FIT
FLOOR
RETURN
& BOX
COCKPIT FLOOR
INSTALL" JIG
233E600037D
U/C BAY UNIT
507E600080D
CENTRE FUSE
TROLLEY UNIT
507E600079D
ASSY PLUG
51E 600068D
SET WING
ASSY JIGS
CENTRE FUSE
TROLLEY CHASSIS
508E600078D
TROLLEY
FROM
101E600006D
WING JIG BOX STRUCTURE
101E600048D
101E600011D
DUMMY RIB 2
520E500004-5D
101E600047D
ASSY JIG
51E600067D
CENTRE BOX
ASSY JIG
51E 500001D
BOXING 6 SITE
101E600009-10D
RIB 2 REF FEMALE
101E600007-8D
TROLLEY
101E 600062D
101E 600063 D
101E6000
101E600603D
101E600604D
101E600606D
101E6000500
MASTER WING BEAM
101E600600D
101E601040 D
101E601471-2D
101E600605D
101E600085D
TO SET WING
ASSY JIG
PYLON BEAM
101E600589D
U/C SUPPORT POST
58E 600064D
101E 600065D
101E 600055D PORT
101E 600057D REAR
101E 600056D STBD
101E200001D
101E600065D
COMPLETE
101E102543D
101E600092D
101E600052D(P)
101E600053D(S)
FLAP FAIRING No. 4
101E D
PLASTER MODEL
167E 600016 D
101E600093D
101E
101E
101E D
101E D
TO SET
WING ASSY
JIG
I.C.Y MED
BUILD S
APPLICAT
CROSS M